Environmental Science and Engineering

Environmental Science

Series editors

More information about this series at http://www.springer.com/series/3234

Sunil Nautiyal · Katari Bhaskar
Y.D. Imran Khan

Biodiversity of Semiarid Landscape

Baseline Study for Understanding the Impact of Human Development on Ecosystems

Springer

Sunil Nautiyal
Centre for Ecological Economics
and Natural Resources
Institute for Social and Economic Change
Bangalore, Karnataka
India

Y.D. Imran Khan
Centre for Ecological Economics
and Natural Resources
Institute for Social and Economic Change
Bangalore, Karnataka
India

Katari Bhaskar
Centre for Ecological Economics
and Natural Resources
Institute for Social and Economic Change
Bangalore, Karnataka
India

ISSN 1863-5520 ISSN 1863-5539 (electronic)
Environmental Science and Engineering
ISSN 1431-6250
Environmental Science
ISBN 978-3-319-15463-3 ISBN 978-3-319-15464-0 (eBook)
DOI 10.1007/978-3-319-15464-0

Library of Congress Control Number: 2015939167

Springer Cham Heidelberg New York Dordrecht London

Printed on acid-free paper

Springer International Publishing AG Switzerland is part of Springer Science+Business Media (www.springer.com)

Foreword

Arid and semiarid regions cover around 40% of terrestrial area of the planet earth and support livelihoods of two billion people, with ~90% people living in developing countries. These regions cover 10 out of 15 agroecological zones in India. The region, though is not as much species-rich and forested as biodiversity hotspots namely Western Ghats and Himalayas, does harbour a large number of unique and crucial species, ecosystems and cultural landscapes that have received only marginal attention of the research community. This volume based on a 3-year-long interdisciplinary study of a vast area (2800 km^2) in the state of Karnataka, India, is a significant addition to the scientific knowledge on diverse dimensions of environment, development and livelihoods in drylands. The volume provides baseline information on the flora, fauna, land cover/use and spatio-temporal dynamics of resource-livelihood linkages. Authors have dealt with multiple components (plants and animals in both aquatic and terrestrial ecosystems) and values of biodiversity in both qualitative and quantitative terms. Data analysis and presentation is such that the volume can serve as a reference point for assessing ecological impacts of development interventions and harmonizing multiple functions of species, ecosystems and cultural landscapes. Deductions on landscape dynamics and resource management are based on an effective integration of modern and conventional ecological monitoring techniques (remote sensing, geographic information system and intensive ground survey/sampling) and scientific knowledge and traditional ecological knowledge on resource management. The contents are thus useful for harmonizing the top-down and bottom-up policies and programmes to sustainable resource management. This contribution from Sunil Nautiyal, K. Bhaskar and Y.D. Imran Khan will be a reference point for future research and development endeavours and will provide a foundation for improving the current conservation-development policies and programmes in drylands.

New Delhi, India
May 2015

K.G. Saxena
School of Environmental Sciences
Jawaharlal Nehru University

Preface

The study on "Biodiversity of Semiarid Landscape: Baseline Study for Understanding the Impact of Human Development on Ecosystems" covers the results compiled from datasets generated from rigorous empirical field studies. The study of biodiversity is meant to provide a reference point against which any future changes associated with any anthropogenic activity can be assessed and to offer information for subsequent monitoring of biodiversity performance. This research aimed to explore, survey, and quantify plant and animal specimens to document the species from aquatic and terrestrial ecosystems. The phytosociological assessment and analysis of diversity indices of different vegetation strata, i.e., trees, shrubs, herbs, climbers, tree saplings, and seedlings across the study region, are part of the objectives of the study. The study and analysis of the conservation status, i.e., to identify and document floral and faunal species (including threatened and endemic), was taken up as a research component. The documentation of traditional knowledge related to use, conservation, and management was also intended to be part of the study. The examination of land-use/land-cover class of the region for vegetation analysis was another objective of this research. The study area is located in North-East Dry Zone of Yadgir and Gulbarga districts in Karnataka, India, and spreads over about 2800 km^2. The central point of the study area is located 16° 43′ 35.40″N latitude and 76° 44′ 40.91″E near Gogi Village in Shahapur taluk (Yadgir district) with average elevation of 460 m (1510 ft.) above mean sea level. The study region falls in Deccan Plateau, mostly covered by dry deciduous plants. Two rivers flow in the study region, and 11 major lakes are also located within the boundary of the studied semiarid landscape. With reference to soil orders of Karnataka State, NBSS & LUP, Nagpur, India, the soil of the study area comprises entisols, vertisols, and inceptisols.

Extreme climatic conditions are the characteristic features of the study region which is hot and receives low rainfall. The average rainfall is less than 650 mm with 40–55 rainy days in a year. The weather of the study area comprises of three seasons. Summer spans from late February to mid-June. It is followed by the southwest monsoon that ranges from late June to late September. It is then followed by dry winter weather until mid-January. The day temperature ranges from 26 °C in winter to 42 °C in summer.

The study area was divided into habitat types and in each habitat 30 quadrats consisting of subgroups—trees, shrubs/climbers, and herbs with sizes of quadrats 10 × 10 m, 5 × 5 m, and 1 × 1 m, respectively were laid down for phytosociological study. When sampling trees and shrubs, the circumference and height of individuals of each species were recorded. Seasonal appearance of plants like flowering, fruiting, and appearance of young leaves was also recorded under phenological study. To study invertebrates, line transects of variable lengths, light traps, pitfall traps, baited traps, and litter collection methods were used. The avian diversity study was done by using line transect and point count methods. Data on big mammals were recorded by sound observation and analysis of pugmarks, scats, pellets, and vocals. The fishes were surveyed and identified by net fishing in major water bodies and exploring the fish markets. Belowground biodiversity was also studied. Litter samples were collected from the field without disturbing the central 10 × 10 cm grid of the different land ecosystems and were transferred to the Berlese funnel for further cleaning and extraction of samples. For all the species, good photographs were taken. The plant species were preserved by preparing herbarium sheets using standard methods given by Botanical Survey of India. The invertebrates and fish species collected were preserved by dry or wet methods for identification and to deposit in regional centers of the Zoological Survey of India. Plants parts such as bark, roots, leaves, and fruits were also collected and preserved in the laboratory. Quantitative data on each species are described in detail. The survey was conducted in the selected villages of the region for documentation of traditional ecological knowledge and conservation and management of biodiversity. Detailed studies were carried out to explore, survey, and collect vertebrates, invertebrates, plants, zooplankton, and phytoplankton. The data collected were tabulated and used for calculating density, abundance, frequency, and importance value index (IVI). The analysis was done for preparing diversity indices such as Shannon–Wiener index, beta-diversity, concentration of dominance (Cd), and Simpson reciprocal index for different vegetation strata.

With the help of the phytosociological study, a total of 376 species of angiosperms (trees, shrubs, herbs, and climbers), 1 bryophyte, 4 pteridophytes, 1 gymnosperm, and 21 phytoplankton were recorded. The number of bryophyte and pteridophyte species recorded is low because the geoclimatic condition of the area is not suitable for those species. Species richness in the study area is dominated by the families belonging to Fabaceae (24), Euphorbiaceae (22), Asteraceae (20), Mimosaceae (18), Poaceae (16), Caesalpiniaceae (16), Convolvulaceae (15), Asclepiadaceae (13), Malvaceae (13), Amaranthaceae (13), Acanthaceae (12), Rubiaceae (10), and Verbenaceae (10). Unique code numbers were given to all the species collected during the study period. Plant biodiversity in this region is sparse and trees grow to limited height, which is a typical semiarid zone characteristic. Phenological data were recorded for tree and shrub species in the study region. The forest of the study region is of the dry deciduous type, and leaf-fall of most of the tree species coincides with the dry season (November to February), and budding and leaf-flushing start from March and continue until April. Majority of plant species were observed for their flowering and fruiting round the year. During field studies, it was found that four species were flowering throughout the year. Seven tree species were flowering for seven months at different times of the year. Some of the species flowered for two months. Changes in ecosystems or habitats are known as beta-diversity (β-diversity). Trees show a high β-diversity value 2.4 however, for shrubs and herbs, the β-diversity values are recorded 1.56 and 2.25, respectively. The Shannon–Wiener index (H^{-}) for the study region is derived to estimate the species diversity . The H^{-} values are in the range of 0.36–1.56.

The study area has several plant species with economic and traditional uses to meet the needs of local people. The species have multiple values in the form of timber wood (9 species), edible fruits (18 species), fuel wood (26 species), fodder (23 species), oral hygiene (11 species), and other uses. The timber requirement is met by tree species such as *Tectona grandis*, *Azadirachta indica*, and *Bauhinia racemosa*. Domestic fuel demand is met by plants namely, *Prosopis juliflora*, *Balanites aegyptiaca*, *Casuarina equisetifolia*, and *Lantana camara*. Apart from plants with traditional and economic values, there are 80 species of medicinal plants that are used for treating 19 ailments such as cold and cough, bone fracture/pain, jaundice, and diabetes. The uses, mode of preparation of medicine, and dosage to cure various ailments were also documented. *Santalum album*, *Acacia ferruginea*, and *Chloroxylon swietenia* are plant species in the study area that have been listed by IUCN under the Red List category as vulnerable species.

The faunal biodiversity study included classes—fishes, amphibians, reptiles, aves, and mammals under vertebrates; and arthropoda, mollusca, and annelida under invertebrates. Among the insects the butterflies and dragonflies were studied extensively during the study period. Around 164 insect species belonging to 13 orders, and 67 families were identified and documented for the study region. As many as 28 butterfly species belonging to six families of order Lepidoptera were recorded in the entire study area. The family Nymphalidae, represented by 13 species (46.43 % of the total butterfly species), was the most dominant one followed by Papilionidae (21.4 3%) with 6 species, Pieridae (17.86 %) with 5, Lycaenidae (7.14 %) with two, Satyridae and Hesperiidae (3.57 %) with one each. Seasonal appearance of butterflies was also studied for different species in different seasons, and it was observed that the Blue Pansy (*Junonia orithya*) and Peacock Pansy (*Junonia almana*) had two seasons—June to August and December to February. Of the 28 species of butterflies observed in the study area, the maximum numbers of butterflies were seen during September to February and peaking during December and January. The least number of butterflies was seen from March to May. A total of 13 species of dragonfly were recorded from the study region.

Eighty-two spider species (individuals including males, females, and juveniles) were collected from different types of habitats. All these species belong to 19 families of the order Araneae in class Arachnida. In addition to this, five species showed new characteristics but not identified so far. The Phylum Mollusca is represented by 17 different species belonging to 9 families collected from the Krishna and Bheema rivers, and lakes of the study region.

The total number of vertebrate species surveyed and listed in the entire study region was 111 belonging to different phyla—fishes (11), amphibians (5), reptiles

(13), aves (71), and mammals (11). The 11 fish species found in the study area come under 4 different families, namely Channidae, Cichlidae, Cyprinidae, and Siluridae. Among those 11 species, six are found only in rivers and the other five are found both in rivers and lakes. The least number of species was represented by the amphibians, which is expected from the semiarid region. Four of the five amphibian species belong to the family Anura. All of the 13 species of reptiles and six families belong to order Squamata. The species found in abundance were the house gecko, keeled Indian mabuya, and Indian monitor lizard. Aves, the most prevalent vertebrates in the region, comprised of 71 species, of which 15 were identified as aquatic, 31 as terrestrial, and four as both aquatic and terrestrial. The most abundant species belonged to the family Ardeidae and was represented by five aquatic and one terrestrial species. Among all the vertebrates, the avian species have a higher density. The region is home to seven migratory birds, five of which visit during April to September when the region gets a considerable amount of rain (29–142 mm). The study area is a habitat for 11 species of mammals of nine different families.

Geographical information system (GIS) and use of satellite data play an important role in understanding landscape dynamics. Landsat and LISS IV data were used for the current study to understand the changes in land use and land cover. The Survey of India (SOI) toposheets on 1:25,000 scale were procured from SOI, Bangalore office, and registered in the UTM Zone-44 N, WGS-84 Projection system. Precisely, georeferenced and radio-metrically calibrated (Level-1G products) satellite images were procured for the same season for the years 1973, 1980, 1992, and 2003/05/06/08. The thematic layers *viz*., rivers, canals, water bodies, and road transportation were prepared for the study region, by sub-setting the toposheets and digitizing the various themes and creating geodatabase by using the ERDAS and ARC-GIS software. The combined unsupervised and supervised classification methods, popularly known as the hybrid classification method, were used in determining the extent of land cover. The land-use/land-cover change study shows that there has been no significant change in the area of water bodies (ponds/lakes), canal and rivers from 1973 to 2008. There has been an increase in the built-up areas/settlements in 2003 compared to 1973. In 2003, a landsat classified image showed that there has been shrinkage in areas with vegetation and barren land due to the conversion of such lands into agricultural land and other purposes, for example, construction of roads and other infrastructures. The degradation of forests in the study region is apparent from this study. This is due to anthropogenic pressure resulting in over-exploitation of forest resources. The land-use/land-cover change analysis does not show any increase in total vegetation cover. In fact, the forest cover has shrunk by 33.5 % in various localities, whereas the human settlement has increased by 66.7 % in some of the pockets of the study region. The area under barren land has decreased drastically, and current analysis showed the decrease for about 70–76 % in various parts of the study region. From the analysis, it is evident that a systematic and scientific approach is required for large-scale plantation of local species and revival of deciduous forests in the semiarid region for ecosystem sustainability, conservation of biodiversity, and sustainable flow of ecosystem services. This work is based on the extensive fieldwork in the semiarid region, and data related to plants and animals were collected and analyzed. The integrated work on plants and animal biodiversity is the unique feature of this book. Therefore, this volume will be highly beneficial not only for the researchers from environmental/ecological sciences but also for the conservationists and policymakers.

We place on records to extend our sincere gratitude to the Environmental Assessment Division (EAD), Board of Research in Nuclear Sciences (BRNS), Bhabha Atomic Research Centre (BARC) for giving us an opportunity and the financial support to undertake the baseline study of flora and fauna at Gulbarga and Yadgir districts which are located in semiarid region, Karnataka, India.

We are deeply thankful to Dr. H.S. Kushwaha (Director, HS&E Group), Dr. P.K. Wattal (Head, BETDD), Dr. Debanik Roy (Programme Officer, BRNS), Dr. R.M. Tripathi (Head, RPS (NF) EAD), Dr. S.K. Jha (EAD), and Dr. S.A. Pandit (Ex-Addl. Director, AMD) from BARC, for their wholehearted support, valuable inputs, suggestions, comments, and cooperation during the study. We are deeply thankful to Dr. M. Sanjappa (Former Director, BSI) and Dr. Ramakrishna (Former Director, ZSI) for their valuable suggestions, inputs, and comments on this study. We are thankful to the Director, Institute for Social and Economic Change (ISEC), for the support providing facilities required for conducting this study. We extend our sincere thanks to Shri. I.V. Saradhi, Principal Collaborator, BARC, Mumbai, for his wholehearted support and cooperation during the entire study period. We are thankful to Dr. Venkateshalu (Head, Department of Entomology, College of Agriculture, Bheemarayanagudi), Dr. K. Ravikumar (Asst. Director, FRLHT, Bangalore), Dr. R.M. Sharma (Officer in charge and Scientist D, ZSI Pune), Dr. Subramaniam (Scientist ZSI Pune), Dr. A.K. Chakravarthy (Principal Scientist & Head, IIHR, Bangalore) and Dr. Prashant Mohanraj (Principal Scientist and Head, NBAII, Bangalore) for the identification of species, Late Prof. B.H. Patel (Professor, Anand Agricultural University (AAU) Anand, Gujarat) for the help in the identification of spider species, Dr. Rengaswamy (Head and Director, Department of Botany, Madras University, Chennai) for the identification of phytoplankton species, Dr. Jayashree Thilak (Scientist ZSI, Chennai) for the identification of zooplankton species, Dr. B. Anjan Kumar Prusty (Scientist SACON, Coimbatore) who helped in the identification of bird species, and Dr. Rama Rao Nidamanuri (Indian Institute of Space Science and Technology (IIST), Thiruvananthapuram) for all the support needed to conduct this study efficiently and successfully. We are thankful to Dr. S.G. Patil (Ex Dean), Dr. B.M. Chittapur (Dean), Dr. Basavaraj Kalmath (Asst. Professor, Department of Entomology), Dr. Shankara Murthy (Asst. Professor, Department of Entomology), Mr. Gajanan, other teaching staff, nonteaching staff, and students of College of Agriculture, Bheemarayangudi. We would like to extend our thanks and appreciations to Ms. Vani Gopi, Christian Witschel, Ms.Kiruthika Poomalai, Springer Project Coordinator (Books) for their kind help, support and assistance during the entire publication process and to Mr. V. Praveen Kumar, Production Editor, Springer's copy editor and series editor for their careful reading of the book manuscript, constructive comments and valuable suggestions. The encouragement and support of Ms. Agata Oelschläger, Springer Book Production, Project Coordinator, Heidelberg, Germany is thankfully acknowledged. We wholeheartedly thank all the field staff in different villages for their support, valuable inputs, and cooperation during the entire period of this study. We thank Mr. Pavan Tikhile, Ms. Ashwini Suryawanshi, Mr. Trivikram Prasad, Ms. Vani Shastry, Mr. G. Boopalakrishnan, Mr. Tamil Selvan, Ms. Padmavathy, and Mrs. B. Akhila for their help and support. We thank Dr. S. Manasi, Associate Professor, CRUA, ISEC, for her support and help extended to us from time to time. We acknowledge with warm gratitude the help we received directly or indirectly from all the officers and administrative and field staff of Forest Department, Government of Karnataka. We are thankful to the people of the study region for their kind support and providing information on traditional ecological knowledge related to use and conservation of biodiversity.

Contents

List of Figures

List of Tables

List of Photoplates

Chapter 1
Introduction

Biodiversity is regarded as a scientific concept, a measurable entity, and a social–political construct (Wilson 1993; Gaston 1996). The 34 global hotspots are considered to be the living laboratories for biodiversity. Myers et al. (2000) argued that most of the species for example 44 % of all plant species and 35 % of the vertebrate species worldwide can be conserved in hotspots with minimal efforts. The hotspots of biodiversity are important not only for the conservation of animal and plant species but also equally important for about 21 % people living in the biodiversity hotspots (Conservation International online). Keeping current environmental problems in view, the hotspots play important role in reducing the ecological footprints. The biodiversity-rich areas have been attracting the attention of the researchers and stakeholders for the decades, and detailed database is developed on the biodiversity which would enable to understand the impact of various driving forces on the resources. The hotspots have attracted the attention of the researchers, and numerous research works have been conducted in the biodiversity-rich areas all across the world. The welfare of mankind is associated with the biodiversity and its sustainable use not only in the hotspots but all across the ecoregions of the planet earth. Thus, the issues related to the conservation of biodiversity are highly crucial since the sustainable flows of services from the ecosystems are important for the livelihood of the people in various ways. Many areas are rich in biodiversity but not able to qualify for the hotspots because of not having 0.5 % species of plants as endemic to that particular region. Various aspects of ecosystems are based on the variability and complexity of organisms that constitute the biodiversity of a given geographic region and, thus, in the world as a whole (Al-Eisawi 2003). The studies related to biodiversity in semiarid and arid regions have not got much attention as comprehensive database based on the empirical field studies is not available in integrated form. The resources demanded to supply the manifold increases in the human population over time have resulted in heavy biomass removal from the ecosystems, which represents one of the primary causes for biodiversity loss and landscape degradation (Sen et al. 2002). Biodiversity provides important ecological, economic, and social/cultural resources that serve as the sustainability basis of any region (Semwal et al. 2004; Nautiyal 2011). This contributes significantly to the quality of human life and has a generally positive effect on the region (Brabyn 2005; Waikato 2003; Nautiyal and Kaechele 2007). Thus balancing approaches are

S. Nautiyal et al., *Biodiversity of Semiarid Landscape*,
Environmental Science, DOI 10.1007/978-3-319-15464-0_1

needed that would support biodiversity conservation and livelihood development goals, rather than putting these goals at odds with each other (Berkes 2004). In addition before planning for developmental activities, there is a need for proper documentation of biodiversity which would be providing baseline data to understand the impact of developmental activities on biodiversity in future. The world population is increasing tremendously, and there is no space on the earth which is not unmolested by the human activities. This has resulted in high biodiversity loss from the earth. There is a need for providing feasible solutions that reduce risks and maximize benefits for biodiversity conservation (Singh 2002), and prior to this, proper documentation of biodiversity of particular region is necessary. The approach should be focused on conducting the biodiversity studies in various agroecological regions. The research on biodiversity documentation and quantification is done in various agroecological regions, however not much work available for dryland regions in India in terms of proper documentation of biodiversity. This is an important region because more than two-thirds of India's geographic area falls within the dryland category that includes arid, semiarid, and dry sub-humid regions, and this is affected by varying kinds of land degradation (Chaudhry et al. 2011). The sustenance of the people living in dryland ecoregion that includes arid, semiarid, and dry sub-humid regions is entirely dependent on the conservation and management of biodiversity. For management and development of biodiversity, there is a need to understand the available biodiversity in regions which are vulnerable not only because of environmental factors, but meantime, the anthropogenic pressure also accelerates the deterioration of the diversity from such ecosystems. The documentation and quantification of the biodiversity (plants and animals) enable the stakeholders for developing appropriate strategies for sustainable social and ecological development. The study related to biodiversity is never-ending process and needs the attention all the times. Unfortunately, the over-exploitation of the biodiversity is leading to fast degradation of the landscapes which is influencing the sustainable flow of the services needed for sustainable livelihood development of the human being. Furthermore, land-use change for food production, the urbanization process, illegal exploitation of organisms for monetary gain, and population pressure are all additional factors involved in the degradation of biodiversity not only in hotspots but all across the ecoregions dominated by human being.

Three years of continuous and rigorous empirical research on biodiversity "phytoplankton to higher plants and zooplankton to higher animals—all flora and fauna" in the semiarid region of India has culminated this volume. The study covers authentic data compiled from field experiments, investigations, rational explanation, and scientific interpretation required to meet its objectives. There are many studies available on the issues related to biodiversity, but they mostly cover either few groups of plants or few groups of animals. Research studies which include all flora and fauna including phyto-zooplanktons in a given ecosystems for understanding the entire structure of biodiversity are not readily available except a few studies. Therefore, this study meant to provide a reference point against any future changes associated with any anthropogenic activity in the semiarid ecosystems. This work will have its own importance as the Govt of India had chosen the area for establishment of nuclear power plant. Therefore, the results of current study will be the baseline information for subsequent monitoring and mapping of biodiversity for conservation and management. Standard protocol was developed for this study. This research aimed to explore, quantify, and survey of plant and animal species from aquatic and terrestrial ecosystems. The phytosociological assessment and quantitative analysis of diversity indices of different vegetation strata are part of the objectives of the study. The conservation status of each species (flora fauna) was investigated as per IUCN category. Study on landscape dynamics using RS and GIS for vegetation analysis was also part of this study. Study also covered traditional ecological knowledge related to use, and conservation and management of biodiversity. The main aim of the current research was to undertake in-depth study on biodiversity of semiarid region of India with following objectives.

1. To document the existing biodiversity of study region conducting various experiments in the field;
2. To conduct phytosociological studies of different vegetation strata (i.e., tree, shrubs, herbs, grasses) in the study region;
3. To explore, survey, and document fauna from different land-cover classes (i.e., forests, main land use, grassland, village commons, shrub land, and open and exposed soil);
4. To prepare diversity indices (Shannon–Wiener Index; beta-diversity; concentration of dominance (cd); and Simpson reciprocal index) for the different vegetation strata;
5. To document endangered/threatened, endemic, and rare species (flora/fauna) from the study region;
6. To document traditional knowledge related to use, and conservation and management of the flora–fauna in the study region; and
7. To study the landscape dynamics of the study region with the help of remote sensing and GIS.

1.1 Semiarid Regions and Socio-Ecological Setup

Arid and semiarid regions come under dry lands which cover approximately 40–42 % of land area of planet earth (Fig. 1.1). As per the estimates from **MEA, 2009 and UN, 2011,** the dry lands support about two billion people of the world and majority of them ($\sim$90 %) live in developing countries. Based on the index developed for aridity, the dry lands can be classified into various sub-habits, for example, hyper-arid, arid, semiarid, and sub-humid. Of the total dry lands, the share of hyper-humid and sub-humid is about 7 and 9 %, respectively. The arid and semiarid regions constitute about 65 % of area of total dry land in the world.

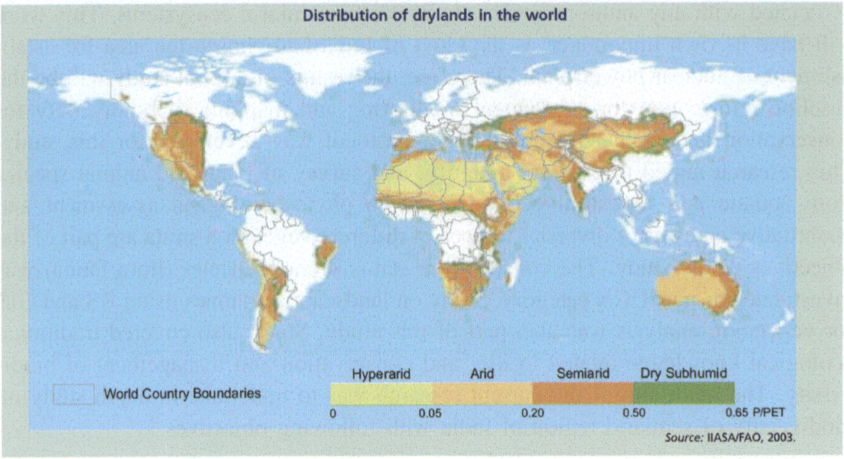

Fig. 1.1 Distribution of dry lands in the world

The share of arid region is about 24 %, and semiarid region occupies about 38 % of area of the total dry lands as per the classification given by United Nations, Environmental Management Group 2011. In semiarid region, about 55 % of land is considered under rangeland and about 35 % is main land use for cultivation of various crops. However, about 10 % land area comes under other category including the urban settlements. At present, there are an estimated 524 million people living in semiarid regions globally (Barakat 2009). The semiarid region spreads all across the continents but largely in central Africa (Niger, Chad, Congo, Cameroon), central Asia (Turkmenistan, Kazakhstan, Kyrgyzstan, Tajikistan, Uzbekistan), south Asia (India, Pakistan, Bangladesh, Sri Lanka), west Africa (Senegal, Mali, Ghana, Benin, Burkina Faso, Sierra Leone etc.), and east and southern Africa (Sudan, South Sudan, South Africa, Swaziland, Uganda, Zambia, Ethiopia, Angola, Botswana, Tanzania etc.). The maximum land in semiarid region comes under the category of rangeland; therefore, the large animal population depends on the vegetation of rangelands. Therefore, the pressure on semiarid region is comparatively high with viewpoints of land categories. A variety of factors influence the process of desertification and land degradation in semiarid region; therefore, the area under this land category is expanding year by year.

In India, about 53 % of land area comprises two important ecoregions: the arid and semiarid regions. In these regions, the livelihood activities of the residing population depend on two sectors—agriculture land use in which the cultivation is restricted to high productive but limited land area and animal husbandry as availability of rangelands provides fodder to large animal population which depends on typical vegetation of arid and semiarid regions. The arid and semiarid regions are highly complex and thus make a sensitive ecosystem in which disturbances at smaller scale would create threats to the sustainability, which seems to be irreversible keeping current environmental and ecological problems in view. The semiarid region in India is home to very significant population which depends on the available land resources, natural as well as cultivable lands. Therefore, the challenge for the stakeholders is to achieve the goals of economic security and environment sustainability. Landscape of semiarid zone comprises of variety of ecosystems which are highly fragile with large risks that is caused due to factors such as increasing population, resource exploitation, unfavorable climate and weather conditions, climate change, increasing pressure on rangelands during excessive grazing by domesticated animals, high intensification of available arable lands, and land degradation and desertification. The implementation of developmental program has acute hurdles. In recent years, natural, anthropogenic factors along with climate variability have contributed to a large extent change in various ecoregions including the semiarid region. A considerable change has been recorded in the area of semiarid region of India over a period of time. The five states of India viz., Madhya Pradesh, Bihar, Uttar Pradesh, Karnataka, and Punjab had shown the increase in semiarid area by 8.45 million ha. However, some states, particularly in the area under semiarid region, have shown decreasing trend by 5 million ha. Overall, there has been a net increase of 3.45 million ha in the semiarid area in the country for two points (1971–2004) of time (Kesava Rao et al. 2013). Asia's 90 % of tropical dryland poor live in India (ICRISAT 2010; Pathak et al. 2013). Cultivable land, rangelands which include common property resources, scrubland, and forest patches are important ecosystems of semiarid regions in India. Majority of farmers are from medium, small and marginal land holding categories. Therefore, the animal husbandry represents important sector which contributes significantly to the livelihood of the people of semiarid region. In India, more than 60 % of the total livestock population is being reared by the people inhabited in semiarid regions (WOT 2013). In semiarid regions, animal husbandry is subsidiary sector of livelihood as elsewhere in the country; however, it becomes main occupation when agriculture fails to withstand the unprecedented climatic conditions and extreme droughts. Therefore,

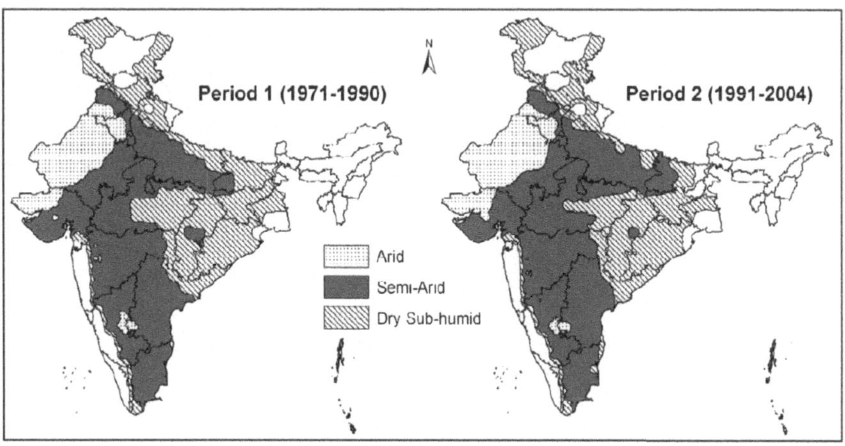

Fig. 1.2 Spatial distribution of semiarid regions in India during two points of time (1971–1990 and 1991–2004). *Source* ICRISAT (Keshav Rao et al. 2013)

documentation, monitoring, and quantification of biodiversity and its conservation become crucial component of the landscape development and management. Integrated management of land, biodiversity including agrobiodiversity, and water could be a potential adaptation strategy for sustainable development of semiarid landscape (Fig. 1.2).

1.2 Geography of the Study Area in Semiarid Region of India

The study region is about 2800 km^2 surrounding the center point geographically located at 16° 43′ 35.40″N latitude and 76° 44′ 40.91″E, near Gogi Village of Yadgir district in Karnataka, India, with an average elevation of 460 m (1510 ft). It is bounded on the west by Bijapur district of Karnataka and Sholapur district of Maharashtra, on the east by Maheboobnagar district of Telangana, on the north by Bidar district of Karnataka and Osmanabad district of Maharashtra, and in the south

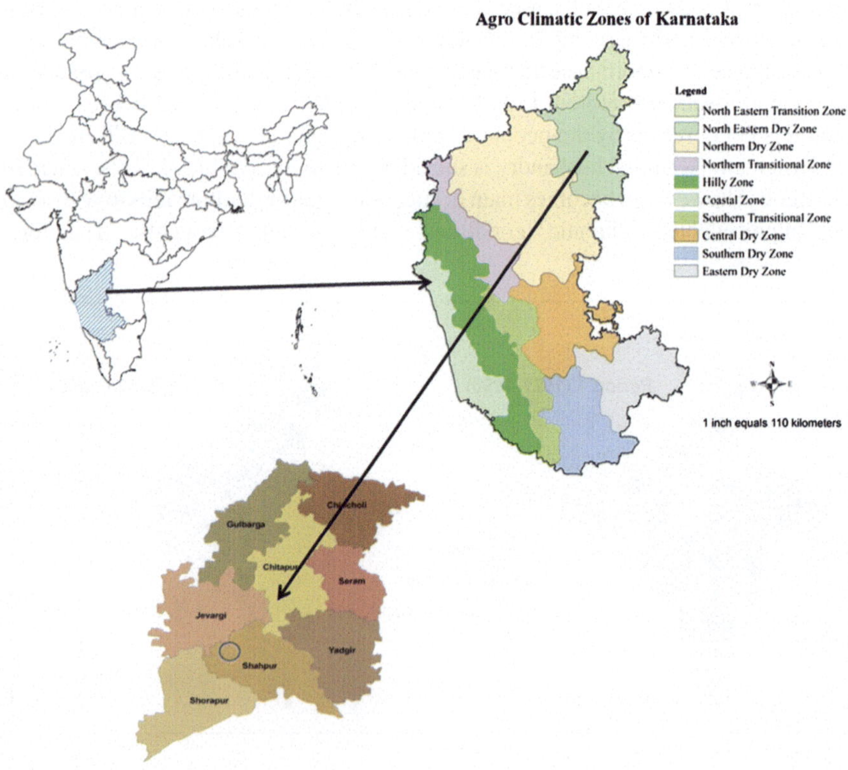

Fig. 1.3 Location of the study area

by Raichur district of Karnataka. A large portion of the study region comes under Yadgir district (Fig. 1.3). The Krishna and Bhima rivers are two main rivers which flow across the sturdy region and provide sustainability to agriculture through the constructions of various canals in the region. Therefore, the cultivable land has increased in the last three decades because the water for irrigation is being supplied from River Krishna through to canals.

1.2.1 Climatic Conditions

The study area is semiarid with low rainfall and high temperature. The weather of the study region experiences three seasons—summer from March to mid-June, followed by the southwest monsoon from late June to late September, and then dry winter weather until mid-February. The average rainfall is less than 650 mm (in Gulbarga district), and the normal rainy days range between 42 and 50 days in a year (personal observations). However, the rainfall during our field investigation in the study area showed only 397 mm annual rainfall. It is very hot during summer, and the temperature during different seasons is as follows. Summer: Min 31 °C to Max 42 °C, Monsoon: Min 28 °C to Max 32 °C, and Winter: Min 15 °C to Max 26 °C.

1.2.2 Land Use in the Study Area

Agriculture forms the primary land use for the study area. The major crops grown in the study area are red gram (*Cajanus cajan*), green gram (*Vigna radiata*), black gram (*Phaseolus mungo*), groundnut (*Arachis hypogaea*), sunflower (*Helianthus annuus*), jowar (*Sorghum vulgare*), and bajra (*Pennisetum glaucum*) during Kharif season. During Rabi season, jowar (*Sorghum vulgare*), bengal gram (*Cicer arietinum*), sunflower (*Helianthus annuus*), cotton (*Gossypium hirsutum*), wheat (*Triticum aestivum*), and paddy (*Oryza sativa*) are grown. Since majority of the area is utilized for cultivation of pulses, the region is known as the "Pulse Bowl of Karnataka." During Kharif (July–November) and Rabi (December–April), the agricultural activities are carried out in the semiarid region. A few decades ago, only traditional crop varieties were grown in the region, for example, sunflower (*Helianthus annuus*), traditional varieties "gajji and dabbi" of groundnut (*Arachis hypogaea*), mungari jola (*Sorghum vulgare*), and bili jola (*Sorghum vulgare*). Recently, cash crops such as Bt cotton (*Gossypium hirsutum*), and hybrid variety of *Arachis hypogaea, Cajanus cajan, Capsicum annum,* etc., were introduced. These major cash crops are sources of income due to ensured irrigation from water canals. Cash crops are exported to various parts of the country.

1.3 Habitat Types in the Arid and Semiarid Regions

The vegetation in semiarid regions is categorized into several land-use/land-cover classes viz., vegetation dominated by shrubs, herbs, and trees, water bodies, agroforestry, and agriculture land. The plant biodiversity is distributed sparsely due to several environmental and demographic factors (Photoplates 1.1, 1.2, 1.3, 1.4, 1.5, 1.6, 1.7, and 1.8). The visual observation indicates that the land cover under scrub vegetation is the most dominant form in the landscape and thus occupies major portions of the landscape followed by other land-cover and land-use types, for example, forests, grasslands, agriculture, water bodies, and barren land.

Photoplate 1.1 Vegetation dominated by shrubs

Photoplate 1.2 Vegetation dominated by annual herbs and shrubs

Photoplate 1.3 Landscape dominated by tree species

Photoplate 1.4 Tree species on the margin agricultural fields

Photoplate 1.5 Aquatic ecosystem in semiarid region

Photoplate 1.6 Introduced tree species in the ecosystem

Photoplate 1.7 Agricultural land under red gram cultivation

Photoplate 1.8 Agricultural land under sunflower cultivation

References

Al-Eisawi D (2003) Effect of biodiversity conservation on arid ecosystem with a special emphasis on Bahrain. J Arid Environ 54:81–90

Barakat HN (2009) Arid lands: challenges and hopes. Earth system: history and natural variability, vol III. EOLSS publishers

Berkes F (2004) Rethinking community based conservation. Conserv Biol 18:621–630

Brabyn L (2005) Solutions for characterising natural landscapes in New Zealand using geographical information systems. J Environ Manag 76(1):23–34

Chaudhry P, Bohra NK, Choudhary KR (2011) Conserving biodiversity of community forests and rangelands of a hot arid region of India. Land Use Policy 28:506–513

Gaston KJ (1996) Biodiversity—a biology of numbers and differences. Blackwell, Oxford

ICRISAT (2010) International crops research institute for the semi-arid tropics (ICRISAT) Strategic plan to 2020: inclusive market-oriented development for smallholder farmers in the tropical drylands. www.icrisat.org/strategic-plan-2020.htm

Kesava Rao AVR, Wani SP, Singh K, Ahmed MI, Srinivas K, Bairagi SD, Ramadevi O (2013) Increased arid and semi-arid areas in India with associated shifts during 1971–2004. J Agrometeorology 15(1):11–18

Myers N, Mittermeier RA, Mittermeier CG, da Fonseca GAB, Kenf J (2000) Biodiversity hotspots for conservation priorities. Nature 403:853–858

Nautiyal S (2011) Can conservation and development interventions in the Indian Central Himalaya ensure environmental sustainability? A socioecological evaluation. Sustain Sci 6(2):151–167

Nautiyal S, Kaechele H (2007) Conserving the Himalayan forests: approaches and implications of different conservation regimes. Biodivers Conserv 16:3737–3754

Pathak P, Chourasia AK, Wani SP, Sudi R (2013) Multiple impact of integrated watershed management in low rainfall semi-arid region: a case study from eastern Rajasthan, India. J Water Resour Prot 5:27–36

Sen KK, Semwal RL, Rana U, Nautiyal S, Maikhuri RK, Rao KS, Saxena KG (2002) Patterns and implications for land use/cover change: a case study in Pranmati watershed (Garhwal Himalaya, India). Mt Res Dev 22:56–62

Semwal RL, Nautiyal S, Sen KK, Rana U, Maikhuri RK, Rao KS, Saxena KG (2004) Patterns and ecological implications of agricultural land-use changes: a case study from central Himalaya, India. Agric Ecosyst Environ 102:81–92

Singh JS (2002) Balancing the approaches of environmental conservation by considering ecosystem services as well as biodiversity. Curr Sci 82:1331–1335

Waikato E (2003) Coastal values and beach use survey report. Technical report no 2003/9, Hamilton, New Zealand

Wilson EO (1993) The diversity of life. Harvard University Press, Cambridge

WOT (Watershed Organisation Trust) (2013) Livestock systems, vulnerability and climate change, pp 23, December 2013

Chapter 2
Methodology for Biodiversity (Flora and Fauna) Study

2.1 Plant Biodiversity and Phytosociological Study

For enumeration and quantification of plant biodiversity, in-depth studies are to be conducted in the region. Considering the potential impact of development on biodiversity, this need to be assessed from a biodiversity viewpoint to indicate the extent to which the disturbance will have impact on biodiversity. Several field studies were undertaken in order to gather authentic information on enumeration, quantification, and distribution of plant biodiversity in the study region. Standard field and laboratory methods of biodiversity studies were followed for data collection and data analysis. In plant biodiversity study, standard field and laboratory methods were followed for data collection and data analysis. However, such a detailed information regarding plant diversity of the study region is not available from the study area, and hence, this study can be considered a baseline study on plant biodiversity of the study area. The photoplates (2.1, 2.2, 2.3, and 2.4) illustrate the procedures for conducting phytosociological study, specimen collection, field observations, and herbarium preparation.

2.1.1 Duration of Survey

The ecological documentation of biodiversity and the survey were started in the month of April 2010 and completed in December 2013. During the field visits, various experiments were conducted. This was followed by surveys, exploration, collection, and preparation of specimens toward building an inventory of floral diversity of the area. Phytosociological studies were conducted to assess the composition, diversity, distribution, and their status in the nature. During the study, the phenological aspects of trees and shrubs were taken into consideration and the annual cycles of these groups recorded based on field observations. This was

S. Nautiyal et al., *Biodiversity of Semiarid Landscape*,
Environmental Science, DOI 10.1007/978-3-319-15464-0_2

Photoplate 2.1 Conducting phytosociology study in the study region

Positioning quadrat for phytosociological study

Recording the GPS points for vegetation analysis

Collection of various plant species for herbarium

Recording taxonomic features of the species

Measuring Tree GBH

Photoplate 2.2 Specimen collection

cross-checked with the traditional knowledge of the people of the study region. A comprehensive study of all the plant species along with their abundance, density, and frequency in the area including diversity indices has been given in this book. The documentation of traditional knowledge relating to the biodiversity forms a very important component of the study. Apart from this, emphasis has been given on the plant species belonging to rare, threatened, and endangered categories included in the IUCN Red Data Book. A comprehensive work on plant biodiversity of the study area is not available as a ready reference barring this study.

Photoplate 2.3 Field observations

2.1.2 Field Equipments

Following tools/equipments were used for conducting phytosociological study.

- Alcohol and mercuric chloride
- Ballpoint pen
- Binoculars
- Camera
- Collecting picks
- Field bags
- Field notebooks

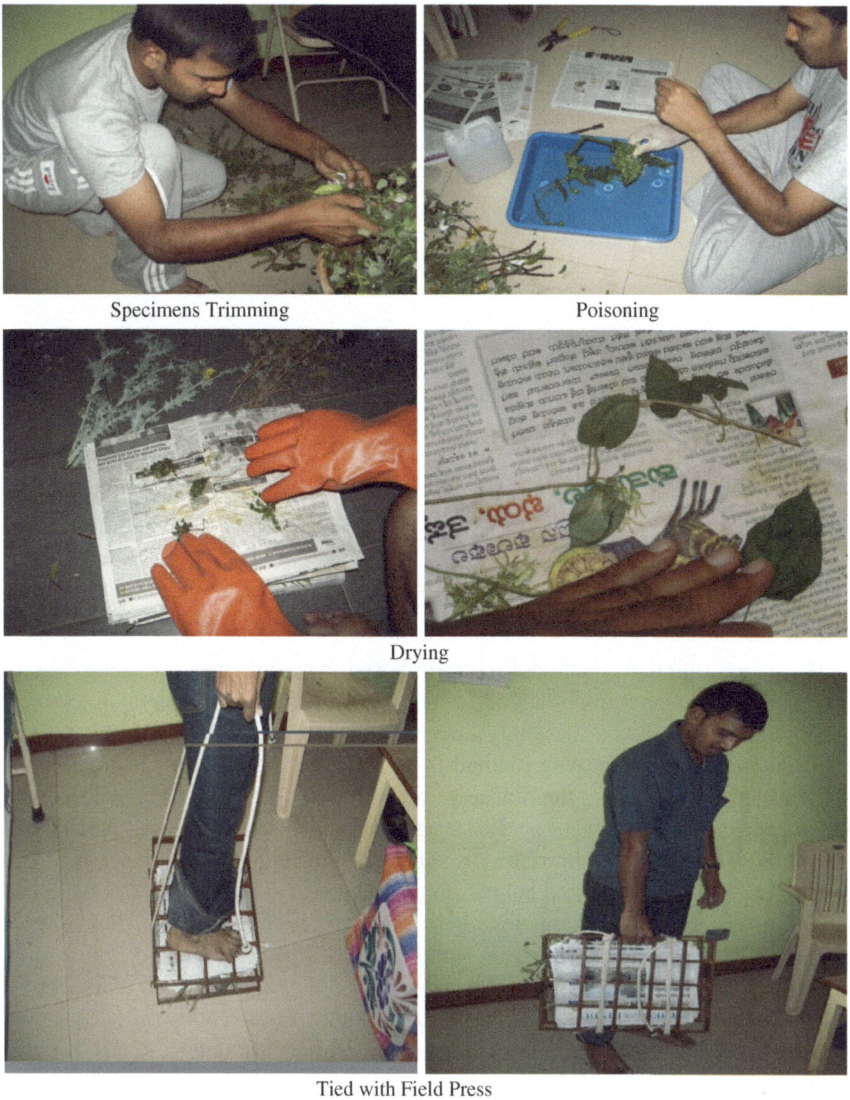

Photoplate 2.4 Herbarium preparation

- Field shoes
- Gloves
- Global positioning system (GPS)
- Identity cards
- Measuring tapes and scales

- Old newspaper
- Plant cutters
- Plant pressers
- Plant tags
- Pocket lens
- Polythene bags for specimen carrying from field to laboratory
- Ropes
- Pruning hooks
- Vasculum

2.1.3 Methods Used

The following groups of plants were surveyed for floral diversity.

- Angiosperms
- Gymnosperms
- Pteridophytes
- Bryophytes

There are different methodologies proposed by ecologists for sampling of angiosperms. The most important and widely used method for a general assessment is belt transect method. The study area was divided according to habitat types followed the random sampling method in the selected area. For plant biodiversity study in the ecosystems, the transect method was followed, and accordingly, transects or straight lines were marked starting from the base of the study area to the end of the vegetation zone in each selected site. The length of a transect was 500 m to 1 km in each of the selected habitat. In each selected site, 30 quadrates were laid down with the size of each quadrate being 10 × 10 m for tree strata, 5 × 5 m for shrubs, and 1 × 1 m for herbs. This is the standard scientific method followed by various workers in respect of phytosociological studies (Cottam and Curtis 1956; Ralhan et al. 1982; Saxena and Singh 1982; Nayak et al. 2000; Lu et al. 2004; Nautiyal 2008). While sampling, circumference at breast height (CBH) of tree species was measured at 1.37 m from ground level, along with the name of the species, phenology (flowering, fruiting, and flushes), and uses. However, in this respect, the circumferences of shrubs were measured at 5 cm above the ground level. The methodology used for conducting the experiments and completing phytosociological studies is given in Fig. 2.1 (Nautiyal and Kaechele 2008).

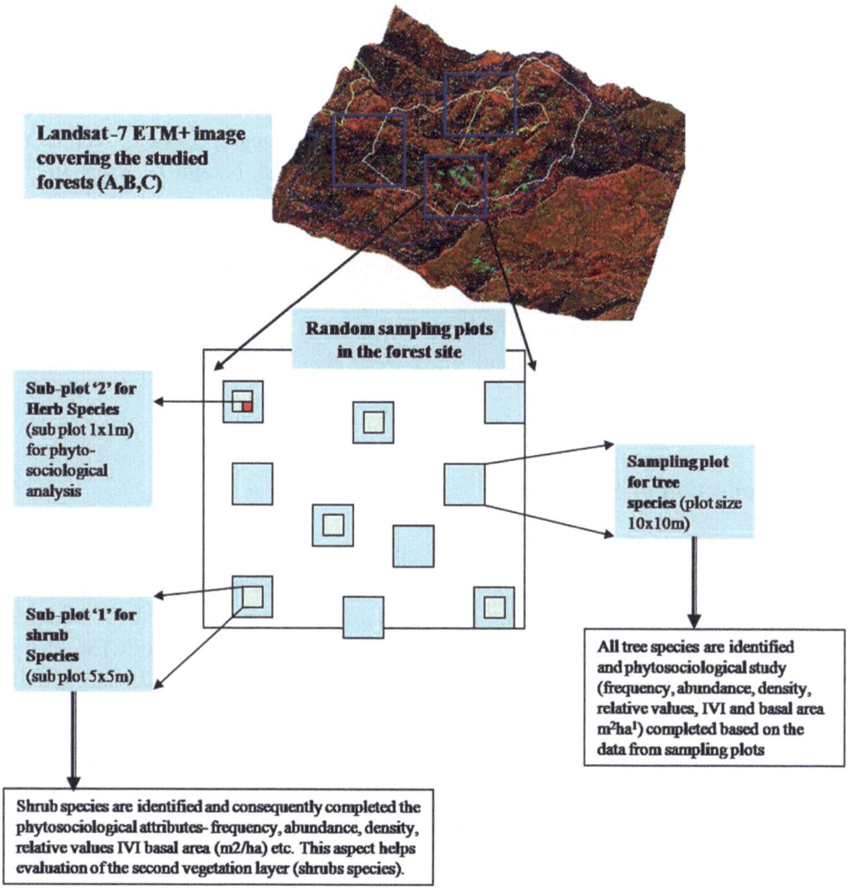

Fig. 2.1 Experimental design for conducting phytodiversity and phytosociological studies

2.1.4 Collection Methods

For studying floral diversity, specimen collection is one of the important aspects. During the specimen collection, field notebooks were carried along for noting down plant-related descriptions. All details related to each plant species were recorded in the field notebooks including the date of collection, local name of species, family, locality of the collection, altitude, habit and habitat, scientific name, vegetation, local use, and collector's name. Such characteristics are recorded for developing proper herbaria. Along with the above characteristics, other special characteristics of each specimen such as latex present or absent and its color, color of the flower, size of the flower and smell, stem shape and size, hairs present or absent, height of the plant, thrones present and size and species association, and root and rhizome character were also recorded.

2.1.4.1 Collection of Trees, Shrubs, and Herbs

The collection of plant specimen is very important for preparing a good herbarium. The size of a herbarium sheet is 28 × 42 cm. The plant collection is always according to the herbarium sheet size, but sometimes, long specimens such as tall grass, creepers, and other large plant parts can bend V, M, N, and V shape. In case of angiosperms (flowering plants) the flowers are the key characters for identifying a specimen's scientific name. During field visits, specimens of trees and shrubs were collected with flowering and fruiting twigs; small herbs were collected with the whole plant and sometimes with roots also; creepers were collected with flowering and fruiting; and grasses were collected with underground parts. Plant parts such as bark, root, leaves, and fruits were also collected and preserved in a laboratory.

2.1.4.2 Collection of Succulents

In the study area, around ten types of succulent plant species were recorded. Generally, the collection of these specimens and making a herbarium is difficult by following general methods. These plants contain more tissues, and their thickness takes a very long time to dry, and hence, they require special methods for drying. After the collection of these plants, their thick parts were cut or made into straight lines with the help of a knife and then boiled in hot water for killing tissues before keeping the specimen on newspapers for up drying with newspapers being continuously changed till they dried up completely.

2.1.4.3 Collection of Minute Plants

The minute plant species required special care for the collection and preservation and the methods for preparing herbaria is different from the higher plants. In view of plants being small, the herbarium of minute plants cannot be prepared through usual methods. After collection, some of the plants were dried and kept in packets which mounted on a herbarium sheet (for example, *Lemna polyrhiza*).

2.1.4.4 Collection of Large Plants like Palms

In the study area, the details of palm trees such as coconut, borassus, and bamboo species were recorded. Generally, these are very large, and the shape of leaves is irregular, and hence, it is difficult to prepare a good herbarium with full characteristics. During the field survey, good photographs of their main parts such as scale trunk, inflorescence, and leaves were taken for the records and identification.

2.1.4.5 Collection of Aquatic Plants

From the aquatic ecosystems, plants were collected. Some specimens were long, and a few of them were small that float on water. The tall specimens were bent into M, N, and V shape to prepare a herbarium. For removing the moisture content from aquatic plants, muslin cloth was used.

2.1.5 Herbarium Preparation

The specimens were collected according to habitat as explained above. After the collection of specimens, press and drying are important to make a herbarium. Plant pressers are of different types: wooden, iron, and book pressers. For pressing and drying, small iron rod pressers and old newspapers were used. The collected specimens were trimmed nicely before poisoning process. The specimens were poisoned with alcohol and mercuric chloride and placed in between the newspapers before being tied with field pressers. Dipping of specimen in mercuric chloride is called dry method. The newspaper needs to be changed every alternative day till the specimen dried up completely. Dried specimens were mounted on a herbarium sheets (28 × 42 cm), while large specimens were mounted on different herbarium sheets shaped like M, N, V, and W identified by professional taxonomists (Allen et al. 1996; Smith 1971; Anderson 1999; Maden 2004; Wondafrash 2008).

2.1.6 Data Analysis

For calculating the species composition, abundance, and diversity indices at the transect level, the following common variables were used: basal area, relative dominance, and relative frequency following Phillips (1959), while the sum of the relative dominance and relative frequency gave the importance value indices (IVI) for various species (Curtis 1959). Species richness, concentration of dominance, evenness, and similarity index were also analyzed. For developing the land-use and land-cover change for the study area, GPS data were recorded.

$$\text{Frequency} = \frac{\text{Number of sampling units (quadrates) in which a species occurs}}{\text{Total number of sampled units studied}} \times 100$$

$$\text{Density} = \frac{\text{Total number of individual in all sampling units}}{\text{Total number of sampled units studied}} \times 100$$

$$\text{Abundance} = \frac{\text{Total number of individuals in all sampling units}}{\text{Total number of sampling units of occurence}}$$

The basal area was calculated using the following formula:
Basal area of a single tree = $\pi \times r^2$
r = radius, $\pi = 3.14$
Basal cover (m^2/ha) for shrub and tree species obtained by adding value of all species together and presented as follows:

$$BC = \frac{\sum_{i=1}^{sh} \text{BASh}}{\text{PA}}, \quad \frac{\sum_{k=1}^{m} \text{BAT}}{\text{PA}}$$

where BC = basal cover or basal area, Sh = shrubs, and m = tree and BASh and BAT are basal area for shrub, tree species respectively, and PA = plot area or quadrat. The total basal cover calculated by the multiplying mean basal cover and density of the species.

$$\text{Relative density (RD)} = \frac{\text{Number of individual of a species}}{\text{Total number of individual of all species}} \times 100$$

$$\text{Relative frequency (RF)} = \frac{\text{Number of occurrences of a species}}{\text{Total number of occurrences of all species}} \times 100$$

$$\text{Relative dominance (RDo)} = \frac{\text{Total basal cover of individual species}}{\text{Total basal cover of all species}} \times 100$$

$$\text{Importance Value Index} = \text{RD} + \text{RF} + \text{RDo}.$$

2.1.7 Diversity Indices

Diversity is a combination of two factors; the number of species present, species richness, and the distribution of individuals among the species are referred to as evenness or equitability. Whittaker distinguishes three types of diversity.

1. *alpha* diversity—diversity within a particular area or ecosystem,
2. *beta*-diversity—the change in diversity between ecosystems, and
3. *gamma* diversity—the overall diversity of a landscape comprising of several ecosystems.

The two most widely used species diversity indices are Shannon and Simpson indices. They are adopted by ecologists to describe the average degree of uncertainty in predicting the species of an individual picked at random from a given community. As the number of species increases, the uncertainty of occurrence also increases along with distribution of individuals, more evenly among the species already present.

The Shannon–Wiener Index or species diversity (Whitt, when properly manipulated, always results in a diversity value (H') ranging between 0, indicating a low community complexity and 4 and above indicating high community complexity.

Species diversity (***H'***) was computed following the Shannon and Weiner (1963) information index as follows

$$H = \sum \frac{n_i}{N} \log_n \frac{n_i}{N}$$

where n_i is the total density value for species, i, and N is the sum of the density values of all the species in that site.

Beta-diversity (***β***) among all the studied forests was calculated following the method given by Whittaker (1975).

$$\beta = \sum \frac{\text{Sc}}{S}$$

Richness: The number of species per sample is a measure of richness. The more species present in a sample, the "richer" the sample.

Evenness: Evenness is a measure of the relative abundance of different species making up the richness of an area. A community dominated by one or two species is considered less diverse than the one in which several different species a similar abundance. Species richness and evenness increase, so does diversity. Simpson diversity index is a measure of diversity which takes into accounts both richness and evenness.

The term "Simpson's Diversity Index" can actually refer to any one of 3 closely related indices.

Simpson's Index (***D***) measures the probability of two individuals randomly selected from a sample belonging to the same species (or some category other than species).

$$\text{Simpson index } D = \frac{\sum n(n-1)}{\sum N(N-1)}$$

where N = total number of species and n = number of species in a given community.

Simpson's Index of Diversity 1 − ***D***: The value of this index also ranges between 0 and 1, but now, the greater the value, the greater the sample diversity. This makes more sense. In this case, the index represents the probability of two individuals randomly selected from a sample belonging to different species.

Simpson Reciprocal Index 1/***D***: The value of this index starts with 1 as the lowest possible figure. This figure represents a community containing only one species. **Simpson reciprocal index increases with an increase in diversity**. The maximum value is the number of species (or other category being used) present in the sample (Simpson's Diversity Index online).

2.2 Phytoplankton Study Materials and Methods

Phytoplankton are single-celled free floating algae and easiest food source for most of the aquatic life form like zooplankton and fishes thus are the basic food producers in any aquatic ecosystem (Suseela 2009). Random sampling method has been applied for the collection of sample from the selected site. Samples were collected in two different seasons from January 2011 to December 2011. The sample were collected using plankton mesh net which consisting of a cylindrical tube with stoppers at each end and closing device, and water sample was collected up to 6 l at each station and passed through the mesh net. Collection of whole water samples from the site, all size classes of plankton can be collected. Different size and categories of plankton were separated by subsequently filtering these whole water samples through netting of the appropriate mesh size. The final volume of water sample was collected 100 ml from the mesh for further study.

Algal samples were also collected small streams and canals surrounding the lake; the algal sample growth was abundant and visible on the surface of the rocks (Pandey and Kashyap 1995).

2.2.1 *Phytoplankton Preservation*

The following methods were used for the preservation of phytoplankton sample.

Lugol's solution: After collection, plankton sample preservation has done using the Lugol's solution. Add 0.3 ml Lugol's solution to 100 ml sample and stored in the dark place. For long time preservation, Lugol's solution add 0.7 ml per 100 ml of sample and buffered formaldehyde add 2.5 % final concentration after 1 h (Grace Analytical Lab 1994). (**Lugol's solution: Dissolve 20 g potassium iodide (KI) and 10 g iodine crystals in 200 ml distilled water containing 20 ml glacial acetic acid)

Formalin: After collection of the sample, 40 ml buffered formalin (20 g sodium borate, $Na_2B_2O_4$, +1 l 37 % formaldehyde) add to 1 l sample immediately (Manickam et al. 2012).

2.2.2 *Brief Procedure for Phytoplankton Study*

- Phytoplankton survey has been carried in two seasons summer (April to May) and rainy season (July to September).
- Four lakes were chosen to study the phytoplankton.
- The lakes are around 200 m length and 65 m breadth.
- The collections were done using phytoplankton net.
- Random sampling method has been applied in the plankton collection procedure. Samples were collected in 20 localities in each lake.
- After collection, the phytoplankton samples were preserved in 4 % of formalin (aqueous solution of formaldehyde).

- After sample collection, it has stored in dark place and later sent to the laboratory for identification.
- Samples are identified with images and documented according to protocol (BARC 2009). The samples were identified in Phycology lab at the Madras University, Gundy campus. Glycerin was used for mounting the material. The centric organism was photographed using a LABOMED microscope with attached SANYO ccd camera (Arulmurugan et al. 2011)

2.3 Animal Biodiversity

2.3.1 Fauna Survey and Study of Insects

Studying the insects is a very big task, and insect taxonomy requires continuous work in selected sampling plots. The numbers of insects are more than 90 % of various life-forms on planet earth; therefore, they are too diverse groups of the kingdom Animalia. And their structure also varies from one species to another species. Collection is the best practice methods for studying the insect taxonomic positions. Not only collection, but insect preservation is also very important process for future references and study. Standard methodologies (Triplehorn and Johnson 2005; Ragumoorthi et al. 2003) were followed for taxonomic study of the insects. The data collection method for fauna is given in Fig. 2.2.

2.3.1.1 Insects: When and Where to Collect

Insects appear in each and every ecosystem (terrestrial or aquatic ecosystems). Some insects are diurnal and some insects are nocturnal. Some insects are active in summer or some are active during winter, but most of the insect will go to hibernate during winter season. Some insect species prefer particular plant for their food purpose. It shows that vegetation also very important for insect collection. Insect may available season wise and habitat wise, but for getting various types of insects, the experiments are to be conducted throughout the year in each and every habitat. Photoplates show the netting method for collection of insects from the study region (Photoplate 2.5), collection and preservation of invertebrates (Photoplate 2.6), below ground diversity survey methods (Photoplate 2.7).

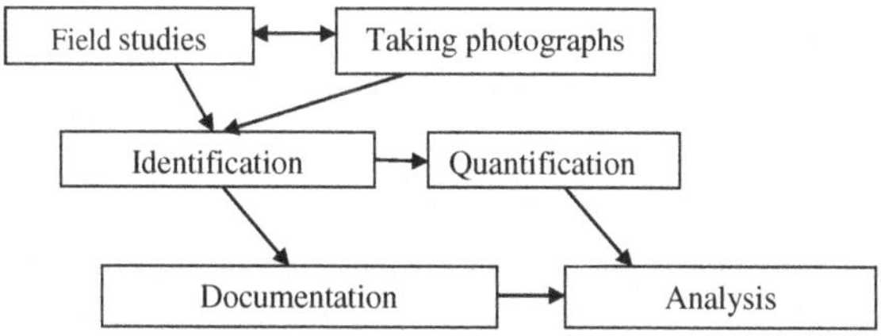

Fig. 2.2 Procedure for data collection in faunal studies

Photoplate 2.5 Netting method for collection of insects from the study region

2.3.1.2 Collecting Equipment

- Insect net
- Killing jars
- Forceps
- Vials
- Envelopes
- Preservatives
- Aspirator
- Beating umbrella/sheet
- Traps
- Knife
- Headlamp
- Box
- 10× lens

2.3.1.3 Collection Methods

Various methods have been used for insect collection. For collecting insects, the methods were (1) hand collection, (2) net collection, (3) traps, (4) aspirator, and (5) beating.

1. **Hand collection**: Large insect's (grasshoppers and beetles) collection can be done by hand. This collection method is unsuitable for poisonous and dangerous (which have stings) insects. In this method, insects were collected by hand and transferred into killing jars for further procedure.

Wet preservation

Collection of insects by using net

Insects preservation in Insect cabinet box

Samples preservation in envelopes

Samples collection near water

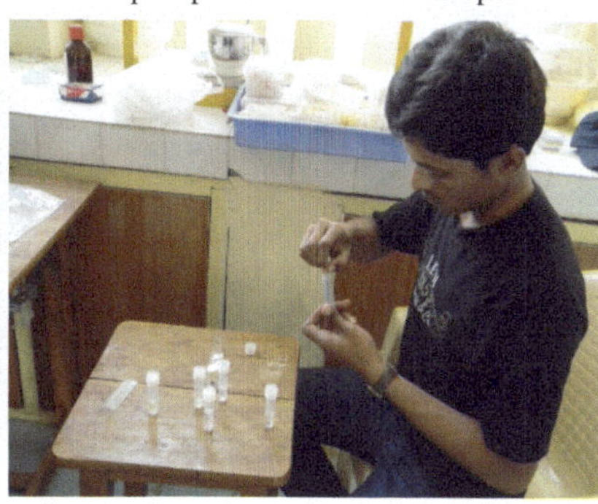

Segregation of samples for identification

Photoplate 2.6 Collection and preservation of invertebrates

2. **Net collection**: Two types of insect nets were used for collection of insects. They are (1) aerial net and (2) sweep net. All flying insects have been collected by using nets, while they are active (mid-morning/late afternoon).

   Aerial net: This net will be used for active fliers such as butterflies, moths, dragonflies, and wasps. This net is very light weight.
   Sweep net: This net is suitable for collecting grasshoppers and leafhoppers and other insects. This net is heavier than aerial net.

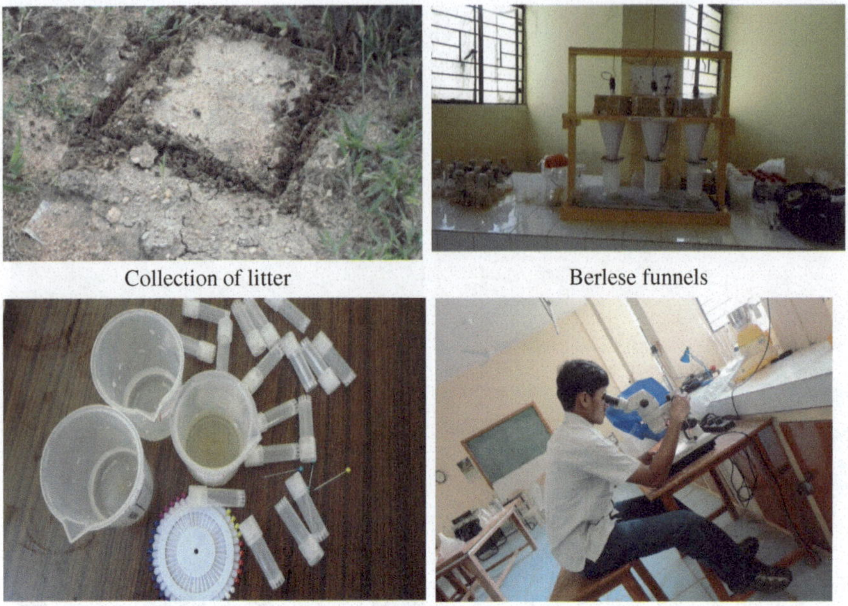

Collection of litter | Berlese funnels

Extraction of samples from funnels | Identification of samples

Photoplate 2.7 Below ground diversity survey methods

Photoplate 2.8 Fishes survey

3. **Traps**: Various types of traps were used for collection of insects. This is very effective and easy method for collecting insects (mainly small insects). Any device, often containing something to which the insects are attracted is a trap. A trap needs to be arranged in a manner so that once the insects get into it, cannot be escaped (Triplehorn and Johnson 2005). Following traps were used for insects collection.

   Bait trap—This method is very effective method for flies sampling. The egg yolk, fried coconut, and honey were used as bait.

Photoplate 2.9 Reptiles survey

Photoplate 2.10 Methods of plankton (phyto and zoo) collection

Light trap—This light trap is effectively useful for sampling nocturnal insects. The high-power florescent or mercury lamp arranged behind the white cloth for trapping the insects. For that light, nocturnal insects (moths

and some another flies and beetles) get attract and they will come to nearby light. Under the light, insects can be collected.
Note: This light trap is effectively works in late evening, and everyday timing should be maintained constantly.
Sticky trap—White flies
Pitfall trap—This is a kind of passive collection, unlke the active collection where the collector catches each animal from the sampling sites. By using this method terrestrial insects and spiders can be collected. However, this method is also useful for sampling herpetofauna (reptiles and amphibians) in addition to collection of invertebrates. For preparing pitfall trap to collect insects and spiders, the cylindrical plastic bottles (11 cm depth and 10 cm diameter) were used (Churchill and Arthur 1999). In this, 69 % water, 30 % ethyl acetate, and 1 % detergent were used as a preservative. Specimens were removed from traps after seven days for further laboratory processing identification (Hore and Uniyal 2008). This pitfall traps could be useful to assess species presence and relative abundance (Bury and Corn 1987) in the study sites.

4. **Berlese funnel**: The berlese funnels are used for collection of soil insect (soil-dwelling insects). This is very effective method for collecting insects from soil and leaf litter. In this method, soil and leaf litter collected from various parts and then transferred into berlese funnel. Funnel contains killing jar or alcohol container below of it. And electric light bulb is placed above the funnel is used to heat the upper part of the soil or litter which makes insects and other animals to move downward and fall into killing jar or alcohol container. After collection of insects, standard procedure was followed for preservation of the insects, spiders and other animals.
5. **Aspirator**: The aspirator was used for capturing small insects.
6. **Beating**: Beating umbrella was used as a sheet beneath the vegetation/bushes where insects/spiders have the dwellings. Upon beating the vegetation/bushes from top the insects and spiders were collected in the beating sheets/umbrella placed underneath. Thereafter, the insect are picked up by hand and transferred into killing jar.
Note: *An umbrella can be used in beating, but it should be made up of muslin or light white color cloth.*

2.3.1.4 Killing of Insects

After collection of insects, killing is one of the most important procedures for preservation of insects. It should be done immediately after capturing of insects. While killing insect, proper handling should be necessary (insect should not get damage and should not break their body parts). Various materials can be used for killing insects such as cyanide, ethyl acetate, and chloroform as a toxic agent in killing jars.

Killing Jars: This killing jar is the equipment where insect can be killed. For killing of insects, traditional killing jar was prepared by using various toxic agents. For preparing killing jar, cyanide and chloroform used as a toxic agent in various size bottles (for various insect sizes). The lepidopteron and other insects were kept in separate killing jars. The reason is that if kept in same containers or killing jars, their wings and other body parts may get damanged thus identification of species would be difficult.

2.3.1.5 Preservation

The preservation of specimens is very important for records in the laboratory. For preservation of specimens two methods namely dry preservation and wet preservation were followed.

(a) ***Dry preservation***: For dry preservation, specimens are preserved by using pins in insect cabinet box. In dry preservation, all hard-bodied insects were preserved by using paper envelopes and pinning of insects. The following materials were used for preservation of insects.

- Paper envelops: Paper envelopes were used for preserving the large winged insects such as butterflies, dragonflies, and moths.
- Spreading board: It is used for spreading the wings of dead insects.
- Pins: For pinning the nickel-plated (rust resistant) pins were used, which are specially prepared for preserving the insects.

Pinning: Pinning is a most common and suitable method for preserving hard-bodied insects. Pinning should be done in a proper way (to identify diagnostic characters clearly). Based on the size of the insect, pins are selected (small size pins for smaller insects and large size pins for lager insects). Place of pinning varies from insects to insects. The pinning region in various groups of insects is depicted in Table 2.1. Once pinning procedure is over, the insects have to be shifted to insect cabinet boxes for long-term preservation.

Table 2.1 Showing pinning position for various insects group

S. no.	Insect group	Pinning region
1	Grasshoppers, crickets, praying mantids, and cockroaches	Pronotum
2	Bugs	Scutellum
3	Stick insects	Metanotum
4	Beetles and weevils	Right elytron
5	Earwigs	Right tegmen
6	Dragonfly, damselfly, green lacewings, moths, butterflies, bees, wasps ants, and true flies	Thorax

Source Ragumoorthi et al. (2003)

(b) ***Wet preservation or liquid preservation***: Soft-bodied insects are nymphs, larvae, caddisflies, and mayflies. Mainly, ethyl alcohol (70–80 %) was used for preservation of soft-bodied specimens. Various solutions such as Hood's solution, Kahle's solution, and alcoholic Bouin's solution can be used for preservation. For this study Hood's solution was used for preservation of insects.

Hood's solution: 95 ml 70–80 % ethyl alcohol + 5 ml Glycerin.

2.3.2 Spider Collection Methods

Hand collection, pitfall trap, beating, and berlese funnels as mentioned above have been used for sampling and collection of spiders in various habitats. After collection, the spiders were placed in Hood's solution for identification and preservation.

2.3.3 Study of Vertebrates

For sampling and monitoring the vertebrates from the study area, the standard methodologies were followed which are given in the "Handbook of Biodiversity methods Survey, Evolution and Monitoring" (Hill et al. 2005) and "Practical Methods in Ecology" (Henderson 2003). The vertebrate species were not preserved, however, along with visual observations, and vocal sounds good photographs were taken for identifying vertebrate species. The standard field guides were referred for proper identification of the species. (Daniel 2002; Daniels 2002; Prater 2005; Manakadan et al. 2012). Survey methods for fishes and reptiles are depicted in photoplates 2.8 and 2.9.

2.3.3.1 Survey and Monitoring of Fishes

Fishes were collected from the water body using locally available fishing gears from different sampling stations. Fishes collected during premonsoon, monsoon, and post-monsoon seasons. The direct count method (visual survey) and netting were used for surveying the fish species.

2.3.3.2 Survey and Monitoring of Amphibians

Survey and sampling was done in both aquatic and terrestrial systems. Sampling required day and night search during all three seasons in specified habitats (under the logs and stones, digging through litter and soil, searching short bushes and tree hollows, and under fallen barks and water-catchments). For surveying and monitoring amphibians, torch count method (aquatic frogs) and pitfall trap methods (terrestrial frogs) were followed. By using vocal sound and photographs, the amphibian species were identified. Continuous surveys are required to obtain reasonably good estimates of amphibian species diversity and abundance in study sites.

2.3.3.3 Survey and Monitoring of Reptiles

Several survey techniques such as standard walk transect, visual encounter survey and pitfall trap methods were used to sampling reptiles in each and every habitat of the study area. While doing this survey, photographs were taken for identification of species. Species identification was done by using standard field guides in consultation with experts.

Visual Encounter Survey method (VES): This method is useful for studying the species richness and abundance in a survey path (Crump and Scott 1994). VES are standard method for inventory of terrestrial herpatofauna (Campbell and Christman 1982; Corn and Bury 1990).

2.3.3.4 Survey and Monitoring of Birds

Birds are sampled by using line transect method, point count method, and opportunistic bird sightings. By using bird vocal sounds and photographs, the species were identified in consultation with ornithologists.

Line transects: In this method, a straight line of 1 km is drawn, and all birds seen or heard till a range of 25 m on either side of the transect were recorded. The transect was worked for one hour.

Point counts: In this method, the observer will stand in a randomly chosen point and birds seen or heard in 50 m radius are recorded for 5 min. This observation is repeated in another point at least 300 m from the first point.

Opportunistic bird sightings: While traveling in study area, many bird species will be detected in survey time. Such species are recorded by their appearance or by their call.

2.3.3.5 Survey and Monitoring of Mammals

Intensive survey has been done by transect method (walking and in vehicle) for all major habitats for surveying of mammals by direct and indirect evidence. Direct observation technique has been used for surveying large and medium sized mammals. But this technique is perfectly suitable for surveying of diurnal mammals. But, for nocturnal mammals, camera traps were used and other evidences (pellets, hair, foot prints, and vocal sounds) also collected for surveying mammals (Martin 2009). Bait traps and visual encounters have been used for small mammals (rodents and squirrels). However, good photographs were also taken for species identification.

2.3.3.6 Identification

The field guides (as cited above) and standard keys were used for species (invertebrates and vertebrates) identification. The experts' laboratories (The Department of Entomology at the University of Agricultural Sciences, Bangalore, National Bureau of Agricultural Insect Resources (NBAIR) Bangalore, regional office of the Zoological Survey of India (ZSI), Pune were visited for the purpose of proper identification of insects. However, frequent visits were made to Anand Agricultural University (AAU) located in Anand, Gujarat for the identification of spiders. The survey methods for various groups of animals are given in Table 2.2.

Table 2.2 Various survey methods used for various groups of animals

Group of animals	Methods used for survey	
Insects	Hand collection	
	Net collection	Aerial net, sweep net
	– Aerial net	
	– Sweep net	
	Traps	Bait trap, light trap, sticky trap, pitfall trap
	– Bait trap	
	– Light trap, sticky trap	
	– Pitfall trap	
	Berlese funnel	
	Beating	
	Aspirator	
Spiders	Pitfall trap	
	Beating	
	Berlese funnel	
Fishes	Direct count method (visual survey)	By using locally available fishing gears
	Netting	
Amphibians	Visual encounter survey	
	Torch count method	
	Pitfall trap	
Reptiles	Visual encounter survey	
	Pitfall trap	
Birds	Line transect method	
	Point count method	
	Opportunistic sightings	
Mammals	Line transect method	All mammals
	Visual encounter survey	Large mammals
	Pellet count method	Large mammals
	Bait traps	Small mammalian groups

2.4 Zooplankton Survey and Monitoring

Zooplankton are micro organisms present in water bodies and play an important role in aquatic ecosystems in energy transfer from primary level to higher level (Tiwari 2011). The role of zooplankton is significant in assessing the water quality and they act as biological indicators of water pollution (Shivashankar and Venkataramana 2013; Gayathri et al. 2014). For collecting zooplankton (20–200 μm) such as protozoa, rotifers, and immature micro-crustacean, the methodology used for phytoplankton collection was followed. The zooplankton usually are sufficiently abundant to yield adequate samples in 5–10 l bottles as described in several research papers. However, during time of sample collection, the experts recommend composite samples over depth and time and bottle samplers are suitable especially for discrete depth samples. If depth-integrated samples are desired, the nets are used. The larger and more robust micro-zooplankters (e.g., loricate forms and crustacea) may be concentrated by passing the water through a 64-μm mesh net. Various methods regarding zooplankton study are described by Pennak (1978), Battish (1992), APHA (1998) and Altaff (2003). The detailed methodology is given in the protocol developed for flora fauna study (BARC 2009). Methods for plankton (phyto and zoo) collection depicted in photoplate 2.10.

References

Allen G, Berg G, Costanzo B, Douglas G, Egan B, Goward T, Kerik J, Lee O, Meidinger D, Nicholson A, Pinder-Moss J, Ruyle G, Ryan M, Warrington P (1996) Techniques and procedures for collecting, preserving, processing and storing botanical specimens. Working paper no 18, Province of British Columbia

Altaff K (2003) A manual of zooplankton. Department of Zoology, The New College, Chennai

Anderson LC (1999) Collecting and preparing plant specimens and producing an herbarium. In: Karcher SJ (ed) Tested studies for laboratory teaching. Proceedings of the 20th workshop/conference of the association for biology laboratory education (ABLE) vol 20, pp 295–300, 399

APHA (1998) Standard methods for the examination of water and waste water, 20th edn. American Public Health Association, Washington, DC

Arulmurugan P, Nagaraj S, Anand N (2011) Biodiversity of fresh water algae from Guindy campus of Chennai, India. J Ecobiotech 3:19–29

BARC (2009). Protocols for surveying flora and fauna. Board of Research in Nuclear Sciences Department of Atomic Energy (unpublished)

Battish SK (1992) Freshwater zooplankton of India. Oxford and IBM Publications

Bury RB, Corn PS (1987) Evaluation of pitfall trapping in northwestern forests: trap arrays with drift fences. J Wildlife Manage 51:112–119

Campbell HW, Christman SP (1982) Field techniques for Herpetofaunal community analysis In: Scott NJ Jr (ed) Herpetological communities, U.S.D.I. fish and wildlife service. Wildlife research report no 13, Washington, D.C., p 239

Churchill TB, Arthur J (1999) Measuring spider richness. Effects of different sampling methods and spatial and temporal scales. J Insect Conserv 3:287–295

Corn PS, Bury RB (1990) Sampling methods for terrestrial amphibians and reptiles. USDA Forest Service, General technical report PNW-GTR-256, 34 pp

Cottam G, Curtis JT (1956) The use of distance measures in phytosociological sampling. Ecology 37(3):451–460
Crump ML, Scott NJ Jr (1994) Visual encounter surveys. In: Heyer WR, Donnelly MA, Mcdiarmid RW, Hayek LAC, Foster MS (eds) Measuring and monitoring biological diversity-standard methods for amphibians. Smithsonian Institution Press, Washington, DC, pp 84–92
Curtis JT (1959) The vegetation of wisconsin; an ordination of plant communities. University of Wisconsin Press, Madison
Daniel JC (2002) The book of Indian reptiles and amphibians. Bombay natural history society. Oxford University Press, Oxford
Daniels RJR (2002) Fresh water fishes of peninsular India. University Press, Hyderabad
Gayathri S, Latha N, Mohan MR (2014) Studies on population dynamics and seasonal abundance of zooplankton community in Doddavoderahallilake, Bangalore. Int J Emerg Trends Eng Dev Issue 4(1):50–55
Grace Analytical Lab (1994) Standard operating procedure for phytoplankton sample collection and preservation", Chicago, IL 60605 http://www.epa.gov/glnpo/lmmb/methods/phy.pdf
Henderson PA (2003) Practical methods in ecology. Wiley, UK
Hill D, Fasham M, Tucker G, Shewry M, Shaw P (2005) Handbook of biodiversity methods, survey, evaluation and monitoring. Cambridge University Press, Cambridge
Hore U, Uniyal VP (2008) Diversity and composition of spider assemblages in five vegetation types of the Terai conservation area, India. J Arachnol 36:251–258
Lu D, Mausel P, Brondizio E, Moran E (2004) Relationship between forest stand parameters and landsat TM spectral responses in the Brazilian amazon basin. For Ecol Manage 198:149–167
Maden K (2004) Plant collection and herbarium techniques. Our Nature 53–57
Manakadan R, Daniel JC, Bhopale N, Dick JH (2012) Birds of the Indian subcontinent a field guide. Bombay natural history society. Oxford University Press, Oxford
Manickam N, Bhavan PS, Vijayan P, Sumathi G (2012) Phytoplankton species diversity in the parambikulam-aliyar irrigational canals (Tamil Nadu, India). Int J Pharma and Bio Sci 3:289–300
Martin CO (2009) Mammalian survey techniques for level II natural resource inventories on corps of engineers projects (Part I). ERDC TN—EMRRP (Ecosystem Management and Restoration Research Programme)-SI-34, July 2009
Nautiyal S (2008) Structure of Central Himalayan forests under different management regimes: an empirical study. Working paper No 206. Institute for Social and Economic Change, Bangalore
Nautiyal S, Kaechele H (2008) Fuel switching from wood to LPG can benefit the environment! Environ Impact Assess Rev 28:523–532
Nayak SNV, Swamy HR, Nagaraj BC, Rao U, Chandrashekara UM (2000) Farmers' attitude towards sustainable management of Soppina Betta forests in Sringeri area of the Western Ghats, South India. Int J For Ecol Manage 132(2):223–241
Pandey KD, Kashyap AK (1995) Diversity of algal flora in six fresh water streams of Schirmacher oasis, Antarctica. Department of Ocean Development, Technical Publication, pp 219–229
Pennak RW (1978) Fresh water invertibrates of the United States, 2nd edn. Wiley, New York
Philips EA (1959) Methods of vegetation study. Henry Holt and Company, New York
Prater SH (2005) The book of Indian animals. Bombay natural history society. Oxford University Press, Oxford
Ragumoorthi KN, Balasubramani V, Srinivasan MR, Natarajan N (2003) Insecta-an introduction. AE Publishers, Coimbatore
Ralhan PK, Saxena AK, Singh JS (1982) Analysis of forest vegetation at and around Nainital in Kumaun Himalaya. Proc Indian Nat Sci Acad B 48:121–137
Saxena AK, Singh JS (1982) A phytosociological analysis of woody species in forest communities of a part of Kumaun Himalaya. Vegeratio 50(1):3–32
Shivashankar P, Venkataramana GV (2013) Zooplankton diversity and their seasonal variations of Bhadra Reservoir, Karnataka, India. Int Res J Environ Sci 2(5):87–91
Shannon CE, Wiener W (1963) The mathematical theory of communication. University of Juionis Press, Urbana. 117

Smith CE (1971) Preparing herbarium specimens of vascular plants. Agriculture information bulletin no. 348, U.S. Government Printing Office

Suseela MR (2009) Conservation and diversity of fresh water algae. In: Anand N (ed) Biology and biodiversity of microalgae. Centre for advanced studies in botany. University of Madras, Chennai, India, pp 41

Tiwari RP, Tripathi K, Sahu R (2011) Studies on monthly population of total zooplanktons and their correlation coefficient with physicochemical factors of Lony dam Theonthar, Rewa (M.P). Indian J Biol Stud Res 1(1):33–38

Triplehorn A, Johnson NF (2005) Borrar and delong's introduction to the study of insects, 7th edn. Brroks/cole, a division of Thomson learning, Inc.

Whittaker RH (1975) Community and ecosystem, 2nd edn. Macmillan Publishing Company, New York 385

Wondafrash M (2008) A preliminary guide to plant collection, identification and herbarium techniques. The National Herbarium AAU, Rustenburg

Chapter 3
Plant Biodiversity

3.1 Phytosociological Study

The phytosociological study was conducted in the study area in all the seasons. The vegetation is distributed in the forests, scrub jungles, agricultural fields, barren lands, streams, and roadsides of the study area. A total of 403 plant species were identified in the study region including angiosperms, bryophytes, pteridophytes, and phytoplankton. The detailed inventory scientific/local names, family, and uses in general are given in Table 3.1. The numbers recorded were angiosperms (376), gymnosperms (1), pteridophytes (4), bryophytes (1), and phytoplankton (21) all belonging to 104 families (Tables 3.1, 3.2 and Fig. 3.1). The various uses of these plant species have also been tabulated emphasizing the traditional knowledge system of the people to understand the human–plants relationship. Figure 3.2 depicts the dominant families in the study region belonging to Fabaceae (24 species), Euphorbiaceae (22), Astaraceae (20), Mimosaceae (18), Poaceae (16), Caesalpiniaceae (16), Convolvulaceae (15), Asclepiadaceae, Malvaceae, and Amaranthaceae (13 species each), Acanthaceae (12), Rubiaceae and Verbenaceae (10 species each). The family Fabaceae is found to be dominating the studied semiarid region. In this section, various attributes of the vegetation strata are given that include frequency, density, dominance, importance value index, and other diversity parameters. However, the details about phytoplankton are given in the separate section.

The detailed analysis based on phytosociological study for tree is depicted in Table 3.3. The top-story vegetation in the semiarid region is dominated by *Prosopis juliflora* (176 individual per ha) followed by *Annona squamosa* (65 individuals per ha), *Leucaena leucocephala* (53 individuals per ha), *Azadirachta indica* (30 individual per ha), and *Acacia nilotica* subsp. *indica* (25 individuals per ha). The least individuals per ha were recorded for 41 species where density was below 1

The erratum of this chapter can be found under DOI 10.1007/978-3-319-15464-0_8

S. Nautiyal et al., *Biodiversity of Semiarid Landscape*,
Environmental Science, DOI 10.1007/978-3-319-15464-0_3

Table 3.1 Floral diversity of angiosperms of semiarid region

S. no.	Scientific name	Vernacular name	Family	Uses
1	*Adhatoda zeylanica* Medic	Adusoge	Acanthaceae	Fd, Mis
2	*Andrographis paniculata* (Burm. f.) Wall. ex Nees	Nelabevu	Acanthaceae	Nu*
3	*Asteracantha longifolia* (L.) Nees	Kalavankabija	Acanthaceae	Nu*
4	*Barleria prionitis* L.	Maragadegida	Acanthaceae	Mis
5	*Crossandra infundibuliformis* (L.) Nees	Kanakambara gida	Acanthaceae	Mis
6	*Dipteracanthus patulus* (Jacq.) Nees	VNA	Acanthaceae	Fd
7	*Justicia* sp.	VNA	Acanthaceae	Fd, M
8	*Lepidagathis cristata* Willd.	Narigoodi	Acanthaceae	Nu*
9	*Peristrophe paniculata* (Forsk.) Brummitt	VNA	Acanthaceae	Nu*
10	*Rostellularia simplex* Wight	VNA	Acanthaceae	Fd
11	*Ruellia* sp.	VNA	Acanthaceae	Fu
12	*Rungia repens* (L.) Nees	VNA	Acanthaceae	Fd
13	*Agave americana* L.	Rakspatti	Agavaceae	Mis
14	*Mollugo lotoides* C.B. Clarke	VNA	Aizoaceae	
15	*Trianthema portulacastrum* L.	Balevadukana soppu	Aizoaceae	Ed, Fd
16	*Alangium salvifolium* (L. f.) Wangerin	Ankaligida	Alangiaceae	Fu
17	*Aloe vera* (L.) N. Burm.	Kathaligidi	Aloaceae	M
18	*Achyranthes aspera* L.	Uttarani	Amaranthaceae	M
19	*Aerva lanata* (L.) Juss. ex Schult.	Bilihindi soppu	Amaranthaceae	M
20	*Alternanthera pungens* Humb., Bonpal. & Kunth	Mullu honagonne	Amaranthaceae	Ed
21	*Alternanthera sessilis* (L.) R. Br	Honagone soppu	Amaranthaceae	Ed
22	*Alternanthera tenella* Colla		Amaranthaceae	Ed
23	*Amaranthus spinosus* L.	Mulludantu	Amaranthaceae	Ed
24	*Amaranthus viridis* L.	Chelakeerae soppu	Amaranthaceae	Ed
25	*Celosia argentea* L.	Anne soppu	Amaranthaceae	
26	*Digera muricata* (L.) Mart.		Amaranthaceae	Ed
27	*Gomphrena celosioides* Mart.	Nelarudrakshi	Amaranthaceae	M, Fd
28	*Psilotrichum* sp.		Amaranthaceae	Fd
29	*Pupalia lappacea* (L.) Juss.	Haridhachhaga	Amaranthaceae	Ed, Fd
30	*Mangifera indica* L.	Mavina mara	Anacardiaceae	Ed, M, Fu, Mis

(continued)

Table 3.1 (continued)

S. no.	Scientific name	Vernacular name	Family	Uses
31	*Annona squamosa* L.	Sithaphal	Annonaceae	Ed, Fu, M
32	*Carissa spinarum* L.	Kaulekayi	Apocynaceae	Fu
33	*Nerium indicum* Mill.	Kangile	Apocynaceae	Mis
34	*Plumeria rubra* L.		Apocynaceae	Nu*
35	*Thevetia neriifolia* Juss. ex Steud.	Haladi kanagile	Apocynaceae	Fu
36	*Vinca rosea* (L.) G. Don		Apocynaceae	M, Mis
37	*Wrightia tinctoria* R. Br.	Kodu muriki	Apocynaceae	M, Fu
38	*Lemna polyrhiza* L.	VNA	Araceae	
39	*Borassus flabellifer* L.	Thatling	Arecaceae	Ed, Fu, Mis
40	*Cocos nucifera* L.	Tengina mara	Arecaceae	Ed, M, Mis
41	*Phoenix sylvestris* (L.) Roxb.	Eachalu	Arecaceae	Ed, Fu, Mis
42	*Aristolochia bracteolata* Lam.	Kuri gida	Aristolochiaceae	M
43	*Aristolochia indica* L.	Eshwari balli	Aristolochiaceae	M
44	*Calotropis gigantea* (L.) R. Br.	Yekkada gida	Asclepiadaceae	M
45	*Calotropis procera* R. Br.	Bili yekkada gida	Asclepiadaceae	M
46	*Caralluma attenuata* Wight	VNA	Asclepiadaceae	M
47	*Cryptolepis buchanani* Roem. and Schult.	VNA	Asclepiadaceae	Nu*
48	*Cryptostegia grandiflora* R. Br.	Rubber hombu	Asclepiadaceae	Nu*
49	*Gymnema sylvestre* (Retz.) R. Br. ex Schult	Kodapathre	Asclepiadaceae	M
50	*Hemidesmus indicus* (L.) Schult	Suganhaberu	Asclepiadaceae	M
51	*Oxystelma esculentum* (L. f.) R. Br. ex Schult.	VNA	Asclepiadaceae	Nu*
52	*Pentatropis capensis* (L. f.) Bullock	VNA	Asclepiadaceae	Nu*
53	*Pergularia daemia* (Forssk.) Chiov.	Kurudigana balli	Asclepiadaceae	Nu*
54	*Telosma pallida* (Roxb.) Craib	VNA	Asclepiadaceae	Nu*
55	*Tylophora indica* (Burm.) Merr.	Adumuttada balli	Asclepiadaceae	M
56	*Wattakaka volubilis* (L. f.) Stapf	VNA	Asclepiadaceae	M
57	*Ageratum conyzoides* L.	Muguthigida	Asteraceae	Fd
58	*Blainvillea acmella* (L.) Philipson.	VNA	Asteraceae	Fd

(continued)

Table 3.1 (continued)

S. no.	Scientific name	Vernacular name	Family	Uses
59	*Blumea eriantha* DC.	VNA	Asteraceae	Fd
60	*Dicoma tomentosa* Cass.	Sannigida	Asteraceae	Nu*
61	*Echinops echinatus* Roxb.	Brahmadande	Asteraceae	Nu*
62	*Eclipta alba* (L.) Hassk.	Bhrungaraja	Asteraceae	M
63	*Erigeron alpinus* L.	VNA	Asteraceae	Fd
64	*Flaveria trinervia* (Spreng.) Mohr.	VNA	Asteraceae	Fd
65	*Grangea maderaspatana* (L.) Poir.	VNA	Asteraceae	Fd
66	*Lagascea mollis* Cav.	VNA	Asteraceae	Fd
67	*Launaea pinnatifida* Cass.	VNA	Asteraceae	Fd
68	*Parthenium hysterophorus* L.	Congress gida	Asteraceae	Nu*
69	*Pentanema indicum* (L.) Ling	VNA	Asteraceae	Nu*
70	*Pulicaria* sp.	VNA	Asteraceae	Fd
71	*Pulicaria wightiana* (DC.) C.B. Clarke	VNA	Asteraceae	Fd
72	*Sonchus oleraceus* L.	VNA	Asteraceae	Fd
73	*Sphaeranthus amaranthoides* Burm. f.	VNA	Asteraceae	Fd
74	*Spilanthes calva* DC.	VNA	Asteraceae	Fd
75	*Tridax procumbens* L.	Kari balli	Asteraceae	M, Fd
76	*Xanthium indicum* DC.	VNA	Asteraceae	Mis
77	*Balanites aegyptiaca* (L.) Delile	Ingalarade	Balanitaceae	M, Fu
78	*Basella alba* L.	Baayi basale	Basellaceae	Mis
79	*Dolichandrone atrovirens* (Heyne) Sprague	VNA	Bignoniaceae	Fu
80	*Dolichandrone falcata* Seem	Oodi mara	Bignoniaceae	Fu, Fd, Mis
81	*Markhamia lutea* (Benth.) K. Schum.	VNA	Bignoniaceae	Fu, Mis
82	*Millingtonia hortensis* L.	Akasha mallige	Bignoniaceae	Fu, Fd, Mis
83	*Spathodea campanulata* Beauv.	Neeru kai	Bignoniaceae	T, Fu
84	*Tecoma stans* (L.) Kunth	VNA	Bignoniaceae	Fu
85	*Cordia dichotoma* G. Forst.	Bankegida	Boraginaceae	Fu, Fd
86	*Cordia wallichii* G. Don	Solle mara	Boraginaceae	Fu, Fd
87	*Coldenia procumbens* L.	Thripakshi gida	Boraginaceae	Fd
88	*Heliotropium indicum* L.	Chelubalada gida	Boraginaceae	Mis
89	*Heliotropium marifolium* Koen. ex Retz.	VNA	Boraginaceae	

(continued)

Table 3.1 (continued)

S. no.	Scientific name	Vernacular name	Family	Uses
90	*Trichodesma indicum* (L.) Lehm.	VNA	Boraginaceae	Fd
91	*Opuntia dillenii* (Ker-Gawl.) Haw.	Papas kalli	Cactaceae	Ed, Mis
92	*Bauhinia racemosa* Lam.	Basavana pada	Caesalpiniaceae	T, Fu
93	*Caesalpinia pulcherrima* L.	Kenjigemara	Caesalpiniaceae	M, Fu, Fd
94	*Cassia fistula* L.	Kakke gida	Caesalpiniaceae	M, Fu, Fd
95	*Cassia javanica* L.	VNA	Caesalpiniaceae	Fu, M, Fd
96	*Cassia leschenaultii* Wall.	VNA	Caesalpiniaceae	Fd
97	*Cassia* sp.	VNA	Caesalpiniaceae	Nu*
98	*Delonix elata* (L.) Gamble	Bile gulmohr	Caesalpiniaceae	M, Fu, Fd
99	*Delonix regia* (Boj. ex Hook.) Raf.	Kempu gulmohr	Caesalpiniaceae	Fu, Fd, Mis
100	*Guilandina bonduc* L.	Gajjaga	Caesalpinaceae	M, Fu
101	*Parkinsonia aculeata* L.		Caesalpinaceae	Fu
102	*Senna auriculata* L.	Honnambre gida	Caesalpinaceae	M, Fu, Fd
103	*Senna italica* (Mill.) Lam.	Neladavare	Caesalpinaceae	M
104	*Senna occidentalis* L.	Tharavadi	Caesalpinaceae	M, Fd
105	*Senna sericea* Sw.	VNA	Caesalpinaceae	M, Fd
106	*Senna siamea* (Lam.) Irwin & Barneby	VNA	Caesalpinaceae	Fd, Fu, M
107	*Tamarindus indica* L.	Hunase mara	Caesalpinaceae	Ed, T, M, Fu
108	*Cadaba fruticosa* (L.) Druce	VNA	Capparaceae	Fu, Fd
109	*Capparis brevispina* DC.	VNA	Capparaceae	M
110	*Capparis divaricata* Lam.	Bhandero	Capparaceae	M, Fu
111	*Capparis zeylanica* L.	Tottilu balli	Capparaceae	Fu
112	*Cleome viscosa* L.	Kaadu saasive	Capparaceae	Fd
113	*Maerua oblongifolia* (Forssk.) A. Rich.	VNA	Capparaceae	Nu*
114	*Carica papaya* L.	Parangi	Caricaceae	Ed
115	*Polycarpaea aurea* Wight & Arn.	VNA	Caryophyllaceae	Nu*
116	*Polycarpaea corymbosa* (L.) Lam.	VNA	Caryophyllaceae	Fd
117	*Casuarina equisetifolia* L.	Survey mara	Casuarinaceae	T, Fu
118	*Maytenus emarginata* (Willd.) Ding Hou	VNA	Celastraceae	Fu, Mis
119	*Chenopodium album* L.	Hunachikka	Chenopodiaceae	Ed
120	*Combretum ovalifolium* Roxb.	Edatiga	Combretaceae	Nu*
121	*Terminalia bellirica* (Gaertn.) Roxb.	Taremara	Combretaceae	M, Fu

(continued)

Table 3.1 (continued)

S. no.	Scientific name	Vernacular name	Family	Uses
122	*Terminalia catappa* L.	Kadu bondami	Combretaceae	M, Fu
123	*Amischophacelus axillaris* (L.) R. S. Rao & Kammathy	Jigale	Commelinaceae	Fd
124	*Commelina benghalensis* L.	Paradene	Commelinaceae	Fd
125	*Commelina diffusa* Burm. f.	VNA	Commelinaceae	Fd
126	*Cyanotis* sp.	VNA	Commelinaceae	Fd
127	*Cyanotis tuberosa* (Roxb.) Schult. & Schult. f.	VNA	Commelinaceae	Fd
128	*Cressa cretica* L.	VNA	Convolvulaceae	Fd
129	*Evolvulus alsinoides* L.	Vishnukanthi	Convolvulaceae	Fd
130	*Ipomoea aquatica* Forssk.	VNA	Convolvulaceae	Fd
131	*Ipomoea cairica* (L.) Sweet.	Bekkina hejje balli	Convolvulaceae	
132	*Ipomoea carnea* Jacq.	Ganesh kaddi	Convolvulaceae	Fd, Fu, Mis
133	*Ipomoea hederifolia* L.	Nakshatra hoovu	Convolvulaceae	Nu*
134	*Ipomoea mauritiana* Jacq.	Nelagumbala	Convolvulaceae	Nu*
135	*Ipomoea obscura* (L.) Ker-Gawl.	VNA	Convolvulaceae	Nu*
136	*Ipomoea* sp.	VNA	Convolvulaceae	Nu*
137	*Ipomoea tricolor* Cav.	VNA	Convolvulaceae	Nu*
138	*Ipomoea quomoclit* L.	VNA	Convolvulaceae	Fd
139	*Jacquemontia pentantha* (Jacq.) G. Don	VNA	Convolvulaceae	Nu*
140	*Merremia emarginata* (Burm. f.) Hallier f.	VNA	Convolvulaceae	Nu*
141	*Merremia tridentata* (L.) Hall. f.	VNA	Convolvulaceae	Fd
142	*Rivea hypocrateriformis* (Desr.) Choisy	VNA	Convolvulaceae	Nu*
143	*Coccinia grandis* (L.) Voigt	Tonde balli	Cucurbitaceae	M, Ed
144	*Corallocarpus epigaeus* (Rottler) C.B. Clarke	VNA	Cucurbitaceae	Nu*
145	*Ctenolepis cerasiformis* (Stocks) C.B Clarke	VNA	Cucurbitaceae	Fu
146	*Cucumis callosus* (Rottler) Cogn.	Mekkekayi balli	Cucurbitaceae	Mis
147	*Diplocyclos palmatus* (L.) Jeffrey	Linge toned balli	Cucurbitaceae	Nu*
148	*Mukia maderaspatana* (L.) Roem.	Gubbi savatikayi	Cucurbitaceae	Nu*
149	*Bulbostylis densa* (Wall. ex Roxb.) Hand.-Mazz.	VNA	Cyperaceae	Fd

(continued)

Table 3.1 (continued)

S. no.	Scientific name	Vernacular name	Family	Uses
150	*Bolboschoenus maritimus* (L.) Palla	Neeru sabbasige	Cyperaceae	Fd, Mis
151	*Cyperus compressus* L.	Vasumani hullu	Cyperaceae	Fd
152	*Cyperus pangorei* Rottb.	VNA	Cyperaceae	Fd
153	*Cyperus alopecuroides* Rottb.	VNA	Cyperaceae	Fd
154	*Cyperus rotundus* L.	Tunge hullu	Cyperaceae	Fd
155	*Pycreus flavidus* (Retz.) T. Koyama	VNA	Cyperaceae	Fd
156	*Cyperus squarrosus* L.	VNA	Cyperaceae	Fd
157	*Cyperus* sp.	VNA	Cyperaceae	Fd
158	*Bergia capensis* L.	VNA	Elatinaceae	Fd
159	*Eriocaulon quinquangular* L.	VNA	Eriocaulaceae	Nu*
160	*Acalypha indica* L.	Kuppe gida	Euphorbiaceae	M
161	*Chrozophora plicata* (Vahl) A. Juss. ex Spreng.	Suttu mare	Euphorbiaceae	Fd
162	*Croton bonplandianum* Baill.	Utti gida	Euphorbiaceae	M
163	*Euphorbia heterophylla* L.	Halalabu	Euphorbiaceae	M, Fd
164	*Euphorbia hirta* L.	Achche gida	Euphorbiaceae	M, Fd
165	*Euphorbia hypericifolia* L.	VNA	Euphorbiaceae	Fd
166	*Euphorbia indica* Lam.	VNA	Euphorbiaceae	Fd
167	*Euphorbia* sp.	VNA	Euphorbiaceae	Mis
168	*Euphorbia tirucalli* L.	Kol kalli gida	Euphorbiaceae	Fu, Mis
169	*Jatropha glandulifera* Roxb.	Totla gida	Euphorbiaceae	M
170	*Kirganelia reticulata* (Poir.) Baill.	Huli gida	Euphorbiaceae	M, Ed, Fu
171	*Phyllanthus amarus* Schum. & Thonn.	Bhoo nelli	Euphorbiaceae	M
172	*Phyllanthus emblica* L.	Nelli kaayi	Euphorbiaceae	Ed, M
173	*Phyllanthus kozhikodianus* Siver. & Mani.	VNA	Euphorbiaceae	M
174	*Phyllanthus maderaspatensis* L.	Madras nalli	Euphorbiaceae	M
175	*Phyllanthus polyphyllus* Willd.		Euphorbiaceae	Fu
176	*Phyllanthus reticulatus* Poir.	Chippullimullu	Euphorbiaceae	M, Fu, Fd
177	*Phyllanthus virgatus* G. Forst.		Euphorbiaceae	M
178	*Ricinus communis* L.	Oudala gida	Euphorbiaceae	M, Fu, Fd
179	*Securinega leucopyrus* (Willd.) Muell.-Arg.	Huli kantae	Euphorbiaceae	Fu
180	*Securinega virosa* (Willd.) Baill.	VNA	Euphorbiaceae	Fu, Mis
181	*Tragia involucrata* L.	VNA	Euphorbiaceae	Nu*

(continued)

Table 3.1 (continued)

S. no.	Scientific name	Vernacular name	Family	Uses
182	*Abrus precatorius* L.	Gulaganji	Fabaceae	M
183	*Alysicarpus vaginalis* (L.) DC.	VNA	Fabaceae	Fd
184	*Canavalia virosa* (Roxb.) Wight &. Arn.	VNA	Fabaceae	
185	*Clitoria ternatea* L.	Shankapushpi gida	Fabaceae	
186	*Crotalaria juncea* L.	Senabu	Fabaceae	Fd, Fu
187	*Crotalaria pallida* Aiton	VNA	Fabaceae	Fd
188	*Crotalaria pusilla* Heyne ex Roth	VNA	Fabaceae	
189	*Crotalaria ramosissima* Roxb.	VNA	Fabaceae	Fd
190	*Crotalaria retusa* L.	VNA	Fabaceae	Fd
191	*Dalbergia sissoo* Roxb.	Sissu	Fabaceae	Fu
192	*Desmodium triflorum* (L.) DC.	Kaadu menthe	Fabaceae	Fd
193	*Erythrina variegata* L.	Mandhara pushpada gida	Fabaceae	Fu, Fd
194	*Indigofera cordifolia* Heyne ex Roth	VNA	Fabaceae	Fd
195	*Indigofera linnaei* Ali	VNA	Fabaceae	Fd
196	*Indigofera tinctoria* L.	Neeligida	Fabaceae	Fd
197	*Pongamia pinnata* (L.) Pierre.	Honge mara	Fabaceae	M, T, Fu
198	*Cullen corylifolium* (L.) Medik Syn *Psoralea corylifolia* L.	Bavanchi gida	Fabaceae	Ed, Fd
199	*Rhynchosia minima* (L.) DC.	VNA	Fabaceae	
200	*Stylosanthes fruticosa* (Retz.) Alston	VNA	Fabaceae	Fd
201	*Tephrosia purpurea* (L.) Pers.	VNA	Fabaceae	Mis
202	*Tephrosia villosa* (L.) Pers	VNA	Fabaceae	Nu*
203	*Teramnus labialis* (L. f.) Spreng	VNA	Fabaceae	Nu*
204	*Vigna aconitifolia* (Jacq.) Marechal	VNA	Fabaceae	Fd, Ed
205	*Vigna trilobat*a (L.) Verdc.	VNA	Fabaceae	Fd, Ed, M
206	*Vigna* sp.	VNA	Fabaceae	Nu*
207	*Canscora diffusa* (Vahl) R. Br. ex Roem. & Schult.	VNA	Gentianaceae	Fd
208	*Enicostemma axillare* (Lam.) Raynal	VNA	Gentianaceae	Fd
209	*Exacum pedunculatum* L.	VNA	Gentianaceae	Nu*
210	*Gyrocarpus americanus* Jacq.	VNA	Hernandiaceae	Fu
211	*Hydrilla verticillata* (L. f.)	VNA	Hydrocharitaceae	Nu*

(continued)

Table 3.1 (continued)

S. no.	Scientific name	Vernacular name	Family	Uses
212	*Hypercium* sp.	VNA	Hypericaceae	Fd
213	*Anisochilus carnosus* (L. f.) Wall.	VNA	Lamiaceae	Fd
214	*Anisomeles malabarica* (L.) R. Br. ex Sims	Gandu beerana gida	Lamiaceae	M
215	*Hyptis suaveolens* (L.) Poit.	Nayitulasi	Lamiaceae	Fd
216	*Leonotis nepetaefolia* (L.) R. Br.	Kadu tumbe gida	Lamiaceae	Fu
217	*Leucas aspera* (Willd.) Spreng.	Thumbe	Lamiaceae	M, Fd
218	*Ocimum americanum* L.	Nayitulasi	Lamiaceae	M
219	*Orthosiphon glabratus* Benth.	VNA	Lamiaceae	M
220	*Cassytha filiformis* L.	Aakaasha balli	Lauraceae	Nu*
221	*Couroupita guianensis* Aubl.	VNA	Lecythidaceae	Nu*
222	*Scilla hyacinthina* (Roth) J.F. Macbr.	Adavi ulla gadde	Liliaceae	Nu*
223	*Ammannia baccifera* L.	Agnivendrapaaku	Lythraceae	Fd
224	*Nesaea brevipes* Koehne		Lythraceae	Nu*
225	*Lawsonia inermis* L.	Mehndi gida	Lythraceae	M, Mis
226	*Abutilon indicum* (L.) Sweet	Mudre gida	Malvaceae	Fu
227	*Bombax ceiba* L.	Kempu buragada mara	Malvaceae	Fu
228	*Hibiscus hispidissimus* Griff.	VNA	Malvaceae	Nu*
229	*Hibiscus micranthus* L. f.	VNA	Malvaceae	Fd
230	*Hibiscus rosa-sinensis* L.	Dasavala	Malvaceae	M, Mis, Fu
231	*Diplopenta odorata* Alef. Syn *Pavonia odorata* Willd.	Bala raakshasi	Malvaceae	Fd
232	*Pavonia zeylanica* Cav.	Chittaamutti gida	Malvaceae	Fd, Fu
233	*Sida acuta* Burm. f.	Bhimana kaddi	Malvaceae	M, Fd
234	*Sida cordata* (Burm. f) Borss. Waalk.	VNA	Malvaceae	M, Fd
235	*Sida cordifolia* L.	Bala	Malvaceae	M, Fd
236	*Sida spinosa* L.	VNA	Malvaceae	M, Fd
237	*Thespesia populnea* (L.) Sol. ex Corr.	Adavibende mara	Malvaceae	T, Fu, Fd
238	*Waltheria indica* L.	VNA	Malvaceae	Fd
239	*Martynia annua* L.	Huli uguru	Martyniaceae	Nu*
240	*Azadirachta indica* A. Juss.	Bevina mara	Meliaceae	M, T, Fu, Fd, Mis
241	*Cocculus hirsutus* (L.) Deils	Dhagadi balli	Menispermaceae	Nu*
242	*Tinospora cordifolia* (Willd.) Hook. f. & Thoms.	Amrutha balli	Menispermaceae	M

(continued)

Table 3.1 (continued)

S. no.	Scientific name	Vernacular name	Family	Uses
243	*Acacia auriculiformis* Benth.	VNA	Mimosaceae	T, Fu, Mis
244	*Acacia catechu* (Roxb.) Willd.	Kachu gida	Mimosaceae	Fu, Fd, M, Mis
245	*Acacia farnesiana* (L.) Willd.	Kasturi jail	Mimosaceae	Fu, Fd, Mis, M
246	*Acacia ferruginea* DC.	Banni gida	Mimosaceae	Fu, Fd, Mis
247	*Acacia horrida* (L.) Willd.	Dodda mullina jali	Mimosaceae	T, Fu, M
248	*Acacia latronum* (L. f.) Willd.		Mimosaceae	Fu, M, Fd
249	*Acacia leucophloea* (Roxb.) Willd.	Bili jalli	Mimosaceae	M, T, Fu, Fd
250	*Acacia nilotica* subsp. indica (L.) Willd.	kari jail	Mimosaceae	T, M, Fu, Fd, Mis
251	*Albizia amara* (Roxb.) Boivin	Tugli mara	Mimosaceae	Fu, T, M, Fd
252	*Albizia lebbeck* (L.) Benth.	Baage mara	Mimosaceae	Fu, T, M, Fd
253	*Dichrostachys cinerea* (L.) Wight & Arn.	Vadavarada gida	Mimosaceae	Fu
254	*Leucaena leucocephala* (Lam.) de Wit	Sanna sogachu	Mimosaceae	Fu, Fd
255	*Mimosa intisia* L.		Mimosaceae	Fu, Fd
256	*Mimosa pudica* L.	Lajja	Mimosaceae	
257	*Pithecellobium dulce* (Roxb.) Benth.	Seeme hunase	Mimosaceae	Ed, Fu, T, M
258	*Prosopis cineraria* (L.) Druce	Shastra mara	Mimosaceae	T, M, Fu, Fd
259	*Prosopis juliflora* (Sw.) DC.	Sarkari jail	Mimosaceae	Fu, Fd, Mis
260	*Samanea saman* (Jacq.) Merr.	VNA	Mimosaceae	Fu, Fd, Mis
261	*Ficus arnottiana* (Miq.) Miq.	Bettada arali	Moraceae	M, Fu, Mis
262	*Ficus benghalensis* L.	Alada mara	Moraceae	M, Fu, Mis
263	*Ficus racemosa* L.	Attimara	Moraceae	M, Fu, Mis
264	*Ficus religiosa* L.	Aralimara	Moraceae	M, T, Mis
265	*Streblus asper* Lour.	VNA	Moraceae	Fu

(continued)

Table 3.1 (continued)

S. no.	Scientific name	Vernacular name	Family	Uses
266	*Moringa oleifera* Lam.	Nugge mara	Moringaceae	Ed, M
267	*Eucalyptus globulus* Labill.	VNA	Myrtaceae	M, T, Fu
268	*Psidium guajava* L.	Seebe gida	Myrtaceae	Ed, M
269	*Syzygium cumini* (L.) Skeels.	Nerale	Myrtaceae	Ed, T, M, Fu
270	*Nelumbo nucifera* Gaertn.	Tavere	Nelumbonaceae	Nu*
271	*Boerhavia diffusa* L.	Punarnava	Nyctaginaceae	Ed, Fd
272	*Boerhavia erecta* L.	VNA	Nyctaginaceae	Ed
273	*Bougainvillea spectabilis* Willd.	VNA	Nyctaginaceae	Nu*
274	*Nymphoides cristata* (Roxb.) Kuntze	VNA	Nymphaeaceae	Nu*
275	*Nymphaea* sp.	VNA	Nymphaeaceae	Nu*
276	*Ximenia americana* L.	VNA	Olacaceae	Fu
277	*Eulophia epidendraea* (J. Koenig ex Retz.) C.E.C. Fischer	VNA	Orchidaceae	Nu*
278	*Eulophia ramentacea* Lindl.	VNA	Orchidaceae	Nu*
279	*Biophytum sensitivum* (L.) DC.	Ghaati horamuni	Oxalidaceae	Nu*
280	*Oxalis corniculata* L.	Pullampurachi soppu	Oxalidaceae	Fd
281	*Argemone mexicana* L.	Daturi gida	Papaveraceae	M
282	*Passiflora foetida* L.	Kukki balli	Passifloraceae	Nu*
283	*Plumbago zeylanica* L.	Chitra moola	Plumbaginaceae	Nu*
284	*Aristida setacea* Retz.	Dodda hanchi hullu	Poaceae	Mis
285	*Arundinella setosa* (L.) Link	VNA	Poaceae	Fd
286	*Arundo donax* L.	VNA	Poaceae	Fd
287	*Bambusa arundinacea* (Retz.) Willd.	Bedru	Poaceae	T
288	*Chloris barbata* Sw.	Manchada kalu hullu	Poaceae	Fd
289	*Cymbopogon martini* (Roxb.) S. Watson	VNA	Poaceae	Fd
290	*Cynodon dactylon* (L.) Pers.	Garike	Poaceae	Fd
291	*Dactyloctenium aegyptium* (L.) Willd.	VNA	Poaceae	Fd
292	*Digitaria bicornis* (Lam.) Roem. & Schult	VNA	Poaceae	Fd
293	*Dinebra retroflexa* (Vahl) Panz.	Naribalada hullu	Poaceae	Fd
294	*Echinochloa colona* (L.)	Kadu haraka hullu	Poaceae	Nu*

(continued)

Table 3.1 (continued)

S. no.	Scientific name	Vernacular name	Family	Uses
295	*Eragrostis japonica* (Thunb.) Trin.	VNA	Poaceae	Fd
296	*Ischaemum rugosum* Salisb.	VNA	Poaceae	Fd
297	*Iseilema hackelii* Shrestha & Gandhi	VNA	Poaceae	Fd
298	*Saccharum spontaneum* L.	Kadu kabbu	Poaceae	Fd
299	*Setaria pumila* (Poir.) Roem. & Schult.		Poaceae	Fd
300	*Antigonon leptopus* Hook. & Arn.	VNA	Polygonaceae	Mis
301	*Persicaria glabra* (Willd.) M. Gómez	Nerru kanagilu	Polygonaceae	Fd
302	*Polygala chinensis* L.	VNA	Polygalaceae	Fd
303	*Eichhornia crassipes* (Mart.) Solms	VNA	Pontederiaceae	Nu*
304	*Monochoria vaginalis* (Burm. f.) C. Presl ex Kunth	Neelothpala	Pontederiaceae	Nu*
305	*Portulaca oleraceae* L.	VNA	Portulacaceae	Fd
306	*Punica granatum* L.	Dalimbre	Punicaceae	Ed
307	*Scutia myrtina* (Burm. f.) Kurz	VNA	Rhamnaceae	Mis
308	*Ziziphus mauritiana* Lam.	Borae hannu	Rhamnaceae	Ed, M
309	*Ziziphus oenoplia* Mill.	VNA	Rhamnaceae	Ed
310	*Ziziphus xylopyrus* (Retz.) Willd	Chuttimullu	Rhamnaceae	Ed, Fu, M
311	*Canthium coramandelicum* (Burm. f.) Alston	Mullukhare	Rubiaceae	Ed, Fu
312	*Catunaregam spinosa* (Thunb.) Tiruv.	Kaarigida	Rubiaceae	Fu, M
313	*Hedyotis puberula* (G. Don) Arn.	Chaaya vaeru	Rubiaceae	Fd
314	*Ixora arborea* Roxb.ex J.E. Sm.	VNA	Rubiaceae	Mis
315	*Meyna laxiflora* Robyns	VNA	Rubiaceae	Fd
316	*Morinda pubescens* J.E. Sm.	Maligana mara	Rubiaceae	Fu, M, T
317	*Pavetta indica* L.	VNA	Rubiaceae	Fu, Fd
318	*Spermacoce hispida* L.	VNA	Rubiaceae	Fd
319	*Spermacoce stricta* L. f.	VNA	Rubiaceae	Fd
320	*Aegle marmelos* (L.) Corr.	Bilvapatre gida	Rutaceae	M, Fu, Ed
321	*Chloroxylon swietenia* DC.	Masivala	Rutaceae	Fu, Fd, M, Mis

(continued)

Table 3.1 (continued)

S. no.	Scientific name	Vernacular name	Family	Uses
322	*Limonia acidissima* L.	Belada gida	Rutaceae	Ed, M, Fu
323	*Murraya koenigii* (L.) Spreng.	Karebevu	Rutaceae	Ed, M, Mis
324	*Salvadora persica* L.	VNA	Salvadoraceae	Nu*
325	*Santalum album* L.	Gandadamara	Santalaceae	T, M
326	*Cardiospermum halicacabum* L.	Agni balli	Sapindaceae	
327	*Dodonaea viscosa* (L.) Jacq.	Bandarike	Sapindaceae	Fu, Fd, M
328	*Sapindus laurifolia* Vahl	Antuvala	Sapindaceae	M, Fu, Mis
329	*Bacopa monnieri* (L.) Wett.	Niru brahmi	Scrophulariaceae	Ed
330	*Buddleia* sp.	VNA	Scrophulariaceae	Nu*
331	*Stemodia viscosa* Roxb.	Antikamini	Scrophulariaceae	Fd
332	*Ailanthus excelsa* Roxb.	Hebbevu	Simaroubaceae	Fu, Mis
333	*Datura metel* L.	Ummatti gida	Solanaceae	M, Fd
334	*Nicandra physalodes* (L.) Gaertn.	VNA	Solanaceae	
335	*Physalis minima* L.	Sanna budda gida	Solanaceae	Ed, M
336	*Solanum nigrum* L.	Kakki hannu	Solanaceae	M
337	*Solanum trilobatum* L.	Mullu mustha	Solanaceae	M
338	*Solanum xanthocarpum* Schrad. & Wendl.	Kantakari	Solanaceae	M
339	*Withania somnifera* (L.) Dunal	Ottatti gida	Solanaceae	M, Fd
340	*Melhania incana* Heyne ex Wight	VNA	Sterculiaceae	
341	*Corchorus aestuans* L.	VNA	Tiliaceae	M, Fd
342	*Corchorus trilocularis* L.	VNA	Tiliaceae	M, Fd
343	*Grewia flavescens* Juss.	VNA	Tiliaceae	Fu
344	*Grewia tenax* (Forssk.) Fiori	VNA	Tiliaceae	Fu
345	*Grewia villosa* Willd.	VNA	Tiliaceae	Fu
346	*Grewia* sp.	VNA	Tiliaceae	
347	*Triumfetta rotundifolia* Lam.	VNA	Tiliaceae	Mis
348	*Turnera ulmifolia* L.	VNA	Turneraceae	Mis
349	*Typha angustifolia* L.	Aapu	Typhaceae	Mis
350	*Holoptelea integrifolia* (Roxb.) Planch.	Tapsi mara	Ulmaceae	Fu
351	*Clerodendrum inerme* (L.) Gaertn.	Vishamadari	Verbenaceae	Mis
352	*Clerodendrum phlomidis* L. f.	VNA	Verbenaceae	Mis

(continued)

Table 3.1 (continued)

S. no.	Scientific name	Vernacular name	Family	Uses
353	*Duranta repens* L.	Duranta kanti	Verbenaceae	Mis
354	*Lantana camara* L.	Hunni gida	Verbenaceae	Fu, M
355	*Phyla nodiflora* (L.) Greene	VNA	Verbenaceae	Fd
356	*Premna latifolia* Roxb.	Nerugala	Verbenaceae	Fu
357	*Priva cordifolia* (L.) Druce	VNA	Verbenaceae	Fd
358	*Stachytarpheta indica* (L.) Vahl	Edurutrani	Verbenaceae	Fd
359	*Tectona grandis* L. f.	Tega mara	Verbenaceae	T, Fu
360	*Vitex negundo* L.	Lakki gida	Verbenaceae	M, Fu, Fd
361	*Hybanthus enneaspermus* (L.) F. Muell.	Ratna purusha	Violaceae	Fd
362	*Cissus glauca* Roxb.	VNA	Vitaceae	Nu*
363	*Cissus quadrangularis* L.	Mangaravalli	Vitaceae	M
364	*Cyphostemma setosum* (Roxb.) Alston	VNA	Vitaceae	Nu*
365	*Fagonia arabica* L.	VNA	Zygophyllaceae	Nu*
366	*Tribulus terrestris* L.	Neggele mullu	Zygophyllaceae	Nu*

Note M Medicinal, *T* Timber, *Ed* Edible, *Fu* Fuel, *Fd* Fodder, *Mis* Miscellaneous, *VNA* Vernacular names not available, *Nu** Not recorded for any use by the local people

Table 3.2 Bryophytes, pteridophytes, and gymnosperms of the studied semiarid region

	Scientific name	Family
Bryophytes		
1	*Marchantia sp.*	Marchantiaceae
Pteridophytes		
1	*Adiantum sp.*	Adiantaceae
2	*Selaginella sp.*	Selaginellaceae
3	*Actiniopteris radiata*	Pteridaceae
4	*Marsilea quadrifolia*	Marsileaceae
Gymnosperms		
1	*Thuja occidentalis* L.	Cupressaceae

individual per ha for example 0.94 for *Acacia latronum* and 0.10 for *Ficus racemosa*. The frequency of the species was recorded high for *Prosopis juliflora* (72.60) followed by *Azadirachta indica* (23.54), *Annona squamosa* (20.52), *Acacia nilotica* (19.58), and least (0.10) for the species, namely *Ziziphus xylopyrus*, *Erythrina*

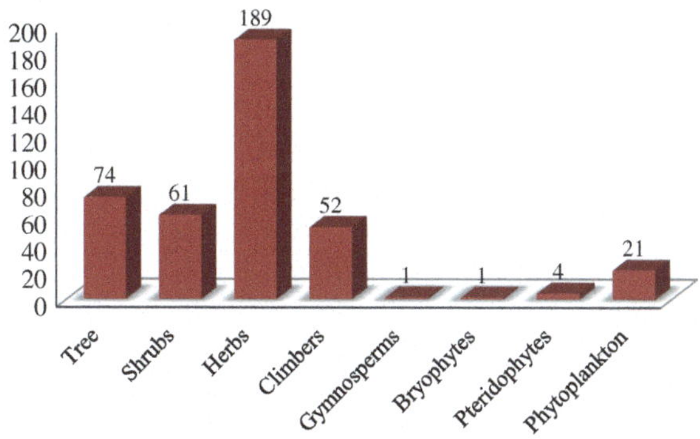

Fig. 3.1 Plant biodiversity in the study region

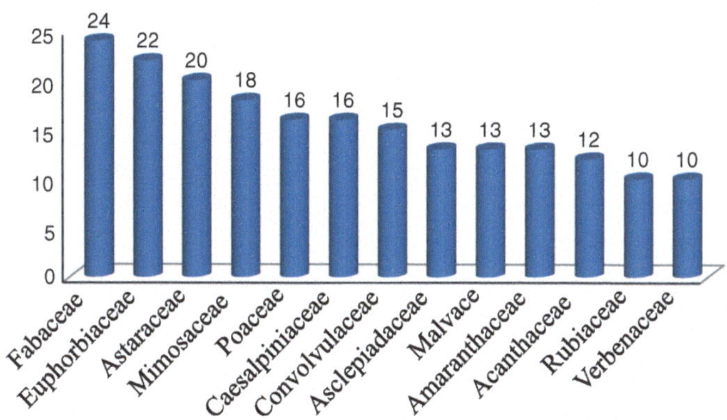

Fig. 3.2 Species richness in the study area

variegata, *Ficus racemosa*, *Terminalia bellirica*. The importance value index (IVI) indicates the overall dominance of a species in a particular ecosystem. Based on IVI values, the *Prosopis juliflora* (88.9) is highly dominant followed by *Azadirachta indica* (66.1) and the least dominant species of the semiarid ecosystem were found to be *Ailanthus excelsa* (0.07) and *Terminalia bellirica* (0.06). The density values for the tree species (468 individuals per ha) from the semiarid region reported to be less in comparison with other ecosystems located in tropical and temperate regions of the country. However, diversity of trees is quite comparable with other rich forested ecosystems.

A total of 60 species of shrubs recorded from the study area. Based on IVI, the most dominant species under the second vegetation layer (shrubs) is found to be *Lantana camara* (IVI 85.8) followed by *Abutilon indicum* (62.2), *Securinega leucopyrus* (21.4), and *Senna auriculata* (19.8). The IVI value indicates that 41 species have IVI value less than 1.0. The lease dominant species are *Dodonaea*

Table 3.3 Phytosociological attributes of trees

S. no.	Scientific name	Density/ha	Frequency	Abundance	IVI
1	*Acacia auriculiformis*	0.10	0.10	1.00	0.1
2	*Acacia farnesiana*	3.13	1.77	1.76	0.76
3	*Acacia ferruginea*	0.83	0.63	1.33	2.14
4	*Acacia horrida*	0.73	0.52	1.40	1.16
5	*Acacia latronum*	0.94	0.31	3.00	0.1
6	*Acacia leucophloea*	4.79	3.96	1.21	2.63
7	*Acacia nilotica subsp. indica*	24.48	19.58	1.25	30.5
8	*Acacia catechu*	0.31	0.31	1.00	0.42
9	*Aegle marmelos*	0.31	0.31	1.00	0.62
10	*Ailanthus excelsa*	0.10	0.10	1.00	0.07
11	*Albizia amara*	0.73	0.31	2.33	0.85
12	*Albizia lebbeck*	1.67	1.35	1.23	7.56
13	*Annona squamosa*	65.42	20.52	3.19	2.6
14	*Azadirachta indica*	29.69	23.54	1.26	66.1
15	*Bauhinia racemosa*	1.77	1.67	1.06	0.87
16	*Bombax ceiba*	0.31	0.21	1.50	0.13
17	*Borassus flabellifer*	0.21	0.21	1.00	0.45
18	*Caesalpinia pulcherrima*	0.31	0.31	1.00	0.7
19	*Carica papaya*	0.31	0.21	1.50	0.2
20	*Cassia fistula*	3.33	2.08	1.60	0.33
21	*Cassia javanica*	2.19	1.56	1.40	0.29
22	*Cassia leschenaultii*	4.48	1.56	2.87	0.16
23	*Casuarina equisetifolia*	1.46	0.63	2.33	0.76
24	*Chloroxylon swietenia*	5.83	2.29	2.55	6.92
25	*Cocos nucifera*	0.73	0.31	2.33	0.6
26	*Cordia dichotoma*	0.21	0.21	1.00	0.23
27	*Cordia wallichii*	0.21	0.21	1.00	0.14
28	*Couroupita guianensis*	0.10	0.10	1.00	0.43
29	*Dalbergia sissoo*	0.94	0.31	3.00	0.24
30	*Delonix elata*	2.08	1.15	1.82	0.62
31	*Delonix regia*	2.81	1.67	1.69	0.86
32	*Dichrostachys cinerea*	0.63	0.42	1.50	0.14
33	*Dolichandrone atrovirens*	0.21	0.21	1.00	0.28
34	*Dolichandrone falcata*	0.10	0.10	1.00	0.16
35	*Erythrina variegata*	0.21	0.10	2.00	0.42
36	*Eucalyptus globulus*	1.67	0.73	2.29	1.19
37	*Ficus arnottiana*	0.10	0.10	1.00	0.28
38	*Ficus benghalensis*	0.31	0.31	1.00	0.44
39	*Ficus racemosa*	0.10	0.10	1.00	0.28
40	*Ficus religiosa*	3.75	3.54	1.06	2.35
41	*Holoptelea integrifolia*	0.31	0.21	1.50	0.17

(continued)

Table 3.3 (continued)

S. no.	Scientific name	Density/ha	Frequency	Abundance	IVI
42	*Leucaena leucocephala*	53.13	11.15	4.77	25.8
43	*Limonia acidissima*	0.73	0.52	1.40	0.7
44	*Mangifera indica*	0.31	0.21	1.50	2.19
45	*Markhamia lutea*	1.15	1.15	1.00	0.92
46	*Millingtonia hortensis*	0.10	0.10	1.00	0.31
47	*Mimosa intisia*	0.31	0.21	1.50	0.1
48	*Morinda pubescens*	4.17	2.60	1.60	3.07
49	*Moringa oleifera*	0.21	0.21	1.00	0.18
50	*Murraya koenigii*	1.35	0.73	1.86	0.35
51	*Phoenix sylvestris*	1.35	0.83	1.63	0.92
52	*Phyllanthus emblica*	0.31	0.31	1.00	0.16
53	*Phyllanthus polyphyllus*	0.52	0.31	1.67	0.19
54	*Pithecellobium dulce*	1.98	1.56	1.27	0.58
55	*Pongamia pinnata*	3.33	2.71	1.23	0.44
56	*Prosopis cineraria*	2.81	2.40	1.17	7.98
57	*Prosopis juliflora*	176.46	72.60	2.43	88.9
58	*Psidium guajava*	1.46	0.42	3.50	0.93
59	*Punica granatum*	0.31	0.21	1.50	0.31
60	*Samanea saman*	2.19	1.88	1.17	1.94
61	*Santalum album*	14.06	8.13	1.73	3.91
62	*Sapindus laurifolia*	0.52	0.31	1.67	0.47
63	*Senna siamea*	3.02	2.19	1.38	0.2
64	*Spathodea campanulata*	0.63	0.52	1.20	0.84
65	*Syzygium cumini*	0.52	0.52	1.00	2.42
66	*Tamarindus indica*	7.08	5.94	1.19	5.47
67	*Tecoma stans*	0.31	0.21	1.50	0.13
68	*Tectona grandis*	0.52	0.21	2.50	0.48
69	*Terminalia bellirica*	0.10	0.10	1.00	0.06
70	*Terminalia catappa*	0.21	0.21	1.00	0.11
71	*Thespesia populnea*	0.42	0.42	1.00	0.61
72	*Wrightia tinctoria*	3.02	1.46	2.07	1.94
73	*Ziziphus mauritiana*	16.67	14.17	1.18	11.8
74	*Ziziphus xylopyrus*	0.63	0.10	6.00	0.34

viscose (0.1), *Euphorbia sp* (0.08) *Maytenus emarginata, Securinega virosa* (0.07 each), and *Alangium salvifolium* (0.03). The shrub density of studied semiarid region is 1295 individuals per ha. The absolute value of density for shrub species is comparable with ecosystems located in the other ecological regions of country. However, only three species of shrubs, viz. *Abutilon indicum, Lantana camara, and Tephrosia purpurea,* have the density value more than 200 individuals per ha. For two species (*Ipomoea carnea* and *Senna occidentalis*), the density value is recorded more than 100 individuals per ha. For nine species, the density value was found ranging between 10.94 and 65.42. However, the density value for 46 species was found to be less than 10 individuals per ha. This indicates that the secondary vegetation layer is dominated by two species *Abutilon indicum* and *Lantana camara.* From the study region, the density of shrubs is 1295 individuals per ha (Table 3.4).

Table 3.4 Phytosociological attributes of shrubs

S. no.	Scientific name	Density/ha	Frequency	Abundance	IVI
1	*Abutilon indicum*	281.15	41.25	6.82	62.2
2	*Adhatoda zeylanica*	1.88	0.625	3.00	0.35
3	*Agave americana*	6.77	3.23	2.10	4.94
4	*Alangium salvifolium*	0.94	0.63	1.50	0.03
5	*Anisomeles malabarica*	14.17	2.92	4.86	4.4
6	*Balanites aegyptiaca*	59.17	19.48	3.04	6.1
7	*Barleria prionitis*	4.79	2.71	1.77	3.1
8	*Cadaba fruticosa*	1.35	0.83	1.63	0.2
9	*Calotropis gigantea*	15.73	11.56	1.36	7.15
10	*Calotropis procera*	2.29	1.88	1.22	0.5
11	*Canthium coramandelicum*	23.02	15.10	1.52	14.5
12	*Capparis brevispina*	0.21	0.10	2.00	0.6
13	*Capparis divaricata*	0.52	0.42	1.25	0.9
14	*Carissa spinarum*	0.83	0.63	1.33	0.4
15	*Clerodendrum inerme*	0.31	0.31	1.00	0.6
16	*Clerodendrum phlomidis*	0.63	0.42	1.50	0.3
17	*Dodonaea viscosa*	17.40	6.77	2.57	0.1
18	*Duranta repens*	0.94	0.42	2.25	0.29
19	*Euphorbia sp.*	4.17	2.40	1.74	0.08
20	*Euphorbia tirucalli*	10.94	5.83	1.88	5.44
21	*Grewia flavescens*	1.56	0.63	2.50	4.1
22	*Grewia tenax*	0.21	0.21	1.00	1.1
23	*Grewia villosa*	2.08	1.56	1.33	0.8
24	*Guilandina bonduc*	1.04	0.83	1.25	0.6
25	*Gyrocarpus americanus*	0.42	0.31	1.33	0.26
26	*Hibiscus hispidissimus*	0.21	0.21	1.00	0.8
27	*Hibiscus micranthus*	1.46	0.83	1.75	0.64
28	*Hibiscus rosa-sinensis*	0.42	0.31	1.33	0.27
29	*Indigofera tinctoria*	0.42	0.21	2.00	0.14
30	*Ipomoea carnea*	125.00	13.44	9.30	25.2
31	*Ixora arborea*	0.31	0.21	1.50	0.17

(continued)

Table 3.4 (continued)

S. no.	Scientific name	Density/ha	Frequency	Abundance	IVI
32	*Jatropha glandulifera*	4.58	2.40	1.91	1.79
33	*Kirganelia reticulata*	1.56	1.04	1.50	0.2
34	*Lantana camara*	245.21	49.58	4.95	85.8
35	*Lawsonia inermis*	0.94	0.52	1.80	0.22
36	*Leonotis nepetaefolia*	0.94	0.42	2.25	0.28
37	*Maytenus emarginata*	0.42	0.31	1.33	0.07
38	*Meyna laxiflora*	1.77	0.73	2.43	1.7
39	*Nerium indicum*	0.21	0.21	1.00	0.5
40	*Opuntia dillenii*	0.63	0.21	3.00	0.7
41	*Parkinsonia aculeata*	0.94	0.63	1.50	0.72
42	*Pavetta indica*	0.21	0.10	2.00	0.9
43	*Pavonia zeylanica*	6.98	2.81	2.48	0.2
44	*Phyllanthus reticulatus*	1.04	0.94	1.11	0.3
45	*Premna latifolia*	1.46	0.52	2.80	0.9
46	*Randia brandisii*	0.10	0.10	1.00	1.9
47	*Ricinus communis*	18.23	3.75	4.86	3.5
48	*Scutia myrtina*	0.83	0.63	1.33	0.3
49	*Securinega leucopyrus*	12.40	8.75	1.42	21.4
50	*Securinega virosa*	1.56	0.73	2.14	0.07
51	*Senna auriculata*	65.42	37.29	1.75	19.8
52	*Senna occidentalis*	125.83	14.17	8.88	4.6
53	*Streblus asper*	0.21	0.10	2.00	0.18
54	*Tephrosia purpurea*	205.63	45.31	4.54	3.54
55	*Thevetia neriifolia*	5.73	1.35	4.23	0.43
56	*Triumfetta rhomboidea*	5.63	2.81	2.00	1.7
57	*Turnera ulmifolia*	0.42	0.10	4.00	0.6
58	*Vitex negundo*	4.38	3.23	1.35	0.8
59	*Ximenia americana*	0.31	0.21	1.50	0.3
60	*Ziziphus oenoplia*	0.63	0.21	3.00	0.34

The characteristic feature of semiarid landscape is the diversity in herbaceous vegetation. This is indicated as well from the study region that 173 species of herbs have been recorded during the study period. Based on the phytosociological study, the 2846 individuals were found in a hectare of land in semiarid region. Species, namely *Parthenium hysterophorus* (398.96 individuals per ha) and *Senna sericea* (29.42 individuals per ha), were found to be most dominant among the herbaceous vegetation followed by *Croton bonplandianum* (189.79), *Bacopa monnieri* (149.27), *Euphorbia hirta* (132.19), *Tridax procumbens* (130.52), *Typha angustifolia* (128.13), and *Achyranthes aspera* (106.46). The density value for 139 species was calculated to be less than 10 individuals per ha (Table 3.5). The density of herbs is calculated to be 2845 individuals per ha in the studied semiarid landscape.

Table 3.5 Phytosociological attributes of herbs

S. no.	Scientific name	Density/ha	Frequency	Abundance
1	*Acalypha indica*	34.06	8.54	3.99
2	*Achyranthes aspera*	106.46	37.08	2.87
3	*Aerva lanata*	39.69	7.19	5.52
4	*Ageratum conyzoides*	7.71	0.42	18.50
5	*Aloe vera*	59.69	3.02	19.76
6	*Alternanthera pungens*	1.25	0.83	1.50
7	*Alternanthera sessilis*	25.73	2.92	8.82
8	*Alternanthera tenella*	9.69	3.44	2.82
9	*Alysicarpus vaginalis*	3.33	1.46	2.29
10	*Amaranthus spinosus*	7.92	1.98	4.00
11	*Amaranthus viridis*	1.88	0.63	3.00
12	*Amischophacelus axillaris*	1.67	0.21	8.00
13	*Ammannia baccifera*	15.63	3.54	4.41
14	*Andrographis paniculata*	0.21	0.21	1.00
15	*Anisochilus carnosus*	1.04	0.63	1.67
16	*Argemone mexicana*	6.98	3.54	1.97
17	*Aristida setacea*	3.96	0.42	9.50
18	*Aristolochia bracteolata*	0.52	0.21	2.50
19	*Arundinella setosa*	2.19	1.35	1.62
20	*Arundo donax*	1.88	0.94	2.00
21	*Asteracantha longifolia*	3.65	0.52	7.00
22	*Bacopa monnieri*	149.27	19.17	7.79
23	*Bambusa arundinacea*	0.63	0.21	3.00
24	*Bergia ammannioides*	0.94	0.31	3.00
25	*Biophytum sensitivum*	1.15	0.42	2.75
26	*Blainvillea acmella*	0.31	0.10	3.00
27	*Blumea eriantha*	0.52	0.31	1.67
28	*Boerhavia diffusa*	18.44	8.85	2.08
29	*Boerhavia erecta*	0.21	0.21	1.00
30	*Bulbostylis densa*	0.94	0.31	3.00
31	*Canscora diffusa*	0.83	0.52	1.60
32	*Caralluma attenuata*	19.69	5.73	3.44
33	*Celosia argentea*	0.31	0.10	3.00
34	*Chenopodium album*	0.31	0.10	3.00
35	*Chloris barbata*	1.25	0.52	2.40
36	*Chrozophora plicata*	0.63	0.21	3.00
37	*Cleome viscosa*	3.96	0.52	7.60
38	*Coldenia procumbens*	2.29	0.52	4.40
39	*Commelina benghalensis*	2.92	0.21	14.00
40	*Commelina diffusa*	4.17	0.94	4.44

(continued)

Table 3.5 (continued)

S. no.	Scientific name	Density/ha	Frequency	Abundance
41	*Corchorus aestuans*	3.33	1.56	2.13
42	*Corchorus trilocularis*	7.29	2.81	2.59
43	*Cressa cretica*	0.83	0.42	2.00
44	*Crossandra infundibuliformis*	0.42	0.31	1.33
45	*Crotalaria juncea*	1.15	0.63	1.83
46	*Crotalaria pallida*	0.63	0.31	2.00
47	*Crotalaria pusilla*	0.52	0.21	2.50
48	*Crotalaria ramosissima*	0.21	0.10	2.00
49	*Crotalaria retusa*	0.83	0.21	4.00
50	*Croton bonplandianum*	189.79	48.44	3.92
51	*Cullen corylifolium*	0.83	0.42	2.00
52	*Cyanotis sp.*	0.94	0.31	3.00
53	*Cyanotis tuberosa*	2.29	1.25	1.83
54	*Cynodon dactylon*	94.06	5.42	17.37
55	*Cyperus alopecuroides*	7.29	1.25	5.83
56	*Cyperus compressus*	2.92	1.15	2.55
57	*Cyperus pangorei*	3.23	0.94	3.44
58	*Cyperus rotundus*	20.10	3.85	5.22
59	*Cyperus sp.*	1.04	0.63	1.67
60	*Cyperus squarrosus*	6.98	1.25	5.58
61	*Dactyloctenium aegyptium*	0.63	0.31	2.00
62	*Datura metel*	27.71	10.52	2.63
63	*Desmodium triflorum*	1.25	0.83	1.50
64	*Dicoma tomentosa*	1.67	1.15	1.45
65	*Digera muricata*	1.04	0.42	2.50
66	*Digitaria bicornis*	0.94	0.63	1.50
67	*Dinebra retroflexa*	3.33	1.46	2.29
68	*Dipteracanthus patulus*	1.25	0.52	2.40
69	*Echinochloa colona*	1.15	0.52	2.20
70	*Echinops echinatus*	0.42	0.31	1.33
71	*Eclipta alba*	73.54	21.04	3.50
72	*Enicostemma axillare*	0.83	0.21	4.00
73	*Eragrostis japonica*	1.67	0.73	2.29
74	*Erigeron alpinus*	1.56	0.42	3.75
75	*Eriocaulon quinquangular*	1.46	0.21	7.00
76	*Eulophia epidendraea*	0.21	0.10	2.00
77	*Eulophia ramentacea*	0.10	0.10	1.00
78	*Euphorbia heterophylla*	62.40	14.90	4.19
79	*Euphorbia hirta*	132.19	24.27	5.45
80	*Euphorbia hypericifolia*	2.92	1.35	2.15

(continued)

Table 3.5 (continued)

S. no.	Scientific name	Density/ha	Frequency	Abundance
81	*Euphorbia indica*	1.98	0.63	3.17
82	*Evolvulus alsinoides*	2.19	0.63	3.50
83	*Exacum pedunculatum*	0.42	0.21	2.00
84	*Fagonia arabica*	0.63	0.42	1.50
85	*Fimbristylis dichotoma*	1.46	0.42	3.50
86	*Flaveria trinervia*	0.73	0.31	2.33
87	*Gomphrena celosioides*	66.04	10.42	6.34
88	*Grangea maderaspatana*	0.63	0.21	3.00
89	*Hedyotis puberula*	1.77	0.52	3.40
90	*Heliotropium indicum*	27.19	6.67	4.08
91	*Heliotropium ovalifolium*	1.15	0.31	3.67
92	*Hemidesmus indicus*	29.69	12.29	2.42
93	*Hybanthus enneaspermus*	44.58	6.88	6.48
94	*Hymenotherum tenniflоum*	1.15	0.52	2.20
95	*Hyptis suaveolens*	3.85	0.94	4.11
96	*Indigofera cordifolia*	1.46	0.73	2.00
97	*Indigofera linnaei*	0.63	0.21	3.00
98	*Ipomoea aquatica*	2.71	0.63	4.33
99	*Ischaemum rugosum*	4.06	1.15	3.55
100	*Justicia sp.*	0.31	0.21	1.50
101	*Lagascea mollis*	0.83	0.42	2.00
102	*Launaea pinnatifida*	1.15	0.63	1.83
103	*Lepidagathis cristata*	1.77	0.73	2.43
104	*Leucas aspera*	42.71	9.79	4.36
105	*Martynia annua*	52.08	21.88	2.38
106	*Melhania incana*	0.83	0.31	2.67
107	*Merremia emarginata*	0.31	0.21	1.50
108	*Mimosa pudica*	0.21	0.10	2.00
109	*Mollugo lotoides*	1.88	0.42	4.50
110	*Monochoria vaginalis*	0.52	0.31	1.67
111	*Nesaea brevipes*	1.15	0.42	2.75
112	*Nicandra physalodes*	0.21	0.10	2.00
113	*Nymphoides cristata*	1.67	0.31	5.33
114	*Ocimum americanum*	79.79	27.81	2.87
115	*Orthosiphon glabratus*	0.42	0.21	2.00
116	*Oxalis corniculata*	14.79	5.52	2.68
117	*Parthenium hysterophorus*	398.96	45.42	8.78
118	*Pavonia odorata*	0.63	0.42	1.50
119	*Pentanema indicum*	0.63	0.21	3.00
120	*Peristrophe paniculata*	15.10	4.27	3.54

(continued)

Table 3.5 (continued)

S. no.	Scientific name	Density/ha	Frequency	Abundance
121	*Persicaria glabra*	0.83	0.10	8.00
122	*Phyla nodiflora*	3.85	0.94	4.11
123	*Phyllanthus amarus*	58.44	19.79	2.95
124	*Phyllanthus kozhikodianus*	0.10	0.10	1.00
125	*Phyllanthus maderaspatensis*	1.15	0.63	1.83
126	*Phyllanthus virgatus*	0.21	0.10	2.00
127	*Physalis minima*	0.94	0.42	2.25
128	*Plumbago zeylanica*	2.19	0.83	2.63
129	*Polycarpaea aurea*	0.83	0.63	1.33
130	*Polycarpaea corymbosa*	18.23	2.92	6.25
131	*Polygala chinensis*	1.25	0.83	1.50
132	*Portulaca oleraceae*	1.77	0.63	2.83
133	*Priva leptostachya*	0.83	0.42	2.00
134	*Psilotrichum sp.*	0.10	0.10	1.00
135	*Pulicaria sp.*	2.92	0.42	7.00
136	*Pulicaria wightiana*	2.92	0.94	3.11
137	*Pupalia lappacea*	0.42	0.21	2.00
138	*Pycreus flavidus*	2.29	0.83	2.75
139	*Rostellularia simplex*	1.25	0.83	1.50
140	*Ruellia sp.*	0.42	0.10	4.00
141	*Rungia repens*	0.52	0.42	1.25
142	*Saccharum spontaneum*	0.83	0.10	8.00
143	*Salvadora persica*	0.10	0.10	1.00
144	*Scilla hyacinthina*	7.60	2.08	3.65
145	*Senna italica*	0.21	0.10	2.00
146	*Senna sericea*	290.42	21.46	13.53
147	*Setaria pumila*	8.96	2.29	3.91
148	*Sida acuta*	64.06	22.92	2.80
149	*Sida cordata*	1.15	0.52	2.20
150	*Sida cordifolia*	3.23	1.77	1.82
151	*Sida spinosa*	2.40	0.94	2.56
152	*Solanum nigrum*	0.10	0.10	1.00
153	*Solanum trilobatum*	0.63	0.21	3.00
154	*Solanum xanthocarpum*	0.31	0.21	1.50
155	*Sonchus oleraceus*	1.15	0.52	2.20
156	*Spermacoce hispida*	0.31	0.31	1.00
157	*Spermacoce stricta*	4.90	1.35	3.62
158	*Sphaeranthus amaranthoides*	0.94	0.10	9.00
159	*Spilanthes calva*	6.46	2.60	2.48
160	*Stachytarpheta indica*	6.15	1.88	3.28

(continued)

Table 3.5 (continued)

S. no.	Scientific name	Density/ha	Frequency	Abundance
161	*Stemodia viscosa*	1.46	0.94	1.56
162	*Stylosanthes fruticosa*	0.42	0.10	4.00
163	*Trianthema portulacastrum*	0.63	0.31	2.00
164	*Tribulas lanuginosus*	12.08	3.54	3.41
165	*Trichodesma indicum*	7.50	2.40	3.13
166	*Tridax procumbens*	130.52	42.81	3.05
167	*Typha angustifolia*	128.13	12.40	10.34
168	*Vigna sp.*	0.63	0.42	1.50
169	*Vigna trilobata*	0.63	0.21	3.00
170	*Vinca rosea*	0.52	0.21	2.50
171	*Waltheria indica*	1.04	0.21	5.00
172	*Withania somnifera*	16.15	7.29	2.21
173	*Xanthium indicum*	31.25	12.08	2.59

A total of 47 species of climbers were recorded from the studied semiarid region with 206 individuals per ha. The climber density in the region was 206 individuals per ha. The dominant species of climbers are *Cryptostegia grandiflora* (40.73 individuals per ha), *Coccinia grandis* (27.60 individuals per ha), *Diplocyclos palmatus* (26.98 individuals per ha), *Tinospora cordifolia* (23.85 individuals per ha), and *Cryptolepis buchanani* and *Cocculus hirsutus* (15 individuals per ha). Forty-one species of climbers have density value less than 10 individuals per ha, out of which the density value for 27 species was calculated less than 1 individual per ha (Table 3.6).

Table 3.6 Phytosociological attributes of climbers of studied semiarid region

S. no.	Scientific name	Density/ha	Frequency	Abundance
1	*Abrus precatorius*	3.85	3.44	1.12
2	*Antigonon leptopus*	0.94	0.63	1.50
3	*Aristolochia indica*	1.15	1.15	1.00
4	*Basella alba*	0.52	0.42	1.25
5	*Bougainvillea spectabilis*	0.31	0.21	1.50
6	*Canvalia africana*	3.13	0.63	5.00
7	*Capparis zeylanica*	0.31	0.31	1.00
8	*Cardiospermum halicacabum*	0.94	0.94	1.00
9	*Cassytha filiformis*	3.85	3.75	1.03
10	*Cissus glauca*	0.21	0.21	1.00

(continued)

Table 3.6 (continued)

S. no.	Scientific name	Density/ha	Frequency	Abundance
11	*Cissus quadrangularis*	0.63	0.21	3.00
12	*Clitoria ternatea*	4.48	3.96	1.13
13	*Coccinia grandis*	27.60	25.63	1.08
14	*Cocculus hirsutus*	15.73	14.69	1.07
15	*Combretum ovalifolium*	0.21	0.21	1.00
16	*Corallocarpus epigaeus*	0.94	0.73	1.29
17	*Cryptolepis buchanani*	15.83	9.58	1.65
18	*Cryptostegia grandiflora*	40.73	22.40	1.82
19	*Ctenolepis cerasiformis*	0.10	0.10	1.00
20	*Cucumis callosus*	0.31	0.21	1.50
21	*Cyphostemma setosum*	0.52	0.42	1.25
22	*Diplocyclos palmatus*	26.98	19.38	1.39
23	*Gymnema sylvestre*	0.42	0.31	1.33
24	*Ipomoea cairica*	0.83	0.73	1.14
25	*Ipomoea hederifolia*	0.73	0.73	1.00
26	*Ipomoea mauritiana*	1.56	1.56	1.00
27	*Ipomoea obscura*	3.13	2.50	1.25
28	*Ipomoea quomoclit*	0.31	0.31	1.00
29	*Ipomoea sp.*	0.42	0.42	1.00
30	*Ipomoea tricolor*	0.94	0.83	1.13
31	*Jacquemontia pentantha*	2.71	2.29	1.18
32	*Maerua oblongifolia*	0.21	0.21	1.00
33	*Merremia tridentata*	0.31	0.31	1.00
34	*Mukia maderaspatana*	4.38	3.23	1.35
35	*Oxystelma esculentum*	1.88	1.77	1.06
36	*Passiflora foetida*	0.31	0.31	1.00
37	*Pentatropis capensis*	0.10	0.10	1.00
38	*Pergularia daemia*	3.13	2.19	1.43
39	*Rhynchosia minima*	2.19	1.67	1.31
40	*Rivea hypocrateriformis*	0.31	0.10	3.00
41	*Telosma pallida*	0.10	0.10	1.00
42	*Teramnus labialis*	0.10	0.10	1.00
43	*Tinospora cordifolia*	23.85	21.46	1.11
44	*Tragia involucrata*	0.21	0.21	1.00
45	*Tylophora indica*	7.19	5.94	1.21
46	*Vigna aconitifolia*	0.42	0.31	1.33
47	*Wattakaka volubilis*	1.04	1.04	1.00

3.2 Phenology (Leaf Flushing, Flowering, Fruiting, Leaf-Fall) of the Species

The study of periodic biological phenomena of plants, for example, budding, leaf flushing, flowering, fruiting, seed dispersal, pollination with relation to the weather and climatic conditions of the area where study, is being conducted. Therefore, study of seasonal appearance and annual cyclic phases of plants is the phenology and each stage in the annual cycle of plants is called pheno-phase (Visser et al. 2010). The entire process is the study of the response of living organisms to climatic changes in the environment in which they live. In studied semiarid region, the phenological observations were made on 66 trees and 53 shrubs species. As described earlier, the phenology and climate changes are closely related; therefore, the study conducted on phenology of the trees and shrubs will have its own significance when similar study is being done on this aspect in the temporal dynamics. In figure, the findings of the study will offer evidence of climate change happening now and help in assessment of the significant effect on plants in future. This data will help in providing the information on the changes that are happening in physical environment of the region. Hence, the study will be useful in providing baseline information on the phonological aspects of the species of the semiarid region. Based on the phytosociological studies, it is being observed that the semiarid landscape has the diversity in the species, but the density of the species is low; therefore, the region appears as sparse and scanty in terms of vegetation structure. The vegetation in the study region is of the dry deciduous type and leaf-fall of most of the tree species coincided with the dry season. A total of 119 species of top layer (tree) and understory layer (shrubs) belonging to 44 families were closely monitored for their pheno-phases (Table 3.7).

3.2.1 Phenology of the Trees

The species have different flowering and fruiting seasons depending on the climatic features such as temperature, rainfall, and humidity (Fig. 3.3). During the field observations, it was found that four species were flowering throughout the year and seven species were flowering for seven months of the year. In some species, flowering was noticed for a period of two months. The detail description about pheno-phases of tree species is given in Table 3.7.

3.2.2 Phenology of Shrubs

Shrubs have a limited flowering and fruiting time when compared to trees. In the study area, there were fewer shrub species than trees and herbs. Shrub species are

Table 3.7 Phenology of trees species of the study region

S. no.	Scientific name	Family	Budding	Flushing	Flowering	Fruiting	Seed dispersal	Leaf-fall
1	*Acacia auriculiformis* A. Cunh. ex Benth.	Mimosaceae	Aug–Sep	May	Jun–Jul	Aug–Oct	Mar–May	Apr–May
2	*Acacia farnesiana* (L.) Willd.	Mimosaceae	Aug/Nov/ Feb–Mar	Apr–May	Mar–May	May–Jul	Aug–Oct	Mar–Apr
3	*Acacia leucophloea* (Roxb.) Willd.	Mimosaceae	Dec/Mar–Apr	May–Jun	Mar–May	Apr–Jun	Sep–Oct	Mar–May
4	*Acacia nilotica subsp. indica (L.)* Willd.	Mimosaceae	Aug/Nov–Dec/Feb	May	Mar–Apr	Apr–Jun	Jul–Aug	Mar–May
5	*Aegle marmelos* (L.) Corr.	Rutaceae	Aug/Nov–Dec	May	Oct–Nov	Dec–Apr	Apr–Jul	Mar–Apr
6	*Ailanthus excelsa* Roxb.	Simaroubaceae	Nov–Jan	May	Feb–Apr	Apr–Jul	Jul–Sep	Apr–May
7	*Albizia amara (*Roxb.) Bovin	Mimosaceae	Dec–Jan	Apr–May	Mar–Apr	May–Jul	Aug–Sep	Feb–Apr
8	*Albizia lebbeck* (L.) Benth.	Mimosaceae	Dec–Feb	May	Mar–Apr	May–Jul	Jul–Aug	Feb–Apr
9	*Annona squamosa* L.	Annonaceae	Jan–Feb	May	May–Jun	Jun–Oct	Oct–Nov	Apr–May
10	*Azadirachta indica* A. Juss.	Meliaceae	Dec–Jan	Apr–May	Mar–Apr	Apr–May	Jun–Aug	Feb–Apr
11	*Bauhinia racemosa* Lam.	Caesalpiniaceae	Nov–Jan/Mar–Apr	May	Mar–May	May–Jul	Aug–Oct	Mar–May
12	*Borassus flabellifer* L.	Arecaceae	Jan	Jan–Dec	Feb–Apr	Apr–Jun	Sep–Nov	Jan–Dec
13	*Carica papaya* L.	Caricaceae	Dec–Jan	Jan–Dec	Apr–May	Jul–Sep	Sep–Nov	Jan–Dec
14	*Cassia fistula* L.	Caesalpiniaceae	Jul–Aug/Nov–Dec	May–Jun	Apr–May	May–Aug	Sep–Nov	Feb–Apr
15	*Cassia javanica* L.	Caesalpiniaceae	Dec–Jan	May–Jun	Jun–Jul	Jul–Sep	Oct–Nov	Mar–May
16	*Chloroxylon swietenia* DC.	Rutaceae	Aug/Nov–Dec	Apr–May	Feb–Apr	Apr–May	May–Jun	Feb–Apr
17	*Cocos nucifera* L.	Arecaceae	Oct–Jan	Jan–Dec	Jan–Dec	Jan–Dec	Jan–Dec	Jan–Dec
18	*Cordia dichotoma* G. Forst.	Boraginaceae	Sep/Dec–Jan	Apr–May	Mar–Apr	Apr–May	Jun–Jul	Feb–Apr
19	*Cordia wallichii* G. Don	Boraginaceae	Sep/Dec–Jan	May	Feb–Mar	Apr–May	Jun–Jul	Mar–Apr
20	*Dalbergia sissoo* Roxb.	Fabaceae	Oct–Dec/Feb	Apr–Jun	Feb–Apr	May–Jul	Aug–Oct	Feb–Mar
21	*Delonix elata* (L.) Gamble	Caesalpiniaceae	Nov–Dec	May–Jun	Mar–May	May–Jul	Jul–Sep	Feb–May

(continued)

Table 3.7 (continued)

S. no.	Scientific name	Family	Budding	Flushing	Flowering	Fruiting	Seed dispersal	Leaf-fall
22	*Delonix regia* (Boj. ex Hook.) Raf.	Caesalpinaceae	Apr–May/Dec	May	Apr–Jun	Jul–Sep	Sep–Nov/ Mar–May	Mar–May
23	*Dichrostachys cinerea* (L.) Wight. & Arn.	Mimosaceae	Sep/Dec–Jan	May	Apr–May	May–Jul	Aug–Sep	Mar–Apr
24	*Dolichandrone atrovirens* (Heyne) Sprague	Bignoniaceae	Oct–Dec	May	Mar–Apr	Apr–May	Jul–Aug	Mar–Apr
25	*Dolichandrone falcata* Seem.	Bignoniaceae	Aug/Dec–Jan	May–Jun	Apr–May	May–Jun	Jul–Sep	Feb–Apr
26	*Eucalyptus globulus* Labill.	Myrtaceae	Aug/Nov–Jan	May	Feb–May	May–Jul	Jul–Aug	Mar–May
27	*Ficus arnottiana* (Miq.) Miq.	Moraceae		May–Jun	Feb–Apr	May–Jul	Aug–Sep	Apr–May
28	*Ficus benghalensis* L.	Moraceae		Mar–May	Feb–Mar	Mar–May	May–Jun	Feb–Mar
29	*Ficus racemosa* L.	Moraceae	Dec–Feb	May–Jun	Apr–May	May–Jul	Jul–Sep	Mar–May
30	*Ficus religiosa* L.	Moraceae	Jan–Feb/Apr	Apr–May	Feb–Mar	Mar–May	May–Jun	Jan–Mar
31	*Holoptelea integrifolia* (Roxb.) Planch.	Ulmaceae	Sep–Nov	May–Jun	Jan–Mar	Mar–May	Jul–Sep	Feb–Mar
32	*Couroupita guianensis* Aubl.	Bignoniaceae		May	Mar–Apr	May–Aug	Oct–Nov	Feb–May
33	*Leucaena leucocephala* (Lam.) de Wit	Mimosaceae	Jul–Aug/ Nov–Dec	May–Jul	Feb–Apr	May–Jun	Aug–Sep	Apr–May
34	*Limonia acidissima* L.	Rutaceae	15th Nov–Jan	Apr–May	Apr–May	Jun–Aug	Sep–Oct	Feb–May
35	*Mangifera indica* L.	Anacardiaceae	Nov–Dec	Apr–May	Dec–Feb	Mar–May	May–Jul	Feb–Apr
36	*Markhamia lutea* (Benth.) K.Schum.	Bignoniaceae	Nov–Jan	May	Mar–May	May–Jun	Jul–Sep	Mar–May
37	*Millingtonia hortensis* L.	Bignoniaceae	Dec–Jan	Feb–Mar	Mar–May	May–Jul	Jul–Aug	Dec–Feb
38	*Mimosa intisia* L.	Mimosaceae	Sep–Nov	May	Apr–May	May–Jul	Aug–Sep	Mar–May
39	*Morinda pubescens* J.E. Sm.	Rubiaceae	Nov–Jan	May	Apr–May	Jun–Jul	Aug–Sep	Mar–May
40	*Moringa oleifera* Lam.	Moringaceae	Dec–Jan	Apr–May	Jun–Jul	Jul–Sep	Sep–Nov	Mar–May
41	*Parkinsonia aculeata* L.	Caesalpinaceae	Nov–Dec	May	Feb–Mar	Apr–Jun	Aug–Oct	Jan–Mar

(continued)

Table 3.7 (continued)

S. no.	Scientific name	Family	Budding	Flushing	Flowering	Fruiting	Seed dispersal	Leaf-fall
42	*Phoenix sylvestris* (L.) Roxb.	Arecaceae	Dec–Jan	Jan–Dec	Jan–Mar	Apr–May	Jun–Aug	Jan–Dec
43	*Phyllanthus emblica* L.	Euphorbiaceae		Apr–May	May–Jun	Aug–Nov	Nov–Jan	Mar–Apr
44	*Pithecellobium dulce* (Roxb.) Benth.	Mimosaceae	Oct–Nov	Apr–May	Feb–Mar	Mar–May	May–Jul	Jan–Apr
45	*Pongamia pinnata* (L.) Pierre.	Fabaceae	Nov/Mar	Mar–Apr	Mar–May	May–Jul	Sep–Nov	Jan–Mar
46	*Prosopis cineraria* (L.) Druce.	Mimosaceae	Nov–Jan	Apr–May	Feb–Mar	Mar–May	May–Jul	Feb–May
47	*Prosopis juliflora* (Sw.) DC.	Mimosaceae	Aug/Nov–Jan/Mar	Apr–May	Mar–Apr	Apr–Jul	Aug–Oct	Mar–May
48	*Psidium guajava* L.	Myrtaceae	Dec–Jan	Apr–May	May–Jun/ Sep–Oct	Aug–Sep/ Jan–Feb	Feb–Sep	Mar–Apr
49	*Samanea saman* (Jacq.) Merr.	Mimosae	Jul–Aug/Dec	Mar–May	Mar–May	May–Jul	Jul–Sep/Mar	Feb–May
50	*Santalum album* L.	Santalaceae	Apr–May/ Oct–Nov	May	May–Jun	Jun–Aug	Aug–Sep	Mar–Apr
51	*Sapindus laurifolia* Vahl	Sapindaceae	Oct–Dec	May–Jun	Jan–Mar	Mar–May	May–Jul	Apr–May
52	*Senna siamea* (Lam.) Irwin & Barneby	Caesalpinaceae	Nov–Jan	Mar–Apr	Apr–May	May–Jul	Aug–Sep	Feb–Apr
53	*Spathodea campanulata* Beauv.	Bignoniaceae	Aug–Sep/Jan	May–Jun	Mar–May	May–Jul	Jul–Aug	Feb–Mar
54	*Syzygium cumini* (L.) Skeels.	Myrtaceae	Jan–Feb	Apr–May	Mar–Apr	May–Jul	Jul–Aug	Mar–Apr
55	*Tamarindus indica* L.	Caesalpinaceae	Aug–Sep/Jan	Feb–Apr	Dec–Feb	Feb–May	Jun–Jul	Jan–Mar
56	*Tecoma stans* (L.) Kunth	Bignoniaceae	Sep–Nov	May	Feb–Apr	Apr–May	May–Jul	Feb–May
57	*Tectona grandis* L. f.	Verbenaceae	Dec–Jan	May–Jun	Jun–Jul	Jul–Sep	Oct–Nov	Feb–Apr
58	*Terminalia bellirica* (Gaertn.) Roxb.	Combretaceae	Jan–Feb	May–Jun	Mar–Apr	May–Aug	Aug–Oct	Feb–May
59	*Thespesia populnea* (L.) Sol. ex Corr.	Malvaceae	Jul–Aug	Mar–Apr	Apr–May	May	Jun–Jul	Feb–Mar
60	*Wrightia tinctoria* R. Br. Mem. Wern.	Apocynaceae	Dec–Jan/Mar	May	Apr–May	May–Jul	Aug–Oct	Feb–Apr
61	*Ziziphus mauritiana* Lam.	Rhamnaceae	Nov–Dec	May–Jun	Jan–Mar	Apr–Jun	May–Jul	Apr–May
62	*Ziziphus oenoplia* Mill.	Rhamnaceae	Dec–Jan	May	May–Jun	Jun–Aug	Aug–Sep	Feb–Apr

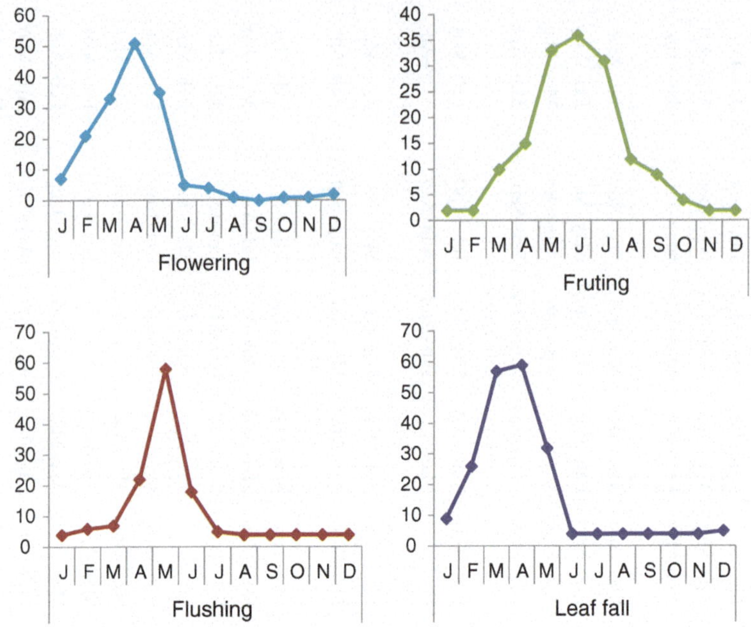

Fig. 3.3 Phenology of tree species

bushy, limited in growth with spines and thorns. The monitoring for flowering and fruiting timings for 53 species of shrubs was done during the study period. Nine species were found to have flowering throughout the year, while 12 species have flowering phase for four to eight months. The detailed description is given in Table 3.8.

3.3 Diversity Study of the Region

3.3.1 Shannon–Wiener Index (Alpha Diversity)

A complex community where high degree of species interaction is possible reveals high species diversity. Therefore, communities with greater diversities have high level of energy transfers (food webs) predation, competition, and niche availability. Low or changing species diversity indicates stress or unstable environment. Some ecologists, however, are of the view that there is no direct correlation between species diversity and community stability or community stress. The theory of species diversity takes into account three different ecological phenomena, i.e., species richness, relative abundance, and community evenness. These parameters can combine to give an excellent overall picture of the community structure. The Shannon–Wiener index for all the species belonging to trees, shrubs, herbs, and

Table 3.8 Flowering and fruiting period of shrub species and weather parameters

			Jan	Feb	Mar	Apr	May	Jun	Jul	Aug	Sep	Oct	Nov	Dec
	Temperature (°C)		26.2	29.2	32.3	34.6	41.1	40.09	38.9	38.1	30.5	35.7	28.5	27.3
	Rain fall (mm)		0	4	0	12	15	18	82	142	54	53	17	0
	Humidity (%)		42.25	39.48	28.3	29	38	56.6	71	71.6	74.1	63	68.9	52.8
S.no	**Scientific Name**	**Months**												
1	*Grewia villosa*	Jan-Feb	Jan	Feb										
2	*Ximenia americana*	Jan-Feb	Jan	Feb										
3	*Premna latifolia*	Jan-Mar	Jan	Feb	Mar									
4	*Turnera ulmifolia*	Jun-Sep						Jun	Jul	Aug	Sep			
5	*Lantana camara*	Jan-Dec	Jan	Feb	Mar	Apr	May	Jun	Jul	Aug	Sep	Oct	Nov	Dec
6	*Crossandra infundibuliformis*	Aug-Dec								Aug	Sep	Oct	Nov	Dec
7	*Jatropha glandulifera*	Jun-Jul						Jun	Jul					
8	*Ricinus communis*	Jan-Dec	Jan	Feb	Mar	Apr	May	Jun	Jul	Aug	Sep	Oct	Nov	Dec
9	*Hibiscus rosa-sinensis*	Jan-Dec	Jan	Feb	Mar	Apr	May	Jun	Jul	Aug	Sep	Oct	Nov	Dec
10	*Duranta repens*	Jan-Dec	Jan	Feb	Mar	Apr	May	Jun	Jul	Aug	Sep	Oct	Nov	Dec
11	*Xanthium indicum*	Jan-Dec	Jan	Feb	Mar	Apr	May	Jun	Jul	Aug	Sep	Oct	Nov	Dec
12	*Clerodendrum inerme*	Jan-Dec	Jan	Feb	Mar	Apr	May	Jun	Jul	Aug	Sep	Oct	Nov	Dec
13	*Calotropis gigantea*	Jan-Dec	Jan	Feb	Mar	Apr	May	Jun	Jul	Aug	Sep	Oct	Nov	Dec
14	*Calotropis procera*	Jan-Dec	Jan	Feb	Mar	Apr	May	Jun	Jul	Aug	Sep	Oct	Nov	Dec
15	*Ipomoea carnea*	Jan-Dec	Jan	Feb	Mar	Apr	May	Jun	Jul	Aug	Sep	Oct	Nov	Dec
16	*Vitex negundo*	Apr-Aug				Apr	May	Jun	Jul	Aug				
17	*Opuntia dillenii*	Feb-Jul		Feb	Mar	Apr	May	Jun	Jul					
18	*Securinega virosa*	Feb-Jul		Feb	Mar	Apr	May	Jun	Jul					
19	*Euphorbia tirucalli*	Mar-May			Mar	Apr	May							
20	*Securinega leucopyrus*	Mar-May			Mar	Apr	May							
21	*Grewia tenax*	Mar-May			Mar	Apr	May							
22	*Kirganelia reticulata*	Mar-Jun			Mar	Apr	May	Jun						
23	*Balanites aegyptiaca*	Mar-Jun			Mar	Apr	May	Jun						
24	*Gyrocarpus americanus*	Mar-Jun				Apr	May							
25	*Canthium coramandelicum*	Apr-May				Apr	May							
26	*Lawsonia inermis*	Apr-Jul				Apr	May	Jun	Jul					
27	*Nerium indicum*	Apr-Aug				Apr	May	Jun	Jul	Aug				
28	*Agave americana*	Jun-Jul						Jun	Jul					
29	*Catunaregam spinosa*	Jun-Sep						Jun	Jul	Aug	Sep			
30	*Randia brandisii*	Jun-Sep						Jun	Jul	Aug	Sep			
31	*Tephrosia purpurea*	Jun-Oct						Jun	Jul	Aug	Sep	Oct		
32	*Ixora arborea*	Jun-Oct						Jun	Jul	Aug	Sep	Oct		
33	*Senna auriculata*	Aug-Oct								Aug	Sep	Oct		
34	*Triumfetta rhomboidea*	Aug-Dec								Aug	Sep	Oct	Nov	Dec
35	*Carissa spinarum*	Aug-Mar	Jan	Feb	Mar					Aug	Sep	Oct	Nov	Dec
36	*Adhatoda zeylanica*	Sep-Nov									Sep	Oct	Nov	
37	*Pavonia zeylanica*	Sep-Nov									Sep	Oct	Nov	
38	*Indigofera tinctoria*	Sep-Nov									Sep	Oct	Nov	
39	*Alangium salvifolium*	Sep-Dec									Sep	Oct	Nov	Dec
40	*Senna occidentalis*	Sep-Dec									Sep	Oct	Nov	Dec
41	*Solanum nigrum*	Sep-Dec									Sep	Oct	Nov	Dec
42	*Clerodendrum phlomidis*	Sep-Dec									Sep	Oct	Nov	Dec
43	*Barleria prionitis*	Sep-Dec									Sep	Oct	Nov	Dec
44	*Leonotis nepetaefolia*	Sep-Jan	Jan								Sep	Oct	Nov	Dec
45	*Abutilon indicum*	Sep-Apr	Jan	Feb	Mar	Apr					Sep	Oct	Nov	Dec
46	*Phyllanthus reticulatus*	Sep-Apr	Jan	Feb	Mar	Apr					Sep	Oct	Nov	Dec
47	*Anisomeles malabarica*	Oct-Nov										Oct	Nov	
48	*Maytenus emarginata*	Oct-Dec										Oct	Nov	Dec
49	*Dodonaea viscosa*	Oct-Dec										Oct	Nov	Dec
50	*Stachytarpheta indica*	Nov-Dec											Nov	Dec
51	*Guilandina bonduc*	Nov-Dec											Nov	Dec
52	*Scutia myrtina*	Dec-Apr	Jan	Feb	Mar	Apr								Dec
53	*Cadaba fruticosa*	Dec-May	Jan	Feb	Mar	Apr	May							Dec

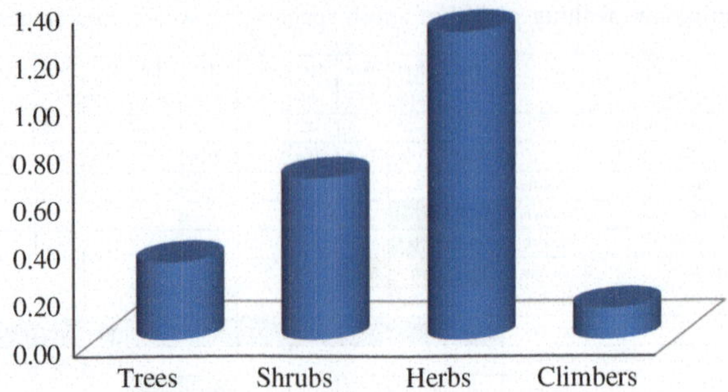

Fig. 3.4 Shannon–Wiener index (Alpha diversity) of the study region

climbers is depicted in Fig. 3.4. Herbs appear to be abundant among all vegetation class as showed high alpha diversity (1.30) followed by shrubs (0.58), trees (0.33), and climbers (0.13). The climbers are relatively less in the studied semiarid region.

3.3.2 Beta-Diversity

In the studied semiarid region, herbs and trees have shown high beta-diversity value 2.03 and 1.71, respectively. However, beta-diversity for shrubs and climbers is calculated to be 1.56 and 1.58, respectively (Fig. 3.5). These findings can help in understanding the diversity between the ecosystems of other semiarid regions of India and elsewhere.

Biological diversity can be quantified in many different ways. It is a measurement that explains the richness and the percentage of each from a biodiversity sample within a zone. The index assumes that the proportion of individuals in an area indicates their importance to biodiversity. Communities with a large number of species that are evenly distributed are the most diverse and communities with few

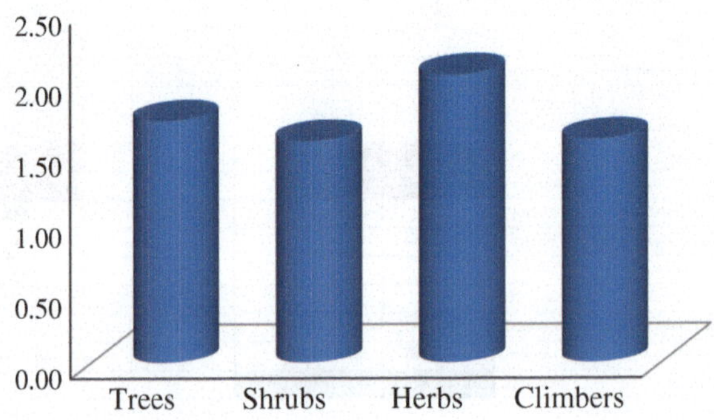

Fig. 3.5 Beta-diversity of the study region

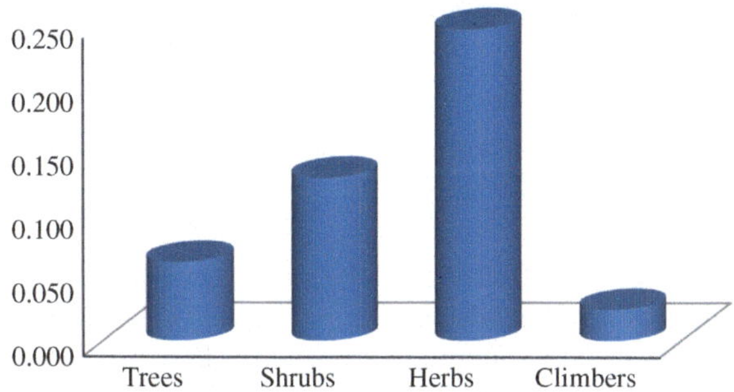

Fig. 3.6 Evenness of the study region

species that are dominated by one species are the least diverse. The two main factors taken into account when measuring diversity are richness and evenness. Figure 3.6 shows the evenness of the species in the study region. The relative abundance of the different species makes the richness of an area is called evenness. As species richness increases, the evenness also increase, so overall diversity increases. Figure 3.6 indicates that among the different vegetation strata the herbs are highly diverse followed by shrubs, trees and least diversity was observed among the climbers of the study region.

3.3.3 Simpson's Diversity Indices

Simpson's Index (D) measures the probability that two individuals randomly selected from a sample will belong to the same species (or some category other than species). There are two versions of the procedure for calculating Simpson's index (D). The value of D ranges between 0 and 1. With this index, 0 represents infinite diversity and 1 represents no diversity.

Simpson's Index of Diversity 1-D The value of this index also ranges between 0 and 1, but now, the greater the value, the greater the sample diversity. In this case, the index represents the probability that two individuals randomly selected from a sample will belong to different species.

Simpson's reciprocal index 1/D The value of this index starts with 1 as the lowest possible figure. This figure would represent a community containing only one species. The higher the value, greater the diversity in the ecosystems and landscapes. The maximum value is the number of species in the sample (http://www.countrysideinfo.co.uk/simpsons.htm).

Table 3.9 indicates Simpson's diversity Indices (D), 1/D, and 1-D for all the vegetation layers (trees, shrubs, herbs, and climbers) of the studied landscape of semiarid region for the total no plant species, which includes trees, shrubs, herbs,

Table 3.9 Simpson's diversity Indices (D), 1/D, and 1-D for plant biodiversity of the studied semiarid region

(D)	1/D	1-D
0.04	2.65	0.29

and climbers. The values of D, 1/D, and 1-D are calculated to be 0.04, 2.65, and 0.29 for the study region. The value of D gives more weight to the more abundant species in an ecosystem. The addition of rare species to a sample causes only small changes in the value of D.

The 1/D values show that trees and herbs have higher diversity in the region. However, shrubs and climbers show the diversity in the region; the values are less in comparison with the trees and herbs. The various types of ecosystems and habitats have not been compared for diversity. Table 3.10 shows the Simpson's index (D), Simpson's reciprocal index 1/D, and Simpson's index of diversity 1-D for vegetation classes (trees, shrubs, herbs, and climbers) of the studied semiarid landscape.

The available relict forest patches in the semiarid region are degrading rapidly due to high anthropogenic pressures. Hence, in situ and ex situ conservation measures are urgently required. These patches are subject to deforestation and fragmentation in many ways due to extraction of timber, fuel wood, fodder, leaves, etc. A systematic and scientific research is required for mass multiplication and revival of deciduous forests in the semiarid region. Efforts are urgently required for plantation of multipurpose tree species (for example *Melia dubia*) with indigenous species, ecologically suitable for the xeric desert environment. Technological interventions are also required to improve the quality and resistance against drought and climate change. The seedlings of various species suited to the semiarid environment need to be distributed among the local people for multiplication of these species in the private and community lands for meeting the requirements of wood, leaf litter, fiber, fodder, etc. Seed orchards and arboretum establishment is suggested for mass multiplication of the existing valuable plant taxa. The traditional ecological knowledge of the people with regard to utilization of a particular plant species also provides valuable information on the status, their use pattern, and dependency of the local communities on the available bioresources. Therefore, the

Table 3.10 Simpson's index (D), 1/D, and 1-D for the vegetation classes (trees, shrubs, herbs, and climbers)

	Simpson's index (D)	Simpson's reciprocal index 1/D	Simpson's index of diversity 1-D
Trees	0.12	11.7	0.88
Shrubs	0.11	9.2	0.89
Herbs	0.11	10.2	0.89
Climbers	0.11	9.3	0.89

documentation of traditional ecological knowledge is important not only for enumeration about the species, but this supports India's initiatives of protecting TEK from patent issues. In-depth studies in traditional ecological knowledge help in a way so that the noncodified information can be codified for development of people and conservation of biodiversity.

3.4 Traditional Ecological Knowledge (TEK) Related to Use of Plant Biodiversity

The survey was done in randomly selected villages of the study area. Data sheets were prepared comprising the village name, locality, name of the informer, age groups, and plant description. The data were collected by approaching the traditional healer and local people, firstly for asking questions about the different uses of plants (medicinal, food, timber, fodder, and fuel) and secondly to make an inventory of species, which includes collection of plant specimens and interviewing the informants for names and uses (Nautiyal et al. 2002). All the information was obtained from knowledgeable people like elders and traditional healers of different communities. All the collected samples prepared for the herbarium were identified with experts and preserved some plant parts such as bark, roots, leaves, dry fruits were also preserved in the institute.

The traditional ecological knowledge related to plant biodiversity of the study region was documented in detail. The detailed phytosociological study has been conducted in this region in different seasons using standard methods. A total of 80 species of medicinal plants belonging to 40 different families are found to be used by traditional healers and local people in the traditional healthcare system to cure various ailments. The survey was done in different seasons in 20 selected villages of the study region. In this region, a few tribal communities, called as Lambadi or Lambanis, live in settlements known as Thanda. Of the total medicinal plants, 35 % are tree species, followed by herbs (31 %), shrubs (19 %), and climbers (15 %) in the study region. The information on medicinal plant resources which includes scientific names, family, and common names is given in Table 3.11. The information on the traditional knowledge of the people with regard to use of species while using various parts, preparation of traditional medicine, amount of dosages, and number of days prescribed for curing the various ailments is given in Table 3.12.

The study reveals that about 21 diseases are cured by the traditional healers while using the local plant species (Fig. 3.7). Maximum numbers of species are used to cure skin-related diseases (25) followed by wounds (15), stomach disorders (15), cold and cough (11), diabetes (8), etc. Some species are used to cure more than one disease following various traditional methods as the knowledge of the local healers. Only 1–2 species are used to cure mouth ulcer, cardiac problems, and earache in the region (Fig. 3.8).

Table 3.11 Medicinal plants of the studied semiarid region

No	Scientific name	Family	Common name
1	*Abrus precatorius* L.	Fabaceae	Gulagangi
2	*Acacia leucophloea* (Roxb.) Willd.	Mimosaceae	Bili jalli
3	*Acacia nilotica* (L.) Willd	Mimosaceae	Kari jali
4	*Acalypha indica* L.	Euphorbiaceae	Kuppe gida
5	*Achyranthes aspera* L.	Amaranthaceae	Uttarani
6	*Adhatoda zeylanica* Medic.	Acanthaceae	Adusoge
7	*Aegle marmelos* (L.) Corr.	Rutaceae	Bilvapatre gida
8	*Agave americana* L.	Amaryllidaceae	Rakspatti
9	*Aloe vera* (L.) N. Burm.	Liliaceae	Kathaligidi
10	*Amaranthus spinosus* L.	Amaranthaceae	Mulludantu
11	*Anisomeles malabarica* (L.) R. Br. ex Sims	Lamiaceae	Gandu beerana gida
12	*Annona squamosa* L.	Annonaceae	Sithaphal
13	*Argemone mexicana* L.	Papaveraceae	Daturi gida
14	*Azadirachta indica* A. Juss.	Meliaceae	Bevina mara
15	*Bacopa monnieri* (L.) Wett.	Scrophulariaceae	Niru brahmi
16	*Balanites aegyptiaca* (L.) Delile	Balanitaceae	Ingalarade
17	*Basella alba* L.	Basellaceae	Baayi basale
18	*Calotropis gigantea* (L.) R. Br.	Asclepidaceae	Yekkada gida
19	*Cardiospermum halicacabum* L.	Sapindaceae	Agni balli
20	*Cassia fistula* L.	Caesalpinaceae	Kakke gida
21	*Cissus quadrangularis* L.	Vitaceae	Mangaravalli
22	*Cleome viscosa* L.	Capparaceae	Nayi sasuve
23	*Clitoria ternatea* L.	Fabaceae	Shankapushpi gida
24	*Coccinia grandis* (L.) Voigt	Cucurbitaceae	Tonde balli
25	*Cocos nucifera* L.	Arecaceae	Tengina mara
26	*Coleus aromaticus* Benth.	Lamiaceae	Thodapatre
27	*Croton bonplandianum* Baill.	Euphorbiaceae	Utti gida
28	*Cyperus rotundus* L.	Cyperaceae	
29	*Datura metel* L.	Solanaceae	Ummatti gida
30	*Delonix elata* (L.) Gamble	Fabaceae	Bile gulmohr
31	*Diplocyclos palmatus* (L.) Jeffrey	Cucurbitaceae	Linge toned balli
32	*Dodonaea viscosa* (L.) Jacq.	Sapindaceae	Bandarike
33	*Dolichandrone falcata* Seem	Bignoniaceae	Oodi mara
34	*Echinops echinatus* Roxb.	Astaraceae	Brahmadande
35	*Eclipta alba* L.	Astaraceae	Bhrungaraja
36	*Eucalyptus globulus* Labill.	Myrtaceae	Neelgiri
37	*Euphorbia hirta* L.	Euphorbiaceae	Achche gida
38	*Ficus racemosa* L.	Moraceae	Attimara
39	*Ficus religiosa* L.	Moraceae	Arali mara
40	*Flaveria trinervia* (Spreng.) Mohr.	Astaraceae	

(continued)

Table 3.11 (continued)

No	Scientific name	Family	Common name
41	*Gymnema sylvestre* (Retz.) R. Br. ex Schult	Asclepiadaceae	Kodapathre
42	*Hemidesmus indicus* (L.) Schult	Asclepidaceae	Suganhaberu
43	*Hibiscus rosa-sinensis* L.	Malvaceae	Dasavala
44	*Jatropha glandulifera* Roxb.	Euphorbiaceae	Totla gida
45	*Lawsonia inermis* L.	Lythraceae	Mehndi gida
46	*Leucaena leucocephala* (Lam.) de Wit	Mimosaceae	Sanna sogachu
47	*Limonia acidissima* L.	Rutaceae	Belada gida
48	*Moringa oleifera* Lam.	Moringaceae	Nugge mara
49	*Mukia maderaspatana* (L.) Roem.	Cucurbitaceae	Gubbi savatikayi
50	*Murraya koenigii* (L.) Spreng.	Rutaceae	Karebevu
51	*Ocimum basilicum* L.	Lamiaceae	Sritulasi
52	*Opuntia dillenii* (Ker-Gawl.) Haw.	Cactaceae	Papas kalli
53	*Pergularia daemia* (Forsk.) Chiov.	Asclepidaceae	Kurudigana balli
54	*Phyllanthus amarus* Schum. & Thonn.	Euphorbiaceae	Bhoo nelli
55	*Phyllanthus emblica* L.	Euphorbiaceae	Bettada nelli
56	*Phyllanthus reticulatus* Poir.	Euphorbiaceae	Chippullimullu
57	*Pongamia pinnata* (L.) Pierre.	Fabaceae	Honge mara
58	*Psidium guajava* L.	Myrtaceae	Seebe gida
59	*Punica granatum* L.	Punicaceae	Dalimbre
60	*Ricinus communis* L.	Euphorbiaceae	Oudala gida
61	*Santalum album* L.	Santalaceae	Gandadamara
62	*Sapindus laurifolia* Vahl.	Sapindaceae	Antuvala
63	*Senna auriculata* L.	Caesalpinaceae	Honnambre gida
64	*Sida spinosa* L.	Malvaceae	
65	*Solanum xanthocarpum* Schrad & Wendl	Solanaceae	Kantakari
66	*Spathodea campanulata* Beauv.	Bignouniaceae	Neru kai
67	*Syzygium cumini* (L.) Skeels.	Myrtaceae	Nerale
68	*Tamarindus indica* L.	Fabaceae	Hunase mara
69	*Tephrosia purpurea* (L.) Pers.	Fabaceae	Koggi gida
70	*Terminalia bellirica* (Gaertn.) Roxb.	Combretaceae	Taremara
71	*Tinospora cordifolia* (Willd.) Hook. f. & Thoms.	Menispermaceae	Amrutha balli
72	*Tridax procumbens* L.	Astaraceae	Kari balli
73	*Tylophora indica* (Burm. f.) Merr.	Asclepidaceae	Adumuttada balli
74	*Typha angustifolia* L.	Typhaceae	Aapu
75	*Vinca rosea* (L.) G. Don	Apocynaceae	Kasikanagila
76	*Vitex negundo* L.	Verbenaceae	Lakki gida
77	*Wrightia tinctoria* R. Br.	Apocynaceae	Kodu muriki
78	*Ziziphus mauritiana* Lam.	Rhamnaceae	Borae hannina mara

Table 3.12 Use of medicinal plants (plant parts used, preparation of local medicine and dosages prescribed and to cure various ailments by the traditional healers)

S. no.	Scientific name	Family	Common name	Parts used	Type of diseases	Preparation and use	Dosage	No. of days
1	*Abrus precatorius* L.	Fabaceae	Gulagangi	Leaves, seeds	Skin diseases	Seeds dried and powdered, then mixed with water/ coconut oil and made into fine paste. This is applied on infected portion	10 g (2)	5
2	*Acacia leucophloea* (Roxb.) Willd.	Mimosaceae	Bili jalli	Bark	Skin diseases	Leaf extract and paste of stem bark is used for treatment of skin disease and boils	10 g (2)	7
3	*Acacia nilotica* (L.) Willd	Mimosaceae	Kari jali	Bark	Wounds	Powder of stem bark is mixed with coconut oil and then applied on wounds	5–10 g (2)	6
4	*Acalypha indica* L.	Euphorbiaceae	Kuppe gida	Leaves	Skin diseases, urinary problems	Leaves crushed with salt and the paste prepared is applied on the body to cure skin diseases. Leaf extract with buttermilk is taken to cure urinary problems	15 g 2 spoons (3)	6

(continued)

Table 3.12 (continued)

S. no.	Scientific name	Family	Common name	Parts used	Type of diseases	Preparation and use	Dosage	No. of days
5	*Achyranthes aspera* L.	Amaranthaceae	Uttarani	Roots, leaves	Rheumatic pain, Foot cracks, and osmosis	The leaf paste is mixed with milk and given to cure the rheumatic pain. Fresh leaves are crushed and applied on foot to treat cracks	5 g (2)	8
6	*Adhatoda zeylanica* Medic.	Acanthaceae	Adusoge	Leaves	Skin diseases, tumors, cough	Leaves are crushed and extract is mixed with ginger to cure the cough. Leaf paste also used to treat skin diseases and tumors	1 spoon (2)	6
7	*Aegle marmelos* (L.) Corr.	Rutaceae	Bilvapatre gida	Leaves and fruits	Dysentery, wounds, stomach ache	Leaf powder is mixed with salt and used for treatment of stomach pain. Leaf extract is used to treat fever. Fruit extract is used to aid digestion	2 spoons (2)	5
8	*Agave americana* L.	Amaryllidaceae	Rakspatti	Roots	Cough	Root extract or powder is mixed with milk to cure cough	1 spoon (2)	2–3

(continued)

Table 3.12 (continued)

S. no.	Scientific name	Family	Common name	Parts used	Type of diseases	Preparation and use	Dosage	No. of days
9	*Aloe vera* (L.) N. Burm.	Liliaceae	Kathaligidi	Leaves	Block spots tumors, coolant	Fleshy part of the leaf is mixed with turmeric and applied on the face to remove the black spots and its consumption reduces heat and weight of the body and also reduces tumors	20 g (1)	10–12
10	*Amaranthus spinosus* L.	Amaranthaceae	Mulludantu	Leaves	Bleeding, digestion	Fried leaves are used by women to reduce excessive bleeding during the menstruation period. Leaf decoction is good for digestion	50 g (3)	4–5
11	*Anisomeles malabarica* (L.) R. Br. ex Sims	Lamiaceae	Gandu beerana gida	Leaves	Cough, cold	Leaf extract mixed with salt is used to relieve the cough and cold. Dry leaf powder is given to infants in case of cough or cold	15 g (2)	4–5

(continued)

Table 3.12 (continued)

S. no.	Scientific name	Family	Common name	Parts used	Type of diseases	Preparation and use	Dosage	No. of days
12	*Annona squamosa* L.	Annonaceae	Sithaphal	Seeds	Headache	Powder of the seeds is used to reduce headache and used in head wash to remove lice	10 g (2)	7–9
13	*Argemone mexicana* L.	Papaveraceae	Daturi gida	Roots	Skin diseases bleeding	Dry roots and bark of mango tree are crushed to prepare powder and taken with sheep milk during menstruation period to reduce excessive bleeding, and whitish fluid. Roots are crushed to prepare paste which is mixed with turmeric and used to treat skin diseases	10–15 g (3)	2–3

(continued)

Table 3.12 (continued)

S. no.	Scientific name	Family	Common name	Parts used	Type of diseases	Preparation and use	Dosage	No. of days
14	*Azadirachta indict* A. Juss.	Meliaceae	Bevina mara	Leaves and stem	Skin diseases, bleeding, snake bite, wounds, diabetes	Neem leaves are given to the victim to confirm snake bite—if the person does not get bitter taste, it implies snake bite. Leaf paste is used to cure skin disease and as a cure to chicken pox. Leaf paste is good for diabetes patients. Parts of the stem are used as toothbrush	5–10 g	10–15
15	*Bacopa monnieri* (L.) Wett.	Scrophulariaceae	Niru brahmi	Stem with leaves	Kidney problem, cough	Whole plant is crushed and extract is taken to cure kidney problems and cough	20 ml (2)	3–5
16	*Balanites aegyptiaca* (L.) Delile	Balanitaceae	Ingalarade	Fruit	Jaundice	Fruit pulp with milk is used to prepare small tablet which is taken to cure jaundice	5 g (1)	7

(continued)

Table 3.12 (continued)

S. no.	Scientific name	Family	Common name	Parts used	Type of diseases	Preparation and use	Dosage	No. of days
17	*Basella alba* L.	Basellaceae	Baayi basale	Leaves	Burns spots, mouth ulcer	Leaf paste is used to treat burn spots and leaf extract is used to reduce mouth ulcers. Ripe fruits are edible	10 g (1)	15–20
18	*Calotropis gigantea* (L.) R. Br.	Asclepidaceae	Yekkada gida	Leaves, latex flower, root	Wounds, cold, head ache	Latex is used to treat wounds. Leaves extract is poured through nose to reduce cold and cough	3–4 drops (2)	2–3
19	*Cardiospermum halicacabum* L.	Sapindaceae	Agni balli	Leaves	Joint pains	Leaf paste with sesamum oil is applied on the joints to relieve the joint pains	15–20 g (2)	7–10
20	*Cassia fistula* L.	Caesalpinaceae	Kakke gida	Leaves, roots	Fever, cardiac problems, scorpion bite, breast diseases	Leaf extract is used to treat fever and cardiac diseases. Root can be used for breast diseases. Fruits are used to treat scorpion bite. Whole plant is used to treat accidental wounds	2 spoon (3)	2

(continued)

Table 3.12 (continued)

S. no.	Scientific name	Family	Common name	Parts used	Type of diseases	Preparation and use	Dosage	No. of days
21	*Cissus quadrangularis* L.	Vitaceae	Mangaravalli	Stem	Wounds, fracture	Stem is crushed to prepare paste and applied on the wound and fracture	15 g (2)	5–7 15–20
22	*Cleome viscosa* L.	Capparaceae	Nayi sasuve	Leaves	Ear ache and skin diseases	Leaf extract is used to treat ear pain. Whole plant is used to treat skin diseases	2 drop (1)	3
23	*Clitoria ternatea* L.	Fabaceae	Shankapushpi gida	Leaves, flower	Tooth ache	Leaves and flowers are chewed to relieve tooth pain	5 g (2)	5–7
24	*Coccinia grandis* (L.) Voigt	Cucurbitaceae	Tonde balli	Leaves, fruits	Skin diseases diabetes	Leaf paste is used to treat skin diseases and fractures. Ripened fruits are edible	10–15 g	5–7
25	*Cocos nucifera* L.	Arecaceae	Tengina mara	Coir, shell	Pain relief in joints, wounds, rash	The coir of coconut is crushed and mixed with sesamum oil and then applied on joints to treat joint pains. The shell is heated, crushed, then mixed with coconut oil, and applied on wounds and rashes	25 g (1)	10–15

(continued)

Table 3.12 (continued)

S. no.	Scientific name	Family	Common name	Parts used	Type of diseases	Preparation and use	Dosage	No. of days
26	*Coleus aromaticus* Benth.	Lamiaceae	Thodapatre	Leaves	Diabetes	Leaf powder with salt to cure diabetes	2 spoons (2)	–
27	*Croton bonplandianum* Baill.	Euphorbiaceae	Utti gida	Leaves, latex	Wounds	Leaf paste is used to treat wounds and ringworm. Latex is used to cure mouth ulcer	5 g (2)	8–10
28	*Cyperus rotundus* L.	Cyperaceae		Rhizome	Hair fall	Dried rhizome powder is mixed with coconut oil and applied to control hair fall	10 g (2)	10–15
29	*Datura metel* L.	Solanaceae	Ummatti gida	Leaves	Scorpion bite, tumors	Leaf extract with salt is used to treat scorpion bite. Burned leaves are used to control tumors. Leaf extract and seed powder is applied on hair to control hair fall and dandruff	6–7 drops 10 ml + 2 spoon	12–15
30	*Delonix elata* (L.) Gamble	Fabaceae	Bile gulmohr	Leaves	Bone fracture	Leaf and bark paste and cold water is used to treat fractures	25–30 g	4–8 weeks

(continued)

Table 3.12 (continued)

S. no.	Scientific name	Family	Common name	Parts used	Type of diseases	Preparation and use	Dosage	No. of days
31	*Diplocyclos palmatus* (L.) Jeffrey	Cucurbitaceae	Linge toned balli	Leaves	Skin diseases, cough	Leaf extract with turmeric is used cure the skin disease and cough	10–15 g (1)	4–7
32	*Dodonaea viscosa* (L.) Jacq.	Sapindaceae	Bandarike	Leaves	Wounds, swelling, burns	Leaf extract is applied to treat wound, swelling, and burns. Leaf paste used to relieve pain in the bones	10 ml (2)	10–12
33	*Dolichandrone falcata* Seem	Bignoniaceae	Oodi mara	Leaves, seeds	Bone fractures	Leaves are crushed with salt and applied on fracture then the affected area is tied with wood stick	20–30 g	4–6 weeks
34	*Echinops echinatus* Roxb.	Astaraceae	Brahmadande	Roots	Anthelminthic in animals	Roots are boiled with water and extract is collected, given to animal for cure the anathematic	50 ml (1)	4–5

(continued)

Table 3.12 (continued)

S. no.	Scientific name	Family	Common name	Parts used	Type of diseases	Preparation and use	Dosage	No. of days
35	*Eclipta alba* L.	Astaraceae	Bhrungaraja	Leaves and stem	Wounds, swellings, cough, skin disease, hair growth, and dandruff	Whole plant extract is used to treat cough. Juice of the plant is used to cure swelling; leaves are crushed and made into paste for wound/dandruff as well as hair growth. Leaves are used as tooth paste	10 g (3)	4–6
36	*Eucalyptus globulus* Labill.	Myrtaceae	Neelgiri	Leaves	Pain relief	Fresh leaves are dipped into hot water and tied with cotton cloth for joint pain relief. Leaf paste and neem leaves paste is applied to treat skin disease	15–20 L	3–4
37	*Euphorbia hirta* L.	Euphorbiaceae	Achche gida	Roots	Urinary disorder	Root powder is mixed with Cuminum cyminum is used for urinary disorder	2 spoon (3)	7

(continued)

Table 3.12 (continued)

S. no.	Scientific name	Family	Common name	Parts used	Type of diseases	Preparation and use	Dosage	No. of days
38	*Ficus racemosa* L.	Moraceae	Attimara	Fruits, bark, latex	Stomach pain	Fruits are edible. The fruits are good digestive and stomach pain reliever. Bark and latex are applied on wounds	–	–
39	*Ficus religiosa* L.	Moraceae	Arali mara	Leaves, bark	Black spots	Leaves are crushed and mixed with coconut oil, and applied on black spots on skin. Extract of bark is used to treat fever	10–15 ml (1)	2–3
40	*Flaveria trinervia* (Spreng.) Mohr.	Astaraceae		Leaves	Wounds	Leaf paste is applied to cure wounds	5 g (3)	8–10
41	Gymnema sylvestre (Retz.) R. Br.ex Schult	Asclepiadaceae	Kodapathre	Leaves	Diabetes, skin diseases	Leaves and jambuva seeds are crushed to make fine paste and that is taken to control diabetes. Leaf paste is used to cure leprosy	2 spoons (2)	–

(continued)

Table 3.12 (continued)

S. no.	Scientific name	Family	Common name	Parts used	Type of diseases	Preparation and use	Dosage	No. of days
42	*Hemidesmus indicus* (L.) Schult	Asclepidaceae	Suganhaberu	Roots	Lactation, reduce body heat,	Roots have sweet odor. Root extract is good for improve lactation in women particularly in pregnancy period. Root powder is used to prepare soft drinks; it helps reduce heat from the body	10 ml (2)	20–30
43	*Hibiscus rosa-sinensis* L.	Malvaceae	Dasavala	Leaves and flowers	Hair growth and dandruff	Crushed leaves and fruits of *Sapindus laurifolia* applied on hair for good growth and to maintain black color. Leaf paste is used to control hair fall and dandruff	15–20 g (1)	–
44	*Jatropha glandulifera* Roxb.	Euphorbiaceae	Totla gida	Latex	Mouth ulcer	Latex is used to cure mouth ulcer	4–5 drop (2)	5–7

(continued)

Table 3.12 (continued)

S. no.	Scientific name	Family	Common name	Parts used	Type of diseases	Preparation and use	Dosage	No. of days
45	*Lawsonia inermis* L.	Lythraceae	Mehndi gida	Leaves	Hair fall, skin diseases	Leaf paste is applied on hair to control hair fall and dandruff. Leaves paste is used to cure skin diseases	15–20 g (1)	–
46	*Leucaena leucocephala* (Lam.) de Wit	Mimosaceae	Sanna sogachu	Pod and bark	Burning micturition	Bark and Pod powder is used to treat burning micturition	10 g (1)	–
47	*Limonia acidissima* L.	Rutaceae	Belada gida	Fruit	Vomiting, digestive	Fruits are edible and good cure for digestive problems	–	–
48	*Moringa oleifera* Lam.	Moringaceae	Nugge mara	Leaves, flower, fruit	Piles, skin diseases, digestive	Fresh leaves and fruits are used as vegetable to prepare as curry/sambar. Leaf fry is good for piles and digestive. Leaf extract is used to treat skin diseases	7–10 ml (3)	7–10
49	*Mukia maderaspatana* (L.) Roem.	Cucurbitaceae	Gubbi savatikayi	Leaves	Skin diseases	Crush the leaves with turmeric prepare paste apply on skin	5–10 g (2)	7–10

(continued)

Table 3.12 (continued)

S. no.	Scientific name	Family	Common name	Parts used	Type of diseases	Preparation and use	Dosage	No. of days
50	*Murraya koenigii* (L.) Spreng.	Rutaceae	Karebevu	Leaves	Burning sensation, skin diseases	Leaves are used as vegetable. Leaf extract and paste are used to treat burning sensation and skin diseases. Leaves are good for digestion	10 ml (2)	5
51	*Ocimum basilicum* L.	Lamiaceae	Sritulasi	Leaves	Cold, cough fever	Leaf decoction used to treat cold, cough, and fever	10 ml (2)	2–3
52	*Opuntia dillenii* (Ker-Gawl.) Haw.	Cactaceae	Papas kalli	Fruit	Cooling	Fruits are edible; fleshy part of leaf is considered as cooling agent	–	–
53	*Pergularia daemia* (Forsk.) Chiov.	Asclepidaceae	Kurudigana balli	Leaves and latex	Bone fractures	Leaf extract and paste are used to treat bone fracture. Latex is used to cure wounds and bone pains	20–25 g	4–8 weeks
54	*Phyllanthus amarus* Schum. & Thonn.	Euphorbiaceae	Bhoo nelli	Whole plant	Jaundice	Whole plant is crushed to prepare paste which is taken with butter milk to reduce jaundice	2–3 Tablet (1)	3 weeks

(continued)

Table 3.12 (continued)

S. no.	Scientific name	Family	Common name	Parts used	Type of diseases	Preparation and use	Dosage	No. of days
55	*Phyllanthus emblica* L.	Euphorbiaceae	Bettada nelli	Fruits	Digestive	Fruits are edible and used to cure digestive problems. Fruits are used to treat cardiac disease, fever, and vomiting	2–3 Spoons (2)	–
56	*Phyllanthus reticulatus* Poir.	Euphorbiaceae	Chippullimullu	Leaves	Control continuous lactation in animals	Leaf paste with asafoetida powder is used to reduce over-lactation in animals	20 g (1)	5–7
57	*Pongamia pinnata* (L.) Pierre.	Fabaceae	Honge mara	Leaves, seeds	Wounds, cardiac problem	Oil from the seeds is used to treat migraine. Leaves and seed paste with turmeric is used to treat skin diseases. Stem parts are used as tooth brush. Crush the leaves along with Vitex *negundo* apply on skin to reduce swellings and skin infections. Seeds' oil is used as pain relief	10 ml (2) 10 g (1)	7 10–15

(continued)

Table 3.12 (continued)

S. no.	Scientific name	Family	Common name	Parts used	Type of diseases	Preparation and use	Dosage	No. of days
58	*Psidium guajava* L.	Myrtaceae	Seebe gida	Fruit and leaves	Diabetes, tooth ache, digestive	Fruits are edible. Tender leaves are eaten with tamarind. Stem parts are used as tooth brush to reduce tooth ache	–	–
59	*Punica granatum* L.	Punicaceae	Dalimbre	Fruit and bark	Dysentery	Fruits and bark powder mix with buttermilk is taken to cure dysentery and diarrhea. Outer layer of fruit chewed to cure stomach problems. Fruits are edible	10 g (3)	2–3
60	*Ricinus communis* L.	Euphorbiaceae	Oudala gida	Leaves	Coolant, wounds, piles	Leaf extract is used for cough, piles, and skin diseases. Leaves and seed oil are used as cooling agent. Seed oil is used as tongue cleaner in infants	–	–

(continued)

Table 3.12 (continued)

S. no.	Scientific name	Family	Common name	Parts used	Type of diseases	Preparation and use	Dosage	No. of days
61	*Santalum album* L.	Santalaceae	Gandadamara	Stem and leaves	Skin diseases	Stem powder and leaf paste are applied externally to treat skin diseases. Santalum oil is used as cooling agent	5–10 g (2)	5–7
62	*Sapindus laurifolia* Vahl.	Sapindaceae	Antuvala	Fruits	To treat poison, hair growth	Crushed fruits mixed with water are taken to treat vomiting	50 ml	–
63	*Senna auriculata* L.	Caesalpinaceae	Honnambre gida	Bark and leaves	Burning sensation and stomach pain, injury, dysentery	Dry bark powder mixed with coconut oil and applied on the burning region. Stem bark is consumed to control dysentery and digestive problems. Leaf paste with coconut oil is used to treat wounds. Stem part are used as tooth brush	10 g (2)	1–2
64	*Sida spinosa* L.	Malvaceae		Leaves	Sexual problems in men	Leaf paste with milk is used to reduce continuous loss of sperm content	5–10 g (1)	7

(continued)

Table 3.12 (continued)

S. no.	Scientific name	Family	Common name	Parts used	Type of diseases	Preparation and use	Dosage	No. of days
65	*Solanum xanthocarpum* Schrad & Wendl	Solanaceae	Kantakari	Fruits	Digestive problems in animals, jaw pain	Fruit juice is used to treat digestive problems in animals. Boiled fruit extract is used to cure jaw pains in humans	50–100 ml (1)	2–3
66	*Spathodea campanulata* Beauv.	Bignouniaceae	Neru kai	Flowers	Bleeding and stomach pain	Soaked flowers are crushed and mixed with milk. This is taken to reduce stomach pain and bleeding	5–10 g (2)	–
67	*Syzygium cumini* (L.) Skeels.	Myrtaceae	Nerale	Leaves and fruit	Allergy, diabetes, digestive	Seeds powder with honey is used to treat diabetes. Leaf paste is used for allergy and vomiting. It will also kill the intestinal worms. Fruits are edible and it will dissolve hairs in intestine. Bark powder is used to treat wounds	10–15 g (2)	–

(continued)

Table 3.12 (continued)

S. no.	Scientific name	Family	Common name	Parts used	Type of diseases	Preparation and use	Dosage	No. of days
68	*Tamarindus indica* L.	Fabaceae	Hunase mara	Leaves, flower, bark	Hyperacidity, burns	Flower soaked in water over night and then eaten with honey for relieving acidity. Fruits contain high vitamin C	15–20 g (1)	7–10
69	*Tephrosia purpurea* (L.) Pers.	Fabaceae	Koggi gida	Roots	Indigestion	Root soup is for good digestion	50 g (1)	2–3
70	*Terminalia bellirica* (Gaertn.) Roxb.	Combretaceae	Taremara	Fruits	Cough and diarrhea	Fruit extract with castor oil or honey are used to cure cough and stomach disorders. Fruit powder with mother's milk is given to children for digestive problems	10 g (2) 5 g (1)	3–5

(continued)

Table 3.12 (continued)

S. no.	Scientific name	Family	Common name	Parts used	Type of diseases	Preparation and use	Dosage	No. of days
71	*Tinospora cordifolia* (Willd.) Hook. f. & Thoms.	Menispermaceae	Amrutha balli	Leaves and stem	Diabetes, fever, cold, jaundice	Leaf paste is applied on forehead to reduce fever. Leaves boiled in hot water and the vapors are inhaled for cold. The leaves and stems are prepared as a curry and eaten for diabetes. Stem and leaf paste is used to treat jaundice	10–15 g (2)	2–3
72	*Tridax procumbens* L.	Astaraceae	Kari balli	Leaves	Wounds	Leaf extract and paste is used to treat wounds and cuts. Leaf paste is used to relieve scorpion bite	5–10 g	3–5
73	*Tylophora indica* (Burm. f.) Merr.	Asclepidaceae	Adumuttada balli	Leaves and roots	Jaundice, whooping cough	Leaves are crushed with pepper and taken for whooping cough	10 g (2)	7–10
74	*Typha angustifolia* L.	Typhaceae	Aapu	Roots	Urinary disorder	Dried root powder is mix with buttermilk to cure urinary disorders	2 spoon (2)	1–2

(continued)

Table 3.12 (continued)

S. no.	Scientific name	Family	Common name	Parts used	Type of diseases	Preparation and use	Dosage	No. of days
75	*Vinca rosea* (L.) G. Don	Apocynaceae	Kasikanagila	Leaves and roots	Wounds, cuts, diabetes	Leaves are powdered and paste is used to treat wounds and cuts. Dry roots and leaves are used to cure diabetes	5–10 g (2)	4–7
76	*Vitex negundo* L.	Verbenaceae	Lakki gida	Leaves	Joint pains, wounds, skin diseases	Leaf paste with milk is applied to reduce bone pains and skin diseases. Leaf paste is also used for fractures	10–15 g (1)	7–10
77	*Wrightia tinctoria* R. Br.	Apocynaceae	Kodu muriki	Leaves	Wounds latex	Leaf is crushed with turmeric and the paste prepared is applied on the wounds. Latex is applied externally to reduce skin diseases	10–15 g (2)	3–5
78	*Ziziphus mauritiana* Lam.	Rhamnaceae	Borae hannina mara	Leaves and fruits	Wounds, scorpion sting	Leaf paste is applied externally to treat wounds, scorpion sting. Fruits are edible	10 g(2)	4–5

Note 1 Once daily, *2* Twice daily, *3* Thrice daily, − till complete cure

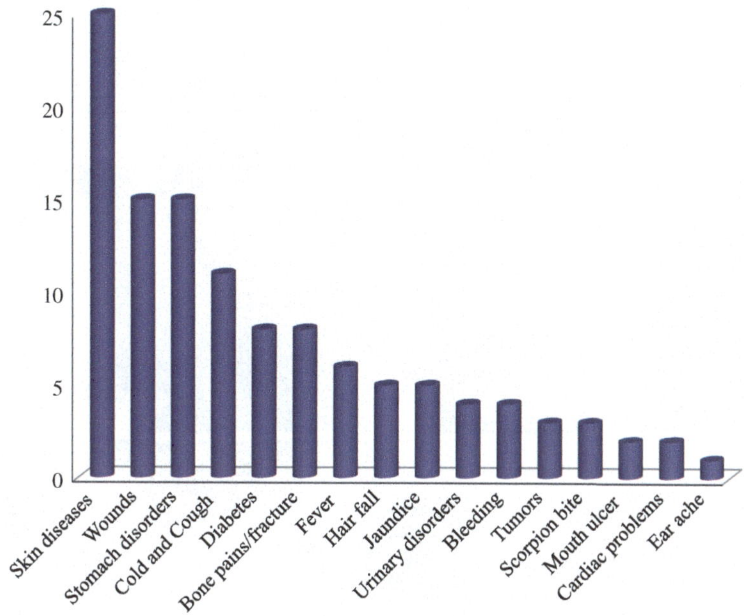

Fig. 3.7 Frequency of use of medicinal plants to cure the various ailments in the study area

Among the plants species used for medicinal purpose (Table 3.11) and the parts used (3.12), the use of leaves is high in preparation of herbal medicine (46 %) followed by fruits (14 %), roots (11 %), bark and stem (6 %), and whole plant (1 %) (Table 3.8).

3.5 Plants with Other Economic Values in the Study Area

The needs of humankind and biodiversity go hand in hand by the very nature of plants, animals, and other living organisms it encompasses. Human beings from time immemorial have been utilizing the bounties provided by planet Earth. There is an abundance of biodiversity in various ecosystems of the semiarid landscape such as scrub jungles, meadows, grass lands, fresh water pond and lakes, forests, wetlands, hillocks, and agricultural ecosystems. A good balance is essential between all these ecosystems with many ecological niches that are directly or indirectly of economic importance to man as well as for the survival of the species. The biodiversity also has great importance as a direct source of income and economic development. The flora of the study area contains many plant species that have been used in the variety of ways apart from the medicinal uses, for example, timber, edible fruits, fuel wood, and fodder (Fig. 3.9). Following sections deal with uses of plant species in social system.

Photoplate 3.1 Documentation of traditional ecological knowledge from the healers and local people

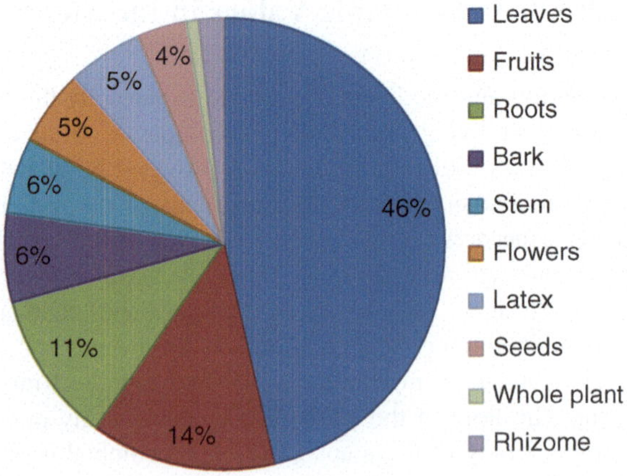

Fig. 3.8 Plant parts used in preparation of herbal medicine by the traditional healers in the study region

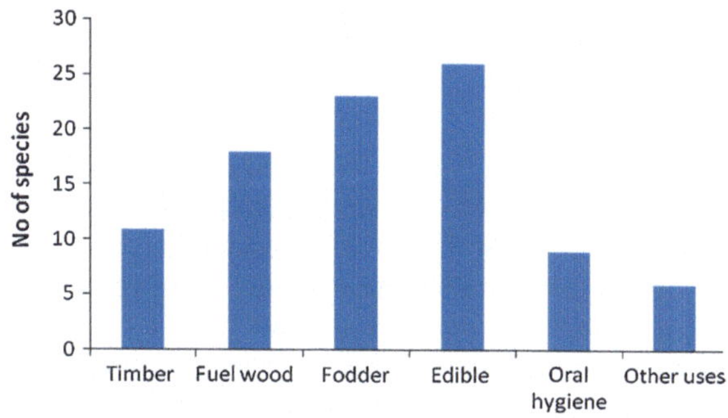

Fig. 3.9 Plant used for various purposes in the study area

Table 3.13 Timber yielding plants of the study region with preference by people

S. no.	Scientific name	Local name	Family	Preference for timber by people (%)
1	*Acacia ferruginea*	Banni gida	Mimosaceae	10
2	*Acacia nilotica*	Kari jali	Mimosaceae	25
3	*Albizia amara*	Tuglimara	Mimosaceae	10
4	*Azadirachta indica*	Bevina mara	Meliaceae	70
5	*Bauhinia racemosa*	Basavana pada	Caesalpiniaceae	25
6	*Ficus benghalensis*	Alada mara	Moraceae	10
7	*Pongamia pinnata*	Honge mara	Fabaceae	5
8	*Prosopis cineraria*	Shastra mara	Mimosaceae	25
9	*Tectona grandis*	Tega mara	Verbenaceae	65

3.5.1 Timber Yielding Plants

Trees are also planted under social forestry program on village common lands, vacant lands, and along roads. Tree species, viz. *Tectona grandis, Azadirachta indica,* and *Bauhinia racemosa,* are found in the margins of agricultural lands; village commons of the study region are used to meet the timber requirement of the people. The names of most commonly tree species used for timber and preference by people of the area are given in Table 3.13. During the study it was noticed that the maximum number of the people prefer Bevina mara (*Azadirachta indica*) and Tega mara (*Tectona grandis*) followed by Sarkari jali (*Prosopis juliflora*) and Shastra mara (*Prosopis cineraria*). Only few people prefer Honge mara (*Pongamia pinnata*) for the purpose of timber.

3.5.2 Plant Species for Fuel Wood

Domestic fuel needs are fulfilled by the use of plants such as *Prosopis juliflora, Balanites aegyptiaca, Casuarina equisetifolia,* and *Lantana camara*. In rural areas, people use tree parts instead of LPG. Generally, in villages parts of trees and shrubs are available in plenty because of the natural vegetation and forests. They prefer to use these as fuel since locally available free of cost. Most of the fuel wood used in this region is from *Prosopis juliflora* around 80 % and it is a highly dominant species found in the area. *Prosopis juliflora* is a native to South America, and it is fast growing, nitrogen-fixing, and tolerant to arid conditions as described by World Agroforestry Centre. Under the right conditions, *Prosopis juliflora* can produce a variety of valuable goods and services such as construction materials, charcoal, soil conservation, and rehabilitation of degraded and saline soils. The lists of species are mainly used by the people in the study region as fuel wood according to their availability. For fuelwood maximum number of the people have preference for Bevina mara (*Azadirachta indica*) and Tega mara (*Tectona grandis*) followed by Sarkari jali (*Prosopis juliflora*) followed by Survey mara (*Casuarina equisetifolia*) and Borae hannu (*Ziziphus mauritiana*). *Typha angustifolia* and *Bauhinia racemosa* are the least preferred species from the study region (Table 3.14).

3.5.3 Plant Species Used as Fodder

India is an agricultural country and about 65 % of the people live in villages. Traditional livestock-based agriculture is the main source of the livelihood of the people living in rural landscape. Agriculture and animal husbandry in India are interwoven with the intricate fabric of the society in cultural, religious, and economical ways as mixed farming and livestock rearing forms an integral part of rural living. Livestock provides rural transport, manure, fuel, milk, and meat. Most often, livestock is the only source of cash income for subsistence farms and serves as insurance in the event of crop failure.

The data on fodder production in the country vary widely. Fodder production and its utilization depend on the cropping pattern, climate, socioeconomic conditions, and type of livestock. The cattle and buffaloes are normally fed on the fodder available from cultivated areas, supplemented to a small extent by harvested from the food crops, grasses, and top feeds. Grazing and harvested grasses are the chief fodder source for equines. The three major sources of fodder supply are crop residues, cultivated fodder, and fodder from common lands resources such as forests, permanent pastures, and grazing lands (ICAR online). The natural grasslands and the cultivable waste and fallow lands provide grazing for the cattle in the monsoon season.

The animal husbandry system of the region comprises of buffalos, sheep, goats, cows, horses, and donkeys that are reared by people for variety of purposes.

Table 3.14 Plant species used as fuel and preference by local people

S. no.	Scientific name	Local name	Family	Preference by people as per their fuel wood requirement (%)
1	*Abutilon indicum*	Mudre gida	Malvaceae	5
2	*Acacia ferruginea*	Banni gida	Mimosaceae	8
3	*Acacia nilotica*	kari jali	Mimosaceae	20
4	*Azadirachta indica*	Bevina mara	Meliaceae	20
5	*Albizia amara*	Tugli mara	Mimosaceae	10
6	*Annona squamosa*	Sithaphal	Annonaceae	10
7	*Balanites aegyptiaca*	Ingalarade	Balanitaceae	40
8	*Bauhinia racemosa*	Basavana pada	Caesalpiniaceae	5
9	*Canthium coromandelicum*	Mulluhare	Rubiaceae	10
10	*Cassia fistula*	Kakke gida	Caesalpiniaceae	15
11	*Casuarina equisetifolia*	Survey mara	Casuarinaceae	45
12	*Eucalyptus globulus*	Niligiri mara	Myrtaceae	20
13	*Euphorbia tirucalli*	Kol kalli gida	Euphorbiaceae	8
14	*Ficus benghalensis*	Alada mara	Moraceae	15
15	*Lantana camara*	Hunni gida	Verbenaceae	30
16	*Leucaena leucocephala*	Sanna sogachu	Mimosaceae	35
17	*Phoenix sylvestris*	Eachalu	Arecaceae	8
18	*Prosopis cineraria*	Shastra mara	Mimosaceae	5
19	*Prosopis juliflora*	Sarkari jali	Mimosaceae	80
20	*Ricinus communis*	Oudala gida	Euphorbiaceae	10
22	*Senna auriculata*	Honnambre gida	Caesalpiniaceae	15
23	*Vitex negundo*	Lakki gida	Verbenaceae	20
24	*Wrightia tinctoria*	Kodu muriki	Apocynaceae	20
25	*Ziziphus mauritiana*	Borae hannu	Rhamnaceae	45

Table 3.15 gives a list of fodder plants available in this region and preference of the people. During monsoon season, most of the herbaceous species and grasses serve as fodder for these animals. Among the fodder species, *Leucaena leucocephala* (85 %) is most preferred followed by *Ipomoea carnea* (65 %). *Cocculus hirsutus* was found to be least preferred species for fodder (5 %).

3.5.4 Edible Species

Forests are endowed with a number of plant species of food value, which yield edible leaves, fruits, seeds, roots, rhizomes, etc. About 17,700 species of flowering plants are present in India; approximately 20 % of the total plants found in the forests have direct utility to mankind (Rathore 2009). Arid and semiarid zone vegetation comprises a wide range of edible fruit-bearing and food-producing species such as *Annona squamosa, Balanites aegyptiaca, Cordia dichotoma,*

Table 3.15 Plant species used for fodder and preference by local people

S. no.	Scientific name	Vernacular name	Family	Preference for fodder (%)
1	*Abutilon indicum*	Mudre gida	Malvaceae	25
2	*Acacia ferruginea*	Banni gida	Mimosaceae	30
3	*Acacia nilotica subsp. indica*	Kari jali	Mimosaceae	40
4	*Acalypha indica*	Kuppe gida	Euphorbiaceae	15
5	*Albizia amara*	Tugli mara	Mimosaceae	25
6	*Albizia lebbeck*	Baage mara	Mimosaceae	20
7	*Coccinia grandis*	Tonde balli	Cucurbitaceae	10
8	*Ficus religiosa*	Aralimara	Moraceae	30
9	*Ipomoea carnea*	Ganesh kaddi	Convolvulaceae	65
10	*Leucaena leucocephala*	Sanna sogachu	Mimosaceae	85
11	*Pithecellobium dulce*	Seeme hunase	Mimosaceae	25
12	*Pongamia pinnata*	Honge mara	Fabaceae	10
13	*Prosopis cineraria*	Shastra mara	Mimosaceae	20
14	*Prosopis juliflora*	Sarkari jali	Mimosaceae	40
15	*Ricinus communis*	Oudala gida	Euphorbiaceae	10
16	*Samanea saman*		Mimosaceae	10
17	*Senna auriculata*	Honnambre gida	Caesalpinaceae	25
18	*Senna occidentalis*	Tharavadi	Caesalpinaceae	20
19	*Syzygium cumini*	Nerale mara	Myrtaceae	10
20	*Tamarindus indica*	Hunase mara	Caesalpinaceae	30
21	*Vitex negundo*	Lakki gida	Verbenaceae	30
22	*Wattakaka volubilis*	Dughdike balli	Asclepiadaceae	5

Ziziphus mauritiana, and *Tamarindus indica.* There are around 30 plant species in arid zone known for their edible use, and of these, around 20 plant species are known for their edible fruits either raw or as a vegetable. Many of the above play multiple roles in the dry agroforestry systems by providing soil cover wind protection, fuel, fodder as well as food (Rathore 2009).

In the study area, 17 edible species were recorded. Leaves of *Moringa oleifera* and *Murraya koenigii* are used as spice and in other plants the fruits are edible. Some species such as *Tamarindus indica, Cocos nucifera, Mangifera indica*, and *Psidium guajava* are cultivated in orchards and plantations.

Annona squamosa and *Limonia acidissima* were growing wild in this area. These wild fruits are collected by the villagers and sold in the market. The collected fruits of other wild species *Phoenix sylvestris, Syzygium cumini, Phyllanthus emblica, Zizyphus mauritiana*, and *Pithecellobium dulce* are also sold in markets and near schools (Table 3.16).

Table 3.16 Edible species recorded in the study area and preference of local people

S. no.	Scientific name	Vernacular name	Family	Part used	Peoples' preference (%)
1	*Annona squamosa*	Sithaphal	Annonaceae	Fruits	30
2	*Canthium coromandelicum*	Mullukhare	Rubiaceae	Fruits	5
3	*Carissa spinarum*	Kaulekayi	Apocynaceae	Fruits	10
4	*Cocos nucifera*	Tengina mara	Arecaceae	Fruit	70
5	*Limonia acidissima*	Belada gida	Rutaceae	Fruits	40
6	*Mangifera indica*	Mavina mara	Anacardiaceae	Fruits	80
7	*Moringa oleifera*	Nugge mara	Moringaceae	Leaves, fruits	40
8	*Murraya koenigii*	Karebevu	Rutaceae	Leaves	55
9	*Opuntia dillenii*	Papas kalli	Cactaceae	Fruits	5
10	*Phoenix sylvestris*	Eachalu	Arecaceae	Fruits	20
11	*Phyllanthus emblica*	Nelli kaayi	Euphorbiaceae	Fruits	45
12	*Pithecellobium dulce*	Seeme hunase	Mimosaceae	Seed aril	15
13	*Psidium guajava*	Seebe gida	Myrtaceae	Fruits	60
14	*Punica granatum*	Dalimbre	Punicaceae	Fruits	50
15	*Syzygium cumini*	Nerale	Myrtaceae	Fruits	45
16	*Tamarindus indica*	Hunase mara	Caesalpiniaceae	Fruits, leaves	80
17	*Ziziphus mauritiana*	Borae hannu	Rhamnaceae	Fruits	70
18	*Ziziphus oenoplia*		Rhamnaceae	Fruits	40

3.5.5 Plants Used in Oral Hygiene

All over the world, native people traditionally make use of disposable toothbrushes made of twigs from plants with therapeutic values, which contain the necessary ingredient for cleaning teeth. Most of the villagers here are accustomed to use the available plant material for their oral hygiene. Moreover, they are available free of cost in and around their settlement. Stems are generally used as a brush and occasionally the leaves of some plants are also used. The leaves of *Eclipta alba, Azadirachta indica, Phoenix sylvestris*, and *Ocimum sanctum* are used in teeth cleaning (Table 3.17).

3.5.6 Other Uses

The other useful products made out of these plant species are ropes and threads for binding, making house roofs and broom sticks (*Typha angustifolia* and *Cocos nucifera* (utensil cleaning fiber). The stems and leaves of *Phoenix sylvestris* are used in making small brooms and threads. *Aristida setacea* is also used in making soft brooms.

Table 3.17 Plants used in oral hygiene in the study region

S. no.	Scientific name	Vernacular name	Family	Part used	Peoples' preference (%)
1	*Acacia nilotica*	Kari jali	Mimosaceae	Twigs	25
2	*Azadirachta indica*	Bevina mara	Meliaceae	Twigs and leaves	80
3	*Pongamia pinnata*	Honge mara	Fabaceae	Twigs	20
4	*Eclipta alba*	Bhrungaraja	Asteraceae	Leaves	10
5	*Phoenix sylvestris*	Eachalu	Arecaceae	Rachis of compound leaves	5
6	*Eucalyptus globulus*	Niligiri mara	Myrtaceae	Twigs	5
7	*Senna auriculata*	Honnambre gida	Caesalpiniaceae	Twigs	10
8	*Achyranthes aspera*	Uttarani	Amaranthaceae	Twigs	5
9	*Prosopis cineraria*	Shastra mara	Mimosaceae	Twigs	5
10	*Ziziphus mauritiana*	Borae hannu	Rhamnaceae	Twigs	5
11	*Ocimum sanctum*	Sri tulasi	Lamiaceae	Leaves	10

3.6 Phytoplankton

The term phytoplankton encompasses all photoautotrophic microorganisms in aquatic food webs. Phytoplanktons are microscopic aquatic plants, occurring as unicellular, colonial, or filamentous forms, without any resistance to currents and are freely floating or suspended in open/pelagic waters. Phytoplanktons are unicellular and free-floating organisms belong to the algae group. Phytoplanktons are the easiest food source for most of the aquatic animals such as zooplanktons, fishes, and other animals of aquatic ecosystems thus are basic food producers in any aquatic ecosystems (Suseela 2009). Phytoplanktons are on the bottom rung of the food chain in any aquatic ecosystem. Many are photosynthetic and are grazed upon by zooplankton and other aquatic organisms. Some species flourish in highly eutrophic waters, while others are very sensitive to organic and chemical wastes. The fresh water ecosystems are differentiated into various types of planktons (free floating), benthoms (attached to the sediments), and epiphytic algae (attaché to the hydrophytes, stones, mud, sand, rock of reservoirs, and lakes) (Arulmurugan et al. 2011; Mahendra Perumal and Anand 2009). Planktonic algae are represented by green algae (chlorophyta), blue green algae (cyanophyta), the diatoms (bacillariophyta), and desmids (chrysophyta).

Total 21 species of phytoplankton were identified in the study area; among which, eight species belong to the class Bacillariophyceae. The rest of the species belong to the classes Chlorophyceae, Cyanophyceae, Charophyceae from different families Cladophoraceae, Scenedesmaceae, Desmidiaceae, Microcystaceae, Bacillariophyceae, Oscillatoriaceae, Bacillariaceae, Nostocaceae, Phormidiaceae, and Zygnemataceae (Fig. 3.10; Photoplate 3.1). All the phytoplankton species observed in the study area are listed according to seasonal availability (Table 3.18).

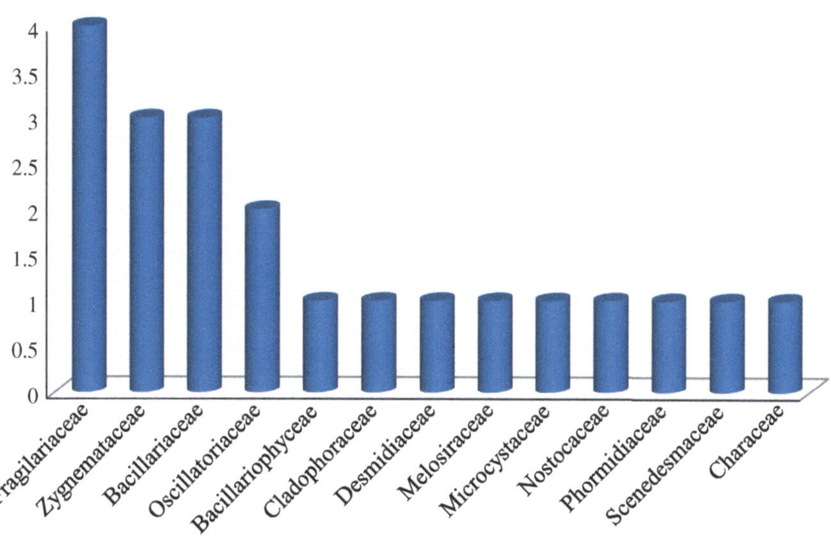

Fig. 3.10 Phytoplankton diversity

Table 3.18 Phytoplankton recorded in the study area

S. No	Scientific name	Family	Class	Season-1							Season-2						
				Mar 2011 to May 2011							Oct 2011 to Nov 2011						
				S-1	S-2	S-3	S-4	S-5	S-6	S-7	S-1	S-2	S-3	S-4	S-5	S-6	S-7
1	*Hantschia amphioxys* (Ehr.) Grun.	Bacillariaceae	Bacillariophyceae								+			+	+	+	
2	*Synedra ulna* (Nitsch) Ehr.	Fragilariaceae	Bacillariophyceae										+				+
3	*Fragilaria fonticola* var. chandolensis	Fragilariaceae	Bacillariophyceae	+	+			+		+		+			+		
4	*Melosira granulata (Kutz.) Ralfs*	Melosiraceae	Bacillariophyceae														
5	*Nitzschia closterium*(Ehr.) Wm. Smith	Bacillariaceae	Bacillariophyceae								+				+	+	
6	*Nitzschia paradoxa* (J.F.Gmelin) Grunow	Bacillariaceae	Bacillariophyceae		+		+										
7	*Synedra ulna* (Nitz.) Ehr. Var. amphirhynchus (Ehr.) Grun.	Fragilariaceae	Bacillariophyceae									+			+		
8	*Synedra ulna* (Nitz.) Ehr. Var. oxyrynchus (Kutz.) Van Heurck	Fragilariaceae	Bacillariophyceae											+		+	
9	*Cladophora glomerata* (L.) Kützing	Cladophoraceae	Chlorophyceae														
10	*Staurastrum gracile forma* Iyengar et Vimala bai	Desmidiaceae	Chlorophyceae									+	+		+		+
11	*Zygnema Khanne* Skuja	Zygnemataceae	Chlorophyceae								+						+
12	*Spirogyra hyalina* Cleve	Zygnemataceae	Chlorophyceae	+		+		+			+	+	+		+	+	
13	*Coelastrum microporum* Naegeli	Scenedesmaceae	Chlorophyceae										+	+			+
14	*Spirogyra condensata* (Vaucher)Kutzing	Zygnemataceae	Chlorophyceae		+	+			+		+	+				+	+
15	*Cosmarium innae*	Desmidiaceae	Chlorophyceae	+		+			+	+							
16	*Lyngbya majuscula* (Dillwyn) Harvey	Oscillatoriaceae	Cyanophyceae										+			+	
17	*Gleocapsa punctata* Nägeli	Microcystaceae	Cyanophyceae			+	+	+		+			+	+			
18	*Nostoc piscinale* Kutz. ex Born. & Flah.	Nostocaceae	Cyanophyceae										+	+			
19	*Oscillatoria tenuis* Ag. Ex Gomont	Oscillatoriaceae	Cyanophyceae								+						+
20	*Phormidium corium* var. captatum Gardner	Phormidiaceae	Cyanophyceae	+	+		+	+		+			+				
21	*Chara*	Characeae	Charophyceae	+	+		+		+		+	+			+	+	

3.7 Description of Plant Species from the Semiarid Region, Karnataka, India, in Alphabetical Family-Wise

1. Family: Acanthaceae

Scientific name: *Adhatoda zeylanica* Medic
Vernacular name: Adusoge
Habit: Shrub

Description: Bushy shrub, branches many. Leaves are elliptic to lanceolate, dark green color above and pale green color underneath. Spike short, flowers white.

Uses: As medicinal uses purpose. Leaf paste is used to cure skin diseases and tumors.

Flowering and Fruiting: September to November
IUCN status: Not Evaluated
Native: India

Scientific name: *Andrographis paniculata* (Burm. f.) Wall. ex Nees
Vernacular name: Nelabevu
Habit: Herb

Description: A medium-sized herb; found fields, roadside and near rocky land. Leaves linear. Flowers small lobed, lower lobe 3 toothed, upper lobed 2 toothed. Fruit is slender capsule.

Flowering and Fruiting: August to December
IUCN status: Not Evaluated
Native: India and Sri Lanka

Scientific name: *Asteracantha longifolia* (L.) Nees
Vernacular name: Kalavankabija
Habit: Herb

Description: Annual herb; found near lakes and moisture places. Stems are sub-quadrangular with thick nodes. Leaves are oblanceolate with yellow or brown color spines. Flowers are clustered, purpled blue.

Flowering and Fruiting: August to January
IUCN status: Not Evaluated
Native: India and Sri Lanka

Scientific name: *Barleria prionitis*
Vernacular name: Mullu jaji
Habit: Shrub

Description: A prickly small shrub; up to 70–80 cm; found in fields, roadside and home gardens. Leaves ovate; flowers yellow in color.

Flowering and Fruiting: September to October
IUCN status: Not Evaluated
Native: Tropical East Africa and Asia

Scientific name: *Crossandra infundibuliformis* (L.) Nees
Vernacular name: Kanakambara gida
Habit: Shrub

Description: A small herb growing up to 2–3 m; found in home gardens as ornamental plant as well as in wild. Leaves lanceolate; flowers appears on spike, orange red color.

Uses: Ornamental plant.

Flowering and Fruiting: January to December
IUCN status: Not Evaluated
Native: India and Sri Lanka

Scientific name: *Lepidagathis cristata* Willd.
Vernacular name: Narigoodi
Habit: Herb

Description: Herb; creeping with stout root stock, recorded in rocky land near hill stations. Branches 20–25 cm long; leaves lanceolate. Flowers arise from the globose head and in pale pink color.

Flowering and Fruiting: July to November
IUCN status: Not Evaluated
Native: Tropical Asia and Indian subcontinent

Scientific name: *Peristrophe paniculata* (Forsk.) Brummitt
Vernacular name: Cheebi gida
Habit: Herb

Description: Herb; found fields and waste land. Stem-angled flowers in panicles, corolla lipped and pink colored.

Flowering and Fruiting: September to November
IUCN status: Not Evaluated
Native: Tropical Asia and India

Scientific name: *Rostellularia simplex* Wight
Vernacular name:
Habit: Herb

Description: Herb up to 5–30 cm tall; recorded in rocky land and hill slopes. Flowers pink in color.

Flowering and Fruiting: July to November
IUCN status: Not Evaluated
Native: India, Nepal, Sri Lanka, and Pakistan

Scientific name: *Rungia repens* (L.) Nees
Vernacular name:
Habit: Herb

Description: Herb up to 20–25 cm high; found near agricultural field. Spike terminal, corolla light blue.

Flowering and Fruiting: July to November
IUCN status: Not Evaluated
Native: India and Sri Lanka

2. Family: Agavaceae

Scientific name: *Agave americana* L.
Vernacular name: Rakspatti
Habit: Succulent Shrub

Description: A large succulent shrub; found margins of the field and waste land. Leaves thick and gray blue color, up to 150–200 cm long 20–25 cm wide and sharp spine margin and tips.

Uses: The plants are growing as boundaries in the field. Very few are using leaves for make fibers. Used as medicine to cure cough.

Flowering and Fruiting: June to February
IUCN status: Not Evaluated
Native: USA and Mexico

3. Family: Aizoaceae

Scientific name: *Mollugo lotoides* C.B. Clarke
Vernacular name:
Habit: Herb

Description: Prostrate herb along the roadsides, cultivated fields, and dried ponds. Leaves are obovate, circular. Flowers are pale green.

Flowering and Fruiting: July to October
IUCN status: Not Evaluated

Scientific name: *Trianthema portulacastrum* L.
Vernacular name: Balevadukana soppu
Habit: Herb

Description: Prostate herb up to 40–50 cm long; identified near cultivated fields. Stem succulent, rounded, and slightly hairy. Leaves are simple, rounded, and glabrous on both the sides. Flowers are whitish.

Uses: The plant is used as vegetable and as fodder to animals.

Flowering and Fruiting: July to December
IUCN status: Not Evaluated
Native: India, Sri Lanka, and Malaysia

4. Family: Alangiaceae

Scientific name: *Alangium salvifolium* (L. f.) Wangerin
Vernacular name: Ankaligida
Habit: Shrub

Description: Shrub or medium-sized deciduous tree. Found near hill slope, leaves are narrowly oblong, glabrous. Flowers are pale white. Fruits are green, thick red when ripe.

Uses: Fruits are edible. Plant parts are used for fuel wood.

Flowering and Fruiting: March to June
IUCN status: Not Evaluated
Native: India and Sri Lanka

5. Family: Aloaceae

Scientific name: *Aloe vera* (L.) N. Burm.
Vernacular name: Kathaligidi
Habit: Succulent herb

Description: Perennial herb up to 100 cm. Leaves are arise from the base of the plant, fleshy, and spiny. Abundantly found in Bheema river basin near Yalwar village.

Uses: Fleshy part is used to cure block spots on face.

Flowering and Fruiting: July to December
IUCN status: Not Evaluated
Native: Mediterranean

6. Family: Amaranthaceae

Scientific name: *Achyranthes aspera* L.
Vernacular name: Uttarani
Habit: Herb

Description: An erect herb up to 90 cm high; spread over all the places. Leaves are obovate; flowers on terminal spike; small, greenish white and numerous.

Uses: The plant is used as medicinal purpose. Roots used as tooth brush. The leaf paste is mixed with mother's milk and given to cure the osmosis problem in children. Fresh leaves are crushed and applied on foot to treat cracks.

Flowering and Fruiting: January to December
IUCN status: Not Evaluated
Native: Southeastern Asia

Scientific name: *Aerva lanata* (L.) Juss. ex Schult.
Vernacular name: Bilihindi soppu
Habit: Herb

Description: Prostate herb; found waste lands, fields. Stem sticky, glabrous; leaves are obovate. Flowers are tiny white and crowded in compact axillary spike.

Flowering and Fruiting: August to December
IUCN status: Not Evaluated
Native: Afrotropics

Scientific name: *Alternanthera pungens* Humb. Bonpal. & Kunth
Vernacular name: Mullu honagonne
Habit: Herb

Description: Prostate branching herb; commonly found in roadside and waste lands. Flowers small and head-like inflorescence.

Uses: Leaves are edible as vegetable.

Flowering and Fruiting: August to November
IUCN status: Not Evaluated
Native: Neotropics

Scientific name: *Alternanthera sessilis* (L.) R.Br
Vernacular name: Honagone soppu
Habit: Herb

Description: Herb; growing in moist places abundantly. Leaves are opposite with variable size; elliptic, ovate, and glabrous. Flowers are small, white color.

Uses: Used as vegetable.

Flowering and Fruiting: August to November
IUCN status: Least Concern
Native: Southern Asia

Scientific name: *Alternanthera tenella* Colla *Syn*
Vernacular name:
Habit: Herb

Description: A small herb; found in near moisture places. Leaves are opposite and elliptic. Flowers are axillary clusters, white color.

Uses: Used for fodder to animals and some local people are using as vegetable.

Flowering and Fruiting: August to November
IUCN status: Not Evaluated
Native: Brazil

Scientific name: *Amaranthus spinosus* L.
Vernacular name: Mulludantu
Habit: Herb

Description: Prickly herb; recorded in agricultural fields and moisture places. Leaves are ovate, elliptic. Flowers are green and whitish.

Uses: As vegetable. Fried leaves are used by women to reduce excessive bleeding during the menstruation period. Leaf decoction is good for digestion.

Flowering and Fruiting: August to December
IUCN status: Not Evaluated
Native: Caribbean and tropical Asia

Scientific name: *Amaranthus viridis* L.
Vernacular name: Chelakeerae soppu
Family: Amaranthaceae
Habit: Herb

Description: Herb; found in agricultural field. Leaves are alternative, lanceolate. Terminal spike.

Uses: Used as vegetable.

Flowering and Fruiting: August to March
IUCN status: Not Evaluated
Native: Southern USA and Mexico

Scientific name: *Celosia argentea* L.
Vernacular name: Anne soppu
Habit: Herb

Description: Annual herb, up to 4–5 ft.; found waste lands and dry agricultural fields. Stem angular; leaves simple, alternate and lanceolate. Spike cylindrical, floral parts are dry; paper-like; and silvery, white, and pink in color.

Flowering and Fruiting: January to December
IUCN status: Not Evaluated
Native: Tropical America and Africa

Scientific name: *Digera muricata* (L.) Mart.
Vernacular name:
Habit: Herb

Description: Herb; spread in the agricultural fields up to 50–70 cm high. Leaves stalk is long up to 5 cm. Flowers appears on spike-like structure; pink in color.

Uses: As vegetable.

Flowering and Fruiting: August to December
IUCN status: Not Evaluated
Native: Asia and Africa

Scientific name: *Gomphrena celosioides* Mart.
Vernacular name: Nelarudrakshi
Habit: Herb

Description: Procumbent herb; distributed most of the places in the study area. Stem greenish white, villose. Leaves are opposite, oblong–obovate. Head inflorescence, silvery.

Flowering and Fruiting: August to December
IUCN status: Not Evaluated
Native: South America

Scientific name: *Pupalia lappacea* (L.) Juss.
Vernacular name: Haridhachhaga
Habit: Herb

Description: Herb, up to 2–3 m tall. Found in near to fields association with *Prosopis juliflora*. Stem branched, swollen at nodes. Leaves are dark green, opposite, ovate–elliptic to oblong. Spike terminal on the stem and branches.

Flowering and Fruiting: August to November
IUCN status: Not Evaluated
Native: Africa and tropical Asia

7. Family: Anacardiaceae

Scientific name: *Mangifera indica* L.
Vernacular name: Mavina mara
Habit: Tree

Description: A large evergreen tree; distributed in the all the places. Leaves are alternative, linear–oblong, lanceolate–elliptical, pointed at both ends, the leaf blades mostly about 20–25 cm long and 5–8 cm wide. Fruits are drupe.

Uses: Fruits are edible. Stem parts are used for fuel wood.

Flowering and Fruiting: February to June
IUCN status: Data Deficient
Native: India

8. Family: Annonaceae

Scientific name: *Annona squamosa* L.
Vernacular name: Sithaphal
Habit: Tree

Description: A small deciduous tree; distributed entire the study area. Flowers are solitary, pale yellow or cream in color. Fruits are globose.

Uses: Fruits are edible.

Flowering and Fruiting: April to September
IUCN status: Not Evaluated
Native: West Indies

9. Family: Apocynaceae

Scientific name: *Carissa spinarum* L.
Vernacular name: Kaulekayi
Habit: Shrub

Description: Shrub 2–3 m tall; found near hill slopes and rocky land. Leaves are opposite, leathery oval shaped, pointed tip. Flowers are small, white. Fruits berries, oval shaped.

Uses: Fruits are edible.

Flowering and Fruiting: April to September
IUCN status: Not Evaluated
Native: Tropical Africa and Arabian Peninsula

Scientific name: *Nerium indicum* Mill.
Vernacular name: Kangile
Habit: Shrub

Description: Shrub; cultivated at home gardens and found in the agricultural fields. Leaves are thick, whorled. Flowers are pink.

Uses: An ornamental plant.

Flowering and Fruiting: January to December
IUCN status: Not Evaluated
Native: Mediterranean region and South Asia

Scientific name: *Plumeria rubra* L.
Vernacular name: Kangile
Habit: Tree

Description: A tree; recorded near hill slopes at Shahapur fort. Stem and branches thick succulent. Leaves are lanceolate–elliptic, 15–20 cm long 7–10 wide. Flowers white with yellow color.

Flowering and Fruiting: November to March
IUCN status: Not Evaluated
Native: Mexico

Scientific name: *Thevetia neriifolia* Juss. ex Steud.
Vernacular name: Haladi kanagile
Habit: Tree

Description: A large shrub; distributed in the waste lands and near agricultural fields. Leaves are linear–elliptic, pointed tip. Flowers yellow, bell shaped.

Flowering and Fruiting: November to February
IUCN status: Not Evaluated
Native: West Indies

Scientific name: *Vinca rosea* (L.) G. Don
Vernacular name:
Habit: Herb

Description: Herb; distributed along roadside, fields and waste lands. Leave obovate–oblong; flowers white color.

Uses: Medicinal plant. Leaves are powdered and paste is used to treat wounds and cuts. Dry roots and leaves are used to cure diabetes.

Flowering and Fruiting: January to December
IUCN status: Not Evaluated
Native: Madagascar

Scientific name: *Wrightia tinctoria* R. Br.
Vernacular name: Kodu muriki
Habit: Tree

Description: A small deciduous tree; found in the waste land and agricultural field. Leaves are opposite, oblong, slightly whitish beneath. Flowers are small, 2–3 cm, white and fragrant. A pair of fruits 20–30 cm long, slender, and hanging.

Used: Leaf is crushed with turmeric and the paste prepared is applied on the wounds. Latex is applied externally to reduce skin diseases. Stem parts uses as fuel wood.

Flowering and Fruiting: April to November
IUCN status: Not Evaluated
Native: India and Burma

10. Family: Araceae

Scientific name: *Lemna polyrhiza* (L.)
Vernacular name:
Habit: Aquatic herb

Description: A small aquatic gregarious herb; found in water bodies, paddy fields.
IUCN status: Least Concern

11. Family: Arecaceae

Scientific name: *Borassus flabellifer* L.
Vernacular name: Thatling
Habit: Tree

Description: A large palm tree up to 15–20 m high; recorded in the boundaries of agricultural field. Leaves are leathery, gray green, fan shaped. Fruits are rounded.
Uses: Fruits are edible. Leaves are used to make traditional house.

Flowering and Fruiting: February to June
IUCN status: Not Evaluated
Native: Indian subcontinent, Southeast Asia, and tropical Africa

Scientific name: *Cocos nucifera* L.
Vernacular name: Tengina mara
Habit: Tree

Description: Tree; commonly cultivated for the fruits. Leaves pinnately compound. Flowers are in a branches spadix. Fruit is drupe.

Uses: The palm tree uses multiple purposes. The coir of coconut is crushed and mixed with sesamum oil and then applied on joints to treat joint pains. The shell is heated, crushed, and then mixed with coconut oil and applied on wounds and rashes.

Flowering and Fruiting: January to December
IUCN status: Not Evaluated
Native: Indo pacific

Scientific name: *Phoenix sylvestris* (L.) Roxb.
Vernacular name: Eachalu
Habit: Tree

Description: A tall palm; distributed entire the study area. Leaves are pinnately compound, terminal cluster forming the crown. Spadix inflorescence. Fruits are one seeded berry.

Uses: Fruits are edible. Used for multipurpose such as to make baskets and mats.

Flowering and Fruiting: January to December
IUCN status: Not Evaluated
Native: India

12. Family: Aristolochiaceae

Scientific name: *Aristolochia bracteolata* Lam.
Vernacular name: Kuri gida
Habit: Herb

Description: Prostrate herb; distributed in the cultivation field and waste land. Stem seriate; leaves are cordate, glabrous, 2–3 cm long, sagittate at base, margin wavy. Inflorescence solitary, axillary. Flowers are bractete, orbicular, tube shaped at base, 1–1.5 cm long, greenish at base purple at tip. Fruit capsule, 1–1.5 cm long, seeds many.

Flowering and Fruiting: January to December
IUCN status: Not Evaluated
Native: Africa, temperate Asia, and tropical Asia

Scientific name: *Aristolochia indica* L.
Vernacular name: Eshwari balli
Habit: Herb

Description: A twinning herb; mainly association recorded with *Prosopis juliflora* trees. Leaves are oblong, 2–3 cm long. Flowers are few in a bunch with pale greenish color, cylindrical tube having trumpet-shaped mouth. Fruit is capsule, brown, and oblong. The dehisced fruit appears like an opened parachute. Seeds are many flat and oval.

Flowering and Fruiting: July to December
IUCN status: Not Evaluated
Native: India, Nepal, Sri Lanka, and Bangladesh

13. Family: Asclepidaceae

Scientific name: *Calotropis gigantea* (L.) R. Br.
Vernacular name: Yekkada gida
Habit: Shrub

Description: A perennial shrub with milky latex; distributed through out the study area. Leaves are simple, opposite decussate; thick, 15–25 cm long, and 7–10 cm wide; powdery on the both surfaces. Flowers are umbellate cyme; 2–4 cm long; white and purple white color. Fruit is a pair of follicles. Seeds many and having long silky hairs at one end.

Uses: A medicinal plant. Latex is used to treat wounds. Leaves are used as an antidote for snake bite. Flora buds along pepper used for shivering fever.

Flowering and Fruiting: January to December
IUCN status: Not Evaluated
Native: Tropical Asia and tropical Africa

Scientific name: *Calotropis procera* R. Br.
Vernacular name: Bili yekkada gida
Habit: Shrub

Description: A shrub with milky latex; found in dry lands, waste lands, and border of agricultural fields. Stem is not much branched and woody. Leaves opposite, simple, sub-sessile, oblong–obovate, apex shortly acuminate, 10–15 cm long and 5–8 cm wide, succulent. Inflorescence umbellate cyme, arise from the nodes, and terminal; flowers white and purple, pedicle 1–3 cm long. Fruit follicles sub-globose up to 10 cm long; seeds numerous, flat, and obovate.

Uses: Use as medicine plant.

Flowering and Fruiting: January to December
IUCN status: Not Evaluated
Native: North Africa, tropical Africa, western Asia, South Asia, and Indochina

Scientific name: *Caralluma attenuata* Wight
Vernacular name:
Habit: Herb

Description: Succulent herb, water latex; found in rocky land, open and dry plain places. Leaves found along the angles, minute spine-like structure. Flowers terminal racemes, brown to dark purple. Fruits are follicles, oblong to linear.

Flowering and Fruiting: July to December
IUCN status: Not Evaluated
Native: India

Scientific name: *Cryptolepis buchanani* Roem. & Schult.
Vernacular name: Haluballi
Habit: Climber

Description: Climber; recorded in near agricultural field. Leaves are simple, elliptic, acuminate, 4–6 cm long and 2–3 cm wide. Flowers are axillary, umbellate racemes, pale yellow colored. Fruits are stout, paired follicles, pointed above. Seeds are many.

Flowering and Fruiting: June to December
IUCN status: Not Evaluated
Native: Indian subcontinent and China

Scientific name: *Cryptostegia grandiflora* R. Br.
Vernacular name: Rubber hombu
Habit: A woody climber

Description: A woody climber; distributed through out the study area. Leaves are opposite, dark green, latex present. Flowers bell shaped with 5 lobes, corolla whitish to pink. Fruits are 2 pods with two angles. Seeds are numerous with bearing silky hairs.

Flowering and Fruiting: November to March
IUCN status: Not Evaluated
Native: Madagascar

Scientific name: *Gymnema sylvestre* (Retz.) R. Br. ex Schult
Vernacular name: Kodapathre
Habit: Climber

Description: Climber; found in near agricultural boundaries. Leaves oblong; flowers axillary, pale yellow, umbel shape.

Uses: Leaf paste with jambuva seeds are used to control diabetes.

Flowering and Fruiting: August to November
IUCN status: Not Evaluated
Native: India

Scientific name: *Hemidesmus indicus* (L.) Schult
Vernacular name: Suganhaberu
Habit: Climber

Description: Creeper; distributed throughout the study area. Leaves are opposite, oblong, narrow. Flowers are crowded, corolla dark purple. Fruits are follicle.

Uses: Roots are aromatic; used for prepare soft drinks called as nannari.

Flowering and Fruiting: August to December
IUCN status: Not Evaluated
Native: India

Scientific name: *Oxystelma esculentum* (L. f.) R. Br. ex Schult.
Vernacular name:
Habit: Climber

Description: A climber; located near water bodies. Leaves are simple, opposite, oblong to linear. Flowers are axillary, solitary, paired, whitish outside, pinkish within. Fruits paired, oblong.

Flowering and Fruiting: March to June
IUCN status: Not Evaluated
Native: Indian subcontinent, China, Malaysia, and Thailand.

Scientific name: *Pentatropis capensis* (L. f.) Bullock
Vernacular name:
Habit: Climber

Description: A twining herb distributed throughout the study area. Leaves are ovate, elliptic, and slightly succulent. Follicles are 3 to 4 cm long.

Flowering and Fruiting: July to November
IUCN status: Not Evaluated
Native: India, Sri Lanka, and Pakistan

Scientific name: *Pergularia daemia* (Forssk.) Chiov
Vernacular name: Kurudigana balli
Habit: Climber

Description: Climber, distributed in all the places. Leaves are thin, ovate, and cordate at base. Flowers light green, white. Fruit is follicles, soft spines on surface. Seeds are hairy, velvety on both sides.

Uses: Leaf extract is used to cure bone fracture.

Flowering and Fruiting: March to May
IUCN status: Not Evaluated
Native: Tropical Asia and Africa

Scientific name: *Telosma pallida* (Roxb.) Craib
Vernacular name:
Habit: Climber

Description: A climber; found in agricultural field and wastelands. Stem is slender, densely pubescent. Leaves are opposite, ovate–oblong, acute, cordate at base. Flowers are many, umbellate cyme, light green, white. Fruits are follicles.

Flowering and Fruiting: July to November
IUCN status: Not Evaluated
Native: Tropical Asia

Scientific name: *Tylophora indica* (Burm.) Merr.
Vernacular name: Adumuttada balli
Habit: Climber

Description: A common climber distributed throughout the study area. Stem pubescent; Leaves opposite, leaf stalk 1–1.5 cm long, oblong–ovate, end with small tip. Flowers are small, green purple. Fruits are follicles 5–8 cm long.

Uses: Crushed leaves with pepper are used to cure whooping cough.

Flowering and Fruiting: January to June
IUCN status: Not Evaluated
Native: India

Scientific name: *Wattakaka volubilis* (L. f.) Stapf.
Vernacular name:
Habit: A woody climber

Description: A large woody climber; found waste lands, agricultural boundaries. Leaves are large, ovate, and acuminate. Flowers are many, umbel cyme, yellowish green. Fruits are large 5–8 cm. Seeds hairy.

Flowering and Fruiting: June to February
IUCN status: Not Evaluated
Native: Indomalayan region

14. Family: Asteraceae

Scientific name: *Ageratum conyzoides* L.
Vernacular name: Muguthigida
Habit: Herb

Description: Erect herb up to 3–4 ft., viscid hairy; found waste lands, agricultural fields. Leaves opposite, ovate, crenate. Head inflorescence, flowers are white.

Flowering and Fruiting: November to February
IUCN status: Not Evaluated
Native: Tropical America and Caribbean

Scientific name: *Blainvillea acmella* (L.) Philipson.
Vernacular name: Muguthigida
Habit: Herb

Description: Herb; distributed waste lands, open fields. Stem erect, branched, and grooved. Leaves are opposite, ovate, crenate. Head inflorescence, solitary, flowers white, turns yellow. Pappus short.

Flowering and Fruiting: August to November
IUCN status: Not Evaluated
Native: India, Sri Lanka, and Pantropis

Scientific name: *Blumea eriantha* DC.
Vernacular name: Muguthigida
Habit: Herb

Description: Herb; found in waste lands. Stem is slightly reddish, hairy. Leaves are obovate, acute, and serrate. Flowers are small, yellow.

Flowering and Fruiting: June to September
IUCN status: Not Evaluated
Native: India

Scientific name: *Dicoma tomentosa* Cass.
Vernacular name: Sannigida
Habit: Herb

Description: Erect annual herb; found in waste lands. Stem stiff, white hairy. Leaves alternate, oblong. Heads solitary, flowers white.

Flowering and Fruiting: December to February
IUCN status: Not Evaluated
Native: Asia and Africa

Scientific name: *Echinops echinatus* Roxb.
Vernacular name: Brahmadande
Habit: Herb

Description: Herb; found near hill stations and wastelands. Stem stiff, branched, stout at base, covered with white hairs. Leaves are lanceolate, end with sharp spines. Flowers solitary, round, end with sharp spines, florets white.

Uses: Boiled root extract is used to treat anathematic in animals.

Flowering and Fruiting: October to April
IUCN status: Not Evaluated
Native: Indian subcontinent

Scientific name: *Eclipta alba* (L.) Hassk.
Vernacular name: Bhrungaraja
Habit: Herb

Description: Ascending herb; distributed entire study area; near by water bodies and paddy field, it is presented abundantly. Stem hairy; leaves opposite, linear to oblong, roughly hairy. Flowers white.

Uses: Plant extract is used to treat cough and swellings. Leaf paste is used to reduce dandruff and hair growth.

Flowering and Fruiting: September to January
IUCN status: Not Evaluated
Native: USA and tropical, subtropical regions of the world

Scientific name: *Erigeron alpinus* L.
Vernacular name:
Habit: Herb

Description: Herb; collected once. Leaves are elliptic, shortly hairy. Flowers are solitary, disk flowers.

Flowering and Fruiting: September to December
IUCN status: Not Evaluated
Native: North America

Scientific name: *Flaveria trinervia* (Spreng.) Mohr.
Vernacular name:
Habit: Herb

Description: Annual herb; found near water bodies. Leaves opposite, lanceolate, crenate. Flowers yellow.

Uses: Leaf paste is used to cure wounds.

Flowering and Fruiting: October to March
IUCN status: Not Evaluated
Native: USA, Mexico, and Caribbean

Scientific name: *Grangea maderaspatana* (L.) Poir.
Vernacular name:
Habit: Herb

Description: Prostate herb with bunch; found in near wet lands. Leaves alternate, wolly, toothed lobes. Flowers yellow, rounded, hairy.

Flowering and Fruiting: November to March
IUCN status: Least Concern
Native: Indian subcontinent, China, and Malaysia

Scientific name: *Lagascea mollis* Cav.
Vernacular name:
Habit: Herb

Description: Annual herb, up to 50–60 cm high, found in waste lands, agricultural fields and near water bodies. Stem is grooved, glabrous. Leaf opposite, crenate, hairy beneath. Head solitary, pubescent, florets are white.

Flowering and Fruiting: November to February
IUCN status: Not Evaluated
Native: Tropical America

Scientific name: *Launaea pinnatifida* Cass.
Vernacular name:
Habit: Herb

Description: Herb; found in open fields. Leaves spathulate, deeply curved, dentate. Flowers white to pale yellow.

Flowering and Fruiting: September to December
IUCN status: Not Evaluated
Native: Africa, Madagascar, India, Sri Lanka, and Southeast Asia

Scientific name: *Parthenium hysterophorus* L.
Vernacular name: Congress gida
Habit: Herb

Description: Annual erect herb, up to 2–3 m high; distributed through out the study area. Stem is hairy, not much branched, grooved. Leaves pale green, hairy. Flowers are small, florets white.

Flowering and Fruiting: January to December
IUCN status: Not Evaluated
Native: Tropical America

Scientific name: *Pentanema indicum* (L.) Ling
Vernacular name:
Habit: Herb

Description: An annual erect herb; found in waste lands and fields. Leaves simple, oblong, alternate, ear shaped at base and hairy. Inflorescence axillary, long stalked, yellow color.

Flowering and Fruiting: August to December
IUCN status: Not Evaluated
Native: Indian subcontinent, Indochina, and tropical Africa

Scientific name: *Pulicaria wightiana* (DC.) C.B. Clarke
Vernacular name:
Habit: Herb

Description: Annual herb; found in grass lands, moisture places and near hill regions. Stem are erect, grooved, and pubescent. Leaves are oblong, toothed, and velvety. Flowers yellow.

Flowering and Fruiting: August to November
IUCN status: Not Evaluated
Native: India

Scientific name: *Sonchus oleraceus* L.
Vernacular name:
Habit: Herb

Description: An erect annual herb; fond near water bodies. Stem is angled. Leaves are thin, dark green, toothed margins, end with soft spines. Flowers heads yellow color.

Flowering and Fruiting: November to February
IUCN status: Not Evaluated
Native: Europe, Asia, and northern Africa

Scientific name: *Sphaeranthus amaranthoides* Burm. f.
Vernacular name:
Habit: Herb

Description: An erect herb up to 30 cm high; distributed in cultivated fields. Stem terete, branches winged. Leaves are spathulate, acute, serrate. Flowers purple.

Flowering and Fruiting: August to February
IUCN status: Least Concern
Native: India

Scientific name: *Spilanthes calva* DC.
Vernacular name:
Habit: Herb

Description: A small annual herb up to 20 cm high. Leaves are opposite, ovate, acute, crenate. Flowers are yellow.

Flowering and Fruiting: August to December
IUCN status: Not Evaluated
Native: South America

Scientific name: *Tridax procumbens* L.
Vernacular name: Kari balli
Habit: Herb

Description: Procumbent herb; distributed throughout the study area. Stem hairy; leaves simple, ovate–lanceolate, toothed, surface rough. Flowers are solitary, edged white and center flowers white.

Uses: Leaf extract and paste are used to treat wounds and cuts. Leaf paste is used to relieve scorpion bite.

Flowering and Fruiting: January to December
IUCN status: Not Evaluated
Native: Central America

Scientific name: *Xanthium indicum* DC.
Vernacular name: Maruluummatti
Habit: Herb

Description: Annual herb; found in waste lands, roadsides and agricultural boundaries. Leaves are broad, palmately veined, and hairy. Fruits end with two beaked hooks and covered with numerous hooks.

Flowering and Fruiting: September to March
IUCN status: Not Evaluated
Native: North America

15. Family: Balanitaceae

Scientific name: *Balanites aegyptiaca* (L.) Delile
Vernacular name: Ingalarade
Habit: Tree

Description: Xerophytic woody tree, growing up to 25–30 ft. tall; branches thorny and distributed in throughout study area mainly dry lands, waste lands, and near agricultural fields. Leaves are compound with two leaflets in different size. Flowers are small, bisexual, greenish white. Fruit ovoid drupe, 2–4 cm long with short thick stalk and ripe fruit is pale brown with brittle coat.

Uses: Used as medicinal purpose and used for fuel wood in local people. Fruit pulp with milk is used to prepare small tablet which is taken to cure jaundice.

Flowering and Fruiting: March to July
IUCN status: Not Evaluated
Native: Tropical Africa and Arabia

16. Family: Bignoniaceae

Scientific name: *Dolichandrone atrovirens* (Heyne) Sprague
Vernacular name:
Habit: Tree

Description: Tree; Leaves are compound, 7–9 cm long. Flowers are white in color, cyme with long stalk. Fruit is capsule up to 25–30 cm long.

Flowering and Fruiting: March to July
IUCN status: Not Evaluated
Native: Tropical Africa

Scientific name: *Dolichandrone falcata* Seem
Vernacular name: Oodi mara
Habit: Tree

Description: A deciduous tree up to 25 ft. high, bark scaly, bluish gray color. Found in Agricultural land boundaries. Compound leaves 3–8 cm long; leaflets 5–7, rounded elliptic, sometimes obovate. Flowers white. Fruits are capsule, curved like a sickle.

Use: Leaf paste is used to treat bone fracture. Wood and branches are used for timber and fuel wood.

Flowering and Fruiting: December to April
IUCN status: Not Evaluated
Native: Tropical Africa

Scientific name: *Markhamia lutea* (Benth.) K. Schum.
Vernacular name:
Habit: Tree

Description: Tree; Leaves compound, leaflets up to 10 cm long, end with small tip. Flowers terminal clusters, yellow in color, bell shaped with small orange red spot inside. Fruit is capsule 30–50 cm long.

Flowering and Fruiting: February to June
IUCN status: Not Evaluated
Native: Tropical Africa

Scientific name: *Millingtonia hortensis* L.
Vernacular name: Akasha mallige
Habit: Tree

Description: Tall tree, up to 30 ft.; pinnate compound leaves. Flowers fragrant, white color, tiny bell-shaped calyx, long slender tube.

Flowering and Fruiting: January to June
IUCN status: Not Evaluated
Native: Myanmar

Scientific name: *Spathodea campanulata* Beauv.
Vernacular name: Neeru kai
Habit: Tree

Description: A deciduous tree; leaves compound. Flowers are terminal, cluster, orange red color. Fruits are woody capsule with pointed tip.

Uses: Soked flowers with milk are used reduce stomach pain and bleeding.

Flowering and Fruiting: January to June
IUCN status: Not Evaluated
Native: Tropical Africa

Scientific name: *Tecoma stans* (L.) Kunth
Vernacular name:
Habit: Tree

Description: Shrub or small tree; recorded in Shahapur mountains. Leaves pinnate; flowers yellow, bell shaped. Fruits are capsule, bean shaped, and 10–15 cm long.

Flowering and Fruiting: July to December
IUCN status: Not Evaluated
Native: South America

17. Family: Cactaceae

Scientific name: *Opuntia dillenii* (Ker-Gawl.) Haw.
Vernacular name: Papas kalli
Habit: Shrub

Description: A succulent shrub, found throughout the study area. Stem is very short. Leaves are modified into spines, stem parts look like leaves, and these are thick green, elliptic, obovate shapes with modified spines. Flowers yellow in color. Fruits are green turn into purple when ripe.

Uses: Fruits are eaten by the local people.

Flowering and Fruiting: February to June
IUCN status: Not Evaluated
Native: South America

18. Family: Caesalpiniaceae

Scientific name: *Bauhinia racemosa* Lam.
Vernacular name: Basavana pada
Habit: Tree

Description: A woody tree; found in near agricultural fields and forest lands. Leaves shapes look like cow's hooves with rough surface. Flowers white. Fruits are pod, falcate.

Flowering and Fruiting: March to September
IUCN status: Not Evaluated
Native: India and Sri Lanka

Scientific name: *Caesalpinia pulcherrima* L.
Vernacular name: Kenjigemara
Habit: Tree

Description: An ornamental tree, recorded near Bheemarayangudi. Leaves are compound, with many small leaflets. Flowers are terminal clusters, yellow to orange red color. Fruit is legume.

Flowering and fruiting: March to November
IUCN status: Not Evaluated
Native: Tropical America and West Indies

Scientific name: *Cassia fistula* L.
Vernacular name: Kakke gida
Habit: Tree

Description: A medium-sized deciduous tree. Leaves are pinnately compound with large leaflets. Flowers are drooping, beautiful, 2–3 cm long. Fruits are long cylindrical pod. Seeds are many.

Uses: All parts are used to cure various diseases such as wounds, cardiac problem. Leaf extract is used to treat fever and cardiac diseases. Root can be used for breast diseases. Fruits are used to treat scorpion bite. Whole plant is used to treat accidental wounds.

Flowering and Fruiting: March to June
IUCN status: Not Evaluated
Native: India, China, and Southeast Asia

Scientific name: *Cassia javanica* L.
Vernacular name:
Habit: Tree

Description: Tree; recorded near at Agriculture college campus, Bheemarayangudi. Leaves pinnately compound, leaflets are elliptic, oblong. Flowers are cluster, pale rose. Pod cylindrical with many seeds.

Flowering and Fruiting: March to June
IUCN status: Not Evaluated
Native: Southeast Asia

Scientific name: *Delonix elata* (L.) Gamble
Vernacular name: Bile gulmohr
Habit: Tree

Description: Tree; distributed entire study area mainly in streams and growing along with agricultural boundaries. Trunk and branches with ash colored bark. Compound leaves, leaflets are many. Flowers are terminal raceme, yellowish white. Pod up to 20 cm long.

Uses: Leaf and bark paste and cold water is used to treat fractures. Stem parts are used for fuel wood.

Flowering and Fruiting: March to July
IUCN status: Not Evaluated
Native: Tropical Africa

Scientific name: *Delonix regia* (Boj. ex Hook.) Raf.
Vernacular name: Kempu gulmohr
Habit: Tree

Description: A medium-sized deciduous tree; trunk large angled toward the base. Leaves are pinnate compound, alternate, 20–40 leaflets. Flowers are orange red. Pods are long up to 60 cm.

Flowering and Fruiting: April to July
IUCN status: Not Evaluated
Native: Madagascar

Scientific name: *Guilandina bonduc* L.
Vernacular name: Gajjaga
Habit: Shrub

Description: A prickly shrub; recorded near roadside. Leaves and rachis together up to 40–50 cm, leaflets 6–8 pairs. Flowers are raceme, greenish yellow. Pod oblong, covered with stiff spines.

Flowering and Fruiting: May to July
IUCN status: Not Evaluated
Native: Tropical America

Scientific name: *Parkinsonia aculeata* L.
Vernacular name:
Habit: Shrub

Description: A deciduous shrub or small tree, up to 20 ft. high. Stems and branches are smooth and have sharp spine. Leaves are pinnate, leaflets small. Flowers are yellow, fragrant. Fruits are pod-like bean.

Flowering and Fruiting: March to June
IUCN status: Not Evaluated
Native: South America

Scientific name: *Senna auriculata* L.
Vernacular name: Honnambre gida
Habit: Shrub

Description: Shrub; distributed throughout the study area. Leaves are unipinnate compound with glands opposite to leaflets. Stipules are leaf like and persistent. Flowers yellow. Pod flat.

Uses: Dry bark powder mixed with coconut oil and applied on the burning region. Stem bark is consumed to control dysentery and digestive problems. Stem parts are used as tooth brush. Used for fuel wood.

Flowering and Fruiting: July to October
IUCN status: Not Evaluated
Native: South America

Scientific name: *Senna italica* (Mill.) Lam.
Vernacular name: Neladavare
Habit: Herb

Description: Herb; collected from rocky land, occasionally found. Leaves are pinnately compound, leaflets elliptic, slightly hairy on both the side. Flowers yellow, pods flat, small, and sickle shape. Seeds are 8–10.

Flowering and Fruiting: June to August
IUCN status: Not Evaluated
Native: Tropical Africa and tropical Asia

Scientific name: *Senna occidentalis* L.
Vernacular name: Tharavadi
Habit: Herb

Description: A glabrous erect herb; distributed throughout the study area. Leaves are alternate, pinnate with small block glands above the pulvinus. Inflorescence is a raceme, flowers yellow. Fruit is flat pod.

Flowering and Fruiting: August to February
IUCN status: Not Evaluated
Native: Central and South America

Scientific name: *Senna sericea* Sw.
Vernacular name:
Habit: Herb

Description: An erect herb with hairy; up to 4–5 ft. high. Leaves are alternate, compound with glands between leaflets. Flowers are axillary, yellow. Pod up to 3–4 cm long, hairy.

Flowering and Fruiting: July to November
IUCN status: Not Evaluated
Native: Tropical America

Scientific name: *Senna siamea* (Lam.) Irwin & Barneby
Vernacular name:
Habit: Tree

Description: A medium-sized tree; found in near conserve land, agricultural fields. Leaves pinnate, alternate, compound, leaflets oblong, end with short tip. Flowers yellow, raceme. Curved pod up to 20 cm long.

Flowering and Fruiting: March to October
IUCN status: Not Evaluated
Native: Southeast Asia

Scientific name: *Tamarindus indica* L.
Vernacular name: Hunase mara
Habit: Tree

Description: Evergreen tree; found in entire study region. Trunk and branches are woody; bark is dark gray, fissured. Leaves are pinnately compound with small leaflets. Flowers are small, many, pale green and pink color. Pods are pale brown, seeds smooth, pulp more acidic, edible.

Uses: Flower soaked in water over night and then eaten with honey for relieving acidity. Fruits are contain high vitamin C. Leaves and fruits are used as vegetable and fruits are used for prepare pickle.

Flowering and Fruiting: February to November
IUCN status: Not Evaluated
Native: Tropical Africa

19. Family: Capparaceae

Scientific name: *Cadaba fruticosa* (L.) Druce
Vernacular name:
Habit: Shrub

Description: Shrub, up to 10–12 ft. high; growing near rocky and barren land. Leaves oval shape, end with short tip. Stem and branches are glandular, pubescent. Flowers pale green to white. Fruits are cylindrical, valvular.

Flowering and Fruiting: January to May
IUCN status: Not Evaluated
Native: Indian subcontinent and Myanmar

Scientific name: *Capparis brevispina* DC.
Vernacular name:
Habit: Shrub

Description: Shrub with thorny stipules; collected from near Surpur mountains. Leaves are thick, short stalk, oblong–lanceolate. Flowers white, axillary. Fruits are berry, smooth and red when ripe.

Flowering and Fruiting: March to July
IUCN status: Not Evaluated
Native: India, Arabia, and North Africa

Scientific name: *Capparis divaricata* Lam.
Vernacular name: Bhandero
Habit: Shrub

Description: A busy shrub with strong spines; recorded near mountain region at Surpur. Leaves are linear to elliptic. Flowers are yellow green, petals velvety. Fruits are berry, almost rounded, ribbed, and red when ripe.

Flowering and Fruiting: March–September
IUCN status: Not Evaluated
Native: India and tropical Africa

Scientific name: *Capparis zeylanica* L.
Vernacular name: Tottilu balli
Habit: Shrub

Description: Climbing shrub; recorded on the way to Bheemarayangudi and adjoining area. Leaves are elliptic–oblong, alternate. Flowers are pink, stamens many.

Flowering: March to April
IUCN status: Not Evaluated
Native: Sri Lanka

Scientific name: *Cleome viscosa* L.
Vernacular name: Kaadu saasive
Habit: Herb

Description: Erect annual herb, slightly hairy and glandular; growing abundantly waste lands, roadside and harvested fields. Leaves are digitately compound with 3–5 leaflets. Flowers yellow; fruits are linear, dark brown, and ribbed.

Uses: Leaf extract is used to treat ear pain. Whole plant is used to treat skin diseases.

Flowering: August to October
IUCN status: Not Evaluated
Native: Tropical Africa, South Arabia, tropical Australia, and Malaysia

Scientific name: *Maerua oblongifolia* (Forssk.) A. Rich.
Vernacular name:
Habit: Shrub

Description: A climbing shrub; rarely found in the study area, collected from near agricultural field at Wandurga. Leaves are linear to elliptic. Flowers are greenish white, stamins 10–20.

Flowering: February to March
IUCN status: Not Evaluated
Native: Mediterranean

20. Family: Caricaceae

Scientific name: *Carica papaya* L.
Vernacular name: Parangi
Habit: Tree

Description: A cultivated tree; found in throughout study area and some places growing wild. Leaves are glabrous up to 2 ft. length, deeply palmately 7-lobed. Flowers are yellow, axillary raceme. Fruits are globose, yellowish color when ripe and edible.

Flowering and Fruiting: February to June
IUCN status: Not Evaluated
Native: Tropical America and Mexico.

Family: Caryophyllaceae

Scientific name: *Polycarpaea aurea* Wight & Arn.
Vernacular name:
Habit: Herb

Description: A small erect herb; abundantly found near rocky places. Branches white in color, tomentose. Leaves opposite, sessile, linear, obtuse at base. Flowers are small, pubescent.

Flowering: July to January
IUCN status: Not Evaluated
Native: Russia and Europe

Scientific name: *Polycarpaea corymbosa* (L.) Lam.
Vernacular name:
Habit: Herb

Description: Annual small herb; up to 25–30 cm high, recorded near mountain region of Shahapur and Surpur. Branches covered with white color, leaves are simple, linear, and hairy. Fruit is capsule.

Flowering: August to November
IUCN status: Not Evaluated
Native: Africa, Asia, and Australia

21. Family: Casuarinaceae

Scientific name: *Casuarina equisetifolia* L.
Vernacular name: Survey mara
Habit: Tree

Description: A deciduous fast growing tree; planted in waste lands, forest lands, and agricultural boundaries. Bark brown color, rough surface. Scale leaves, whorl shape. Whole plant is used as timber and fuel wood.

Uses: A large timber wood plant; used for multipurpose.

Flowering and Fruiting: November to January
IUCN status: Not Evaluated
Native: Australia, Southeast Asia, India, and Pacific Islands

22. Family: Celastraceae

Scientific name: *Maytenus emarginata* (Willd.) Ding Hou
Vernacular name:
Habit: Tree

Description: Shrub; recorded near Yelwar village, at Shahapur. Stem and branches reddish brown, leaves round shape with small emarginated apex. Fruits are capsule.

Flowering and Fruiting: December to February
IUCN status: Not Evaluated
Native: India, Sri Lanka, Indonesia, Malaysia, Australia, and Taiwan.

23. Family: Chenopodiaceae

Scientific name: *Chenopodium album* L.
Vernacular name: Hunachikka
Habit: Herb

Description: Herb; presented entire study area. Leaves are nonaromatic, ovate to lanceolate, margins dentate. Flowers are cluster, seeds compressed.

Flowering and Fruiting: January to February
IUCN status: Not Evaluated
Native: North America

24. Family: Combretaceae

Scientific name: *Combretum ovalifolium* Roxb.
Vernacular name: Edatiga
Habit: Climber

Description: A woody climber; occasionally distributed, recorded only once. Leaves are simple, oblong, opposite decussate, turn red before fall. Flowers are axillary, greenish yellow.

Flowering: February to March
IUCN status: Not Evaluated
Native: India

Scientific name: *Terminalia bellirica* (Gaertn.) Roxb.
Vernacular name: Taremara
Habit: Tree

Description: Tree; occasionally found, recorded near Shahapur port. Leaves are large 12–20 cm long, 8–15 cm width, thick, leathery, dotted, end with pointed tip. Flowers are creamy white. Fruits are obovoid, covered with minute pale pubescent.

Uses: Fruit extract with castor oil or honey is used to cure cough and stomach disorders. Fruit powder with mother's milk is given to children for digestive problems.

Flowering and Fruiting: March to August
IUCN status: Not Evaluated
Native: India

Scientific name: *Terminalia catappa* L.
Vernacular name: Kadu bondami
Habit: Tree

Description: A tree; recorded near coconut forms at Surpur range. Leaves are simple and large, turn red before fall. Flowers are small. Fruits are ellipsoidal, slightly bilaterally compressed.

Flowering and Fruiting: March to June
IUCN status: Not Evaluated
Native: India

25. Family: Commelinaceae

Scientific name: *Amischophacelus axillaris* (L.) R.S. Rao and Kammathy
Vernacular name: Jigale
Habit: Herb

Description: Annual slender prostate herb; collected from near rocky land. Rooting at nodes; leaves linear–lanceolate, succulent, acute. Flowers purple.

Flowering and Fruiting: August to October
IUCN status: Not Evaluated
Native: Indo-Malaysia

Scientific name: *Commelina benghalensis* L.
Vernacular name: Paradene
Habit: Herb

Description: Annual prostate herb with diffuse branches; found in through out study region. Leaves are ovate to oblong, obtuse, hairy on both surfaces. Flowers are small, blue.

Flowering: August to December
IUCN status: Least Concern
Native: India

Scientific name: *Commelina diffusa* Burm. f.
Vernacular name:
Habit: Herb

Description: A small herb; distributed through out the study region. Roots from nodes, leaves alternate, sessile, lanceolate with sharp tip. Flowers are small, axillary.

Flowering: March to August
IUCN status: Least Concern
Native: Asia

Scientific name: *Cyanotis tuberosa* (Roxb.) Schult. & Schult.f.
Vernacular name:
Habit: Herb

Description: A succulent herb; recorded near rocky lands. Stalk less sheathed leaves, coming from the roots. Flowers are bluish purple. Stamins six with densely bearded with long blue hairs.

Flowering: August to September
IUCN status: Not Evaluated
Native: India and Sri Lanka

26. Family: Convolvulaceae

Scientific name: *Cressa cretica* L.
Vernacular name:
Habit: Herb

Description: A diffuse herb; up to 30 cm high, rarely found, recorded only once near Gogi village of semiarid region. Leaves are many, stalk less, small, ovate, acute tipped, hairy. Flowers are small, pink.

Flowering: March to April
IUCN status: Not Evaluated
Native: India, Australia, and Mediterranean region

Scientific name: *Evolvulus alsinoides* L.
Vernacular name: Vishnukanthi
Habit: Herb

Description: Small herb; distributed through out study area. Branches are hairy; leaves covered with white silky hairs, lanceolate. Flowers are pale blue color.

Flowering: July to August
IUCN status: Not Evaluated
Native: Tropical America

Scientific name: *Ipomoea aquatica* Forssk.
Vernacular name:
Habit: Herb

Description: Semiaquatic herb; recorded only once at Kancham kavi village. Stem is hollow, roots at the nodes. Leaves are lanceolate 5–12 cm long. Flowers are trumpet shaped, white color with a purple in center.

Flowering: April to May
IUCN status: Not Evaluated
Native: Tropical Africa, tropical Asia, India, and China

Scientific name: *Ipomoea cairica* (L.) Sweet.
Vernacular name: Bekkina hejje balli
Habit: Herb

Description: A large climber; growing near garden and roadside. Leaves are palmate, lobed. Flowers are purple. Fruit is capsule.

Flowering: January to December
IUCN status: Not Evaluated
Native: Tropical Africa, and tropical Asia

Scientific name: *Ipomoea carnea* Jacq.
Vernacular name: Ganesh kaddi
Habit: Shrub

Description: Shrub; distributed throughout study region and abundantly found in marshy places. Leaves are simple, alternate, heart shaped with pointed tip. Flowers are cluster, funnel shaped, pale pink color. Fruit is capsule.

Flowering and Fruiting: Throughout year
IUCN status: Not Evaluated
Native: South America

Scientific name: *Ipomoea hederifolia* L.
Vernacular name: Nakshatra hoovu
Habit: Climber

Description: A climber; leaves broad, alternate, simple, heart shaped. Flowers are pretty red 2–3 cm across. Fruit is globular.

Flowering: September to December
IUCN status: Not Evaluated
Native: North Mexico

Scientific name: *Ipomoea mauritiana* Jacq.
Vernacular name: Nelagumbala
Habit: Climber

Description: A climber; occasionally recorded. Leaves palmately divided. Flowers are many, pink in color.

Flowering: September to December
IUCN status: Not Evaluated
Native: Tropical America

Scientific name: *Ipomoea obscura* (L.) Ker-Gawl.
Vernacular name:
Habit: Climber

Description: A slender twining climber, recorded near agricultural field. Leaves simple, heart shaped. Flowers are solitary, up to 3 cm across.

Flowering: August to March
IUCN status: Not Evaluated
Native: India, and Sri Lanka

Scientific name: *Ipomoea tricolor* Cav.
Vernacular name:
Habit: Climber

Description: A twining climber; recorded near agricultural fields and gardens. Leaves are heart-shaped end with tip. Flowers are trumpet shaped and blue color with a white yellow at center.

Flowering: January to December
IUCN status: Not Evaluated
Native: Southern Mexico

Scientific name: *Ipomoea quomoclit* L.
Vernacular name:
Habit: Climber

Description: Climber; rarely recorded once at Bheemarayangudi. Leaves are unipinnate compound, linear. Flowers are small up 3–4 cm long, red in color.

Flowering: August to September
IUCN status: Not Evaluated
Native: South America

Scientific name: *Jacquemontia pentantha* (Jacq.) G. Don
Vernacular name:
Habit: Climber

Description: An annual twiner, found along with the streams and agricultural lands. Leaves are ovate–cordate. Flowers are pale blue.

Flowering: September to January
IUCN status: Not Evaluated
Native: South America

Scientific name: *Merremia emarginata* (Burm. f.) Hallier f.
Vernacular name:
Habit: Herb

Description: Prostate herb, occasionally recorded. Stems rooting at nodes, leaves broadly ovate, margin crenate, glabrous. Flowers are yellow, tubular–campanulate.

Flowering: September to January
IUCN status: Not Evaluated
Native: Tropical Africa, Indian subcontinent, China, and Australia

Scientific name: *Merremia tridentata* (L.) Hall. f.
Vernacular name:
Habit: Climber

Description: A twining herb, found in waste lands, rocky land, agricultural fields. Stem is slender, long and angular. Leaves sub-sessile, oblong, acute, and dentate at the base. Flowers are axillary cymes. Corolla pale creamy yellow and dark purplish brown at the center.

Flowering: August
IUCN status: Not Evaluated
Native: Myanmar, Micronesia, Malaysia, and Australia

27. Family: Cucurbitaceae

Scientific name: *Coccinia grandis* (L.) Voigt
Vernacular name: Tonde balli
Habit: Climber

Description: A tendril climber; the tendril are long, elastic, coil-like spring shaped; distributed throughout the study region. Leaves are simple, lobed, and alternate. Flowers are solitary, white color, bell shaped. Fruits are berry, fleshy with many seeds. Ripe fruits are edible.

Uses: Leaf paste is used to treat skin diseases and fractures. Ripened fruits are edible.

Flowering and Fruiting: January to December
IUCN status: Not Evaluated
Native: India and Sri Lanka

Scientific name: *Corallocarpus epigaeus* (Rottler) C.B. Clarke
Vernacular name:
Habit: Climber

Description: A small creeper; rarely recorded. Stout tuberous roots, leaves are lobed, cordate and vary in shapes, rough surface. Flowers are yellow. Fruits are small, berry, fleshy, red when ripe and beak green.

Flowering and Fruiting: December to March
IUCN status: Not Evaluated
Native: Tropical Africa, Indian subcontinent, and Arabian Peninsula

Scientific name: *Ctenolepis cerasiformis* (Stocks) C. B. Clarke
Vernacular name:
Habit: Climber

Description: A climber, spread in large space; distributed in entire study area mainly agricultural fields. Stem sub-filiform, elongated, and much branched. Tendrils simple, elongated. Leaves are deeply lobed, margins dentate, acute, and ciliate with hairs as long as the breadth of bract. Flowers are small, pale yellowish. Fruits are globose.

Flowering and Fruiting: August to March
IUCN status: Not Evaluated
Native: Tropical Africa

Scientific name: *Cucumis callosus* (Rottler) Cogn.
Vernacular name: Mekkekayi balli
Habit: Climber

Description: A prostrate climber; found throughout study region, mainly agricultural fields. Stem is angular, hairy. Leaves are ovate, rough surface on both sides, dentate. Flowers yellow, small, fruits are berry, ellipsoid.

Flowering and Fruiting: September to December
IUCN status: Not Evaluated
Native: Tropical Africa, tropical Asia, and Arabian Peninsula

Scientific name: *Diplocyclos palmatus* (L.) Jeffrey
Vernacular name: Linge toned balli
Habit: Climber

Description: A slender vine, with angular stems; tendrils opposite to leaves and bifid, distributed total study area and associated with tree species. Leaves palm-like 5 lobed, base deeply cordate, acute, margin toothed. Flowers cream color, axillary clusters. Fruits berries, brick red with white streaks when ripe. Seeds resemble miniature "Shiva ling."

Uses: Leaf extract with turmeric is used cure the skin disease and cough.

Flowering and Fruiting: November to March
IUCN status: Not Evaluated
Native: Australia, Malaysia, tropical Africa, and New Guinea

Scientific name: *Mukia maderaspatana* (L.) Roem.
Vernacular name: Gubbi savatikayi
Habit: Herb

Description: A tendril climbing herb, with strongly spreading bristly hairs; tendrils simple. It is mainly distributed in agricultural field boundaries, along with streams and waste land. Leaves alternate, broadly triangular, 3–5 lobed with rough hairs. Flowers are small, pale yellow. Fruits are berry, globose, red when ripe.

Uses: Leaf paste is applied to cure skin problems.

Flowering and Fruiting: August to December
IUCN status: Not Evaluated
Native: Tropical Africa, tropical Asia, Australia, and temperate Asia

28. Family: Cyperaceae

Scientific name: *Bulbostylis densa* (Wall. ex Roxb.) Hand.-Mazz.
Vernacular name:
Habit: Grass

Description: A grass up to 35 cm tall; distributed in moisture places and near water bodies. Rhizome is absent, slender, and smooth. Leaf sheath are membranous, mouth long pilose and leaf blade filiform. Inflorescence is 4–7 spikelets, rays 3–5 cm long. Spikelets are solitary or clustered.

Flowering and Fruiting: April to December
IUCN status: Least Concern
Native: Bhutan, Ethiopia, Guinea, Indian subcontinent, and Indochina.

Scientific name: *Bolboschoenus maritimus* (L.) Palla
Vernacular name: Neeru sabbasige
Habit: Grass

Description: An aquatic grass up to 2 m tall. Leaves sheath membranous at mouth, apex convex. Inflorescence is spiked with sessile, and flowers are bristle.

Flowering and Fruiting: April to December
IUCN status: Not Evaluated
Native: North America

Scientific name: *Cyperus compressus* L.
Vernacular name: Vasumani hullu
Habit: Grass

Description: Stem tufted, erect, and sometimes much reduced. Leaves are shorter, linear, and narrow. Inflorescence simple umbel, sessile, and spikelts are linear. Fruit is nut and dark brown.

Flowering and Fruiting: July to December
IUCN status: Least Concern
Native: Eastern and central USA, West Indies, Mexico, Central America, South America, and the Old World.

Scientific name: *Cyperus pangorei Rottb.*
Vernacular name:
Habit: Grass

Description: Annual or perennial grass, rhizome is short. Leaves reduced, basal sheath up to 20 cm, greenish gray, soft mouth margin, leaf blade 15 cm long. Inflorescence is cluster of spike.

Flowering and Fruiting: August to December
IUCN status: Least Concern
Native: Indian subcontinent, Indochina

Scientific name: *Cyperus alopecuroides* Rottb.
Vernacular name:
Habit: Grass

Description: A large grass, perennial, rhizomatous up to 3 m tall, glabrous; stem is triangular. Leaves are large up to 120 cm long. Inflorescence is spike 1–4, cylindrical; spikelets many linear. Achenes are dark brown.

Flowering and Fruiting: August to December
IUCN status: Least Concern
Native: Tropical Africa, Madagascar, and Indo-Malaysia

Scientific name: *Cyperus rotundus* L.
Vernacular name: Tunge hullu
Habit: Grass

Description: A perennial rhizomatous, tuberiferous grass; common all the moisture places and in water bodies. Stem is triangular and tubers are aromatic. Leaves arise from the stem base, long, and linear. Flowers are cluster on elongated stalk, reddish brown.

Uses: Dried rhizome powder is mixed with coconut oil and applied to control hair fall.

Flowering and Fruiting: January to December
IUCN status: Least Concern
Native: Africa

Scientific name: *Pycreus flavidus* (Retz.) T. Koyama
Vernacular name:
Habit: Grass

Description: A perennial herb, roots are fibrous, rhizome are short. Clums tufted, 3–15 cm tall, 3-angled, smooth few leaves at base. Leaves are shorter, sheath reddish brown, linear and flat. Inflorescence is simple anthela, rays 1–6 and unequal length. Spikelets congested at apex.

Flowering and Fruiting: July to December
IUCN status: Not Evaluated
Native: Madagascar

Scientific name: *Cyperus squarrosus* L.
Vernacular name:
Habit: Grass

Description: A small herb, annual. Roots are fibrous. Clums 3–20 cm, trigonous, and glabrous. Leaves are 1–3, flat. Inflorescence is spike. Achenes light brown.

Flowering and Fruiting: August to February
IUCN status: Least Concern
Native: Tropical Africa, Arabian Peninsula, China, Indian subcontinent, Australia, South America, and North America.

29. Family: Elatinaceae

Scientific name: *Bergia capensis* L.
Vernacular name:
Habit: Herb

Description: Annual herb, up to 30 cm high; recorded near water bodies and wet lands. Stem is prostrate, rooting at lower portion, glabrous, reddish, and fleshy. Leaves are elliptic–lanceolate, serrate, acute at apex. Flowers axillary clusters, sepals 3, entire at margin, obtuse at apex. Stamens 5, carpels 5, ovary globose.

Flowering and Fruiting: September to December
IUCN status: Not Evaluated
Native: South America and Africa

Family: Euphorbiaceae

Scientific name: *Acalypha indica* L.
Vernacular name: Kuppe gida
Habit: Herb

Description: An erect herb, growing up to 50–70 cm tall; distributed all cultivated fields and waste lands. Leaves are rhomboid–ovate, toothed margin. The basal leaves have long petiole and upper leaves with short petiole. Flowers are axillary spikes. Fruits are schizocarpic.

Uses: Leaf paste is used to cure skin diseases. Leaf extract with buttermilk is taken to cure urinary problems.

Flowering and Fruiting: August to December
IUCN status: Not Evaluated
Native: Tropical Africa and tropical Asia.

Scientific name: *Chrozophora plicata* (Vahl) A. Juss. ex Spreng.
Vernacular name: Suttu mare
Habit: Herb

Description: A prostrate herb; recorded in agricultural field at Kancham kavi village. Stem is slightly hairy. Leaves are ovate, crenate, rounded at apex, 3–5 nerved. Flowers are axillary, opposite to leaves. Fruits are capsule, pubescent.

Flowering and Fruiting: March to May
IUCN status: Not Evaluated
Native: Tropical Africa, tropical Asia, and Arabian Peninsula

Scientific name: *Croton bonplandianum* Baill.
Vernacular name: Utti gida
Habit: Herb

Description: An herb up to 1–1.5 m high, branching from the base, scabrid; abundantly distributed throughout the study area in all the places. Leaves are simple, alternate, lanceolate, toothed margin. Terminal inflorescence, flowers are small, stamens many. Fruits are 3 lobed, rough surfaces, dehiscing into single seeded bits.

Uses: Leaf paste is used to treat wounds and ringworm. Latex is used to cure mouth ulcer.

Flowering and Fruiting: January to December
IUCN status: Not Evaluated
Native: Tropical America

Scientific name: *Euphorbia heterophylla* L.
Vernacular name: Halalabu
Habit: Herb

Description: An erect herb; recorded at agricultural fields, waste lands and roadsides. Leaves alternate below and whorled at above, show variegation, red patch at the base. Inflorescence is terminal cluster. Fruits are capsule.

Flowering and Fruiting: Throughout the year
IUCN status: Not Evaluated
Native: Tropical America

Scientific name: *Euphorbia hirta* L.
Vernacular name: Achche gida
Habit: Herb

Description: An annual prostrate herb; up to 30–40 cm tall with milky sap. Stem is densely clothed, covered with hairs, slightly woody. Leaves opposite, toothed margin, color varying green to copper red. Flowers numerous, minute, greenish yellow, look like a small globose. Fruits capsules, very small.

Uses: Root powder is mixed with Cuminum cyminum is used for urinary disorder.

Flowering and Fruiting: Throughout the year
IUCN status: Not Evaluated
Native: India, Nepal, and Sri Lanka

Scientific name: *Euphorbia hypericifolia* L.
Vernacular name:
Habit: Herb

Description: A small prostrate herb; up to 40 cm tall; occasionally found in the study region and recorded near agricultural field. Latex is present. Leaves are small, opposite, dentate, oblanceolate. Flowers are greenish pink.

Flowering and Fruiting: August to March
IUCN status: Not Evaluated
Native: South America

Scientific name: *Euphorbia tirucalli* L.
Vernacular name: Kol kalli gida
Habit: Tree

Description: A succulent milky tree; found in throughout study area, mainly agricultural field boundaries, rocky land, waste land, barren land, and roadside. Leaves are small, deciduous, alternate, oblanceolate, and acute at tip. Flowers are yellow at nodes.

Flowering and Fruiting: November to January
IUCN status: Least Concern
Native: Africa

Scientific name: *Jatropha glandulifera* Roxb.
Vernacular name: Totla gida
Habit: Shrub

Description: A shrub with glandular hairs; latex is present; found toward roadside at Bheemarayangudi. Leaves simple, large, alternate, lobed, and margin glandular. Flowers are small, brick red color. Fruits are schizocarpic.

Uses: Latex is used to cure mouth ulcer.

Flowering and Fruiting: June to September
IUCN status: Not Evaluated (NE)
Native: Brazil

Scientific name: *Kirganelia reticulata* (Poir.) Baill.
Vernacular name: Huli gida
Habit: Shrub

Description: A large shrub; recorded in entire study region—agricultural fields, roadside and waste lands. Leaves compound, distichous, glabrescent, and oblong. Flowers small, fruits globose, black when ripe and edible.

Flowering and Fruiting: February to June
IUCN status: Not Evaluated
Native: Tropical Africa, China, Indian subcontinent, Indochina, and Malaysia

Scientific name: *Phyllanthus amarus* Schum. & Thonn.
Vernacular name: Bhoo nelli
Habit: Herb

Description: A small, erect, annual herb, up to 50–70 cm tall; found throughout the study region, mainly agricultural fields abundantly available. Leaves are many, small, glaucous, and pinnately compound. Flowers are small, greenish yellow. Fruits are small, capsule, globose, and dehiscent.

Uses: The plant paste is used to cure jaundice.

Flowering and Fruiting: March to November
IUCN status: Not Evaluated
Native: America

Scientific name: *Phyllanthus emblica* L.
Vernacular name: Nelli kaayi
Habit: Tree

Description: A large deciduous tree; bark is smooth, greenish gray color, recorded near agricultural fields. Leaves are small, linear, compound. Flowers small, greenish yellow, numerous. Fruits are spherical, fleshy, yellowish green and 3–6 lobed. Fruits are medicinal and edible.

Uses: Fruits are edible and used for medicinal purpose.

Flowering and Fruiting: March to August
IUCN status: Not Evaluated
Native: India and Southeastern Asia

Scientific name: *Phyllanthus kozhikodianus* Siver. & Mani.
Vernacular name
Vernacular name:Habit: Herb

Description: An erect herb; rarely found in the study area, recorded only once at Shahapur mountains. Leaves are elliptic–obovate, hairless, pointed tip. Flowers are small, greenish white. Fruit is capsule, rounded.

Flowering and Fruiting: August to October
IUCN status: Not Evaluated
Native: India

Scientific name: *Phyllanthus maderaspatensis* L.
Vernacular name: Madras nalli
Habit: Herb

Description: An herb up to 50 cm tall; found in mainly agricultural fields, waster lands and near wet lands. Leaves are alternate, arranged in 2 ranks, obovate, tip is rounded with sharp point. Stipules peltate, perianth six lobed. Flowers are small greenish yellow. Fruits are capsule and smooth.

Flowering and Fruiting: July to December
IUCN status: Not Evaluated
Native: India, Sri Lanka and Old World Tropics

Scientific name: *Phyllanthus polyphyllus* Willd.
Vernacular name:
Habit: Tree

Description: A small deciduous tree, branches, and trunk brownish gray; recorded once at Bheema River basin. Leaves are distichous, simple, alternate, stipule persistent, apex obtuse with tip, oblong. Flowers are axillary, whitish. Fruits are capsule, smooth, trilocular, purplish black.

Flowering and Fruiting: March to August
IUCN status: Not Evaluated
Native: India

Scientific name: *Phyllanthus virgatus* G. Forst.
Vernacular name:
Habit: Herb

Description: A small prostrate herb branched, glabrous; found in agricultural fields, waste lands, and wet places. Leaves are elliptic, linear alternate, sub-sessile. Flowers are hanging from slender stalks, greenish white. Fruit is globose capsule.

Flowering and Fruiting: March to September
IUCN status: Not Evaluated
Native: China

Scientific name: *Ricinus communis* L.
Vernacular name: Haralu
Habit: Shrub

Description: A Shrub with soft trunk and branches with nodes; occasionally recorded in the study area. Leaves alternate, large, peltate, and palmately lobed, margins toothed and stalk with many glands. Flowers are yellowish green. Fruits are capsule, coved with soft spines, 3-lobed. Seeds smooth.

Uses: Leaf extract is used to treat piles, cough, and skin diseases.

Flowering and Fruiting: March to August
IUCN status: Not Evaluated
Native: Tropical Africa

Scientific name: *Securinega leucopyrus* (Willd.) Muell. Arg.
Vernacular name: Huli kantae
Habit: Shrub

Description: A large deciduous shrub with angular branches; recorded mainly roadside and waste lands. Leafy terminate in sharp spines. Leaves are simple, alternate, and ovate. Flowers are many, axillary clusters, white color. Fruits are smooth, globose, white in color.

Flowering and Fruiting: July to September
IUCN status: Not Evaluated
Native: India, Sri Lanka, and Myanmar

Scientific name: *Securinega virosa* (Willd.) Baill.
Vernacular name:
Habit: Shrub

Description: A deciduous smooth shrub; rarely recorded. Leaves are elliptic–ovate, obovate, rounded, pointed at the tip. Axillary flowers, fruits small, globose, white in color.

Flowering and Fruiting: June to September
IUCN status: Not Evaluated
Native: Tropical Africa and Asia

Scientific name: *Tragia involucrata* L.
Vernacular name:
Habit: Shrub

Description: A twining herb; covered with hairs; found in waste lands and along with streams. Leaves are ovate–lanceolate, margin serrate. Flowers are axillary, small, pale yellow. Fruits are capsule, 3 lobed covered with hairs.

Flowering and Fruiting: September to January
IUCN status: Not Evaluated
Native: India, Sri Lanka, and Nepal

30. Family: Fabaceae

Scientific name: *Abrus precatorius* L.
Vernacular name: Gulaganji
Habit: Climber

Description: A perennial climber; occasionally found in the study area. Leaves pinnately compound, up to 8–10 cm long. Flowers are axillary raceme, small, pinkish white, crowded and curved. Fruits pods, rectangular, thick, hard, found in cluster, each pod with 5 seeds. Seeds are usually red and black color, ovoid.

Flowering and Fruiting: Throughout the year
IUCN status: Not Evaluated
Native: India

Scientific name: *Alysicarpus vaginalis* (L.) DC.
Vernacular name:
Habit: Herb

Description: A prostrate herb; found in waste places and dry lands. Leaves are alternate, oblong–obovate; Flowers are raceme, cluster at terminal, pink in color. Fruit is pod, lightly hairy, cylindrical up to 2–3 cm long. Seeds are dark red.

Flowering and Fruiting: August to November
IUCN status: Not Evaluated
Native: India, Asia, and North America

Scientific name: *Canavalia virosa* (Roxb.) Wight &. Arn.
Vernacular name:
Habit: Climber

Description: A climber; found in near agricultural field and cultivated in some places. Stem is pubescent when young. Leaves are trifoliate, petiole 5–10 cm long, leaf lets are broad, ovate, obtuse to acuminate, pubescent on both the surface. Inflorescence is raceme, light pink color. Fruit is 12–15 cm long, oblong.

Flowering and Fruiting: January to December
IUCN status: Not Evaluated
Native: North America

Scientific name: *Clitoria ternatea* L.
Vernacular name: Shankapushpi gida
Habit: Herb

Description: A twinning herb; recorded throughout the study area. Leaves with 5 leaflets, leaflets elliptic, entire, papery. Flowers are solitary, white, blue in color. Fruit pod, oblong, seeds 10–12.

Uses: Leaves and stem parts are used to relive tooth pain.

Flowering and Fruiting: Throughout year
IUCN status: Not Evaluated
Native: India

Scientific name: *Crotalaria juncea* L.
Vernacular name: Senabu
Habit: Herb

Description: An erect herb up to 2–4 m tall, found in cultivated fields. Stem are ribbed, cylindrical. Leaves simple, alternate and spirally arranged along with stem, oblong–lanceolate. Inflorescence terminal raceme up to 20 cm, flowers are yellow, sepals 5 hairy. Pod is cylindrical, light brown, tomentose.

Uses: The plant is used as green manure for live stock and improving for the soil nitrate quality.

Flowering and Fruiting: September to March
IUCN status: Not Evaluated
Native: India, Pakistan, and Southeast Asia

Scientific name: *Crotalaria pallida* Aiton
Vernacular name:
Habit: Herb

Description: Herb; occasionally recorded. Leaves are trifoliate, leaflets are elliptic, membranous. Inflorescence raceme, flowers are many, yellow and orange color. Pod is oblong.

Flowering and Fruiting: August to April
IUCN status: Not Evaluated
Native: Africa

Scientific name: *Crotalaria pusilla* Heyne ex Roth
Vernacular name:
Habit: Herb

Description: A small herb; recorded near mountain regions. Leaves are small, linear, elliptic, and acute at apex, base cuneate. Flowers are terminal and lateral, 10–15 cm long, yellow in color with purple. Pods are oblong.

Flowering and Fruiting: August to December
IUCN status: Not Evaluated
Native: Asia

Scientific name: *Crotalaria ramosissima Roxb.*
Vernacular name:
Habit: Herb

Description: A small hairy herb; found in near rocky land, waste land and barren land. Leaves are small, linear, hairy, and acute. Flowers are yellow. Fruit is pod, covered with hairs.

Flowering and Fruiting: August to December
IUCN status: Not Evaluated
Native: Western Australia

Scientific name: *Crotalaria retusa* L.
Vernacular name:
Habit: Herb

Description: A branched herb, branches are white hairy; occasionally recorded form the study area. Leaves are simple, obovate, retuse at apex, glabrous above. Terminal inflorescence with many flowers, flowers is attractive, yellow orange color. Fruit is pod with many seeds.

Flowering and Fruiting: August to April
IUCN status: Not Evaluated
Native: Africa, Asia, and Australia

Scientific name: *Dalbergia sissoo* Roxb.
Vernacular name: Sissu
Habit: Tree

Description: A tall deciduous tree; found along with agricultural field boundaries and roadside. Leaves are compound, 3–5 leaflets, entire, sub-orbicular, acuminate, and leathery. Flowers are small, pale yellow, stamens are 9 united. Fruits are pod, compressed, 3–5 seeds.

Flowering and Fruiting: February to August
IUCN status: Not Evaluated
Native: India

Scientific name: *Erythrina variegata* L.
Vernacular name: Mandhara pushpada gida
Habit: Tree

Description: A tree; rarely found in the study region, recorded in agricultural boundaries. Leaves are compound, trifoliate, leaflets broad, 2-are opposite, acute. Flowers are bright crimson color, 2–3 cm long, terminal cluster. Fruit pod, cylindrical, up to 20 cm long. Seeds are black.

Flowering and Fruiting: February to June
IUCN status: Not Evaluated
Native: Tropical Asia and India

Scientific name: *Indigofera cordifolia* Heyne ex Roth
Vernacular name:
Habit: Herb

Description: A prostrate, pilose herb; recorded in waste places and roadside. Leaves are simple, cordate shape at base, apex acute. Flowers are raceme, axillary, short, pink in color. Pods are cylindrical, grayish white, pubescent.

Flowering and Fruiting: September to December
IUCN status: Not Evaluated
Native: Tropical Africa, India, Malaysia, and Australia

Scientific name: *Indigofera linnaei* Ali
Vernacular name:
Habit: Herb

Description: A prostrate, branching herb; recorded at agricultural field and roadside. Leaves are pinnately compound with 7–9 leaflets. Flowers are small, attractive, clusters. Fruit is small pod, lightly covered with hairs.

Flowering and Fruiting: July to December
IUCN status: Not Evaluated
Native: Northern China, Korea, and Japan

Scientific name: *Indigofera tinctoria* L.
Vernacular name: Neeligida
Habit: Herb

Description: A small dry deciduous shrub; found in roadside. Leaves are compound, leaflets 4–7 pairs. Flowers are raceme, pink in color. Pods are linear, slightly curved.

Flowering and Fruiting: July to November
IUCN status: Not Evaluated

Scientific name: *Pongamia pinnata* (L.) Pierre.
Vernacular name: Honge mara
Habit: Tree

Description: A medium-sized, semievergreen tree, up to 20–25 m tall. Bark grayish green, smooth. Leaves are alternate, pinnate, shining. Flowers are axillary drooping raceme, white and pink color. Fruits are pod, compressed, woody cover, yellowish gray when mature, small curved beak. Seeds are kidney shaped, usually one.

Uses: The plant is used for multiple purposes. Oil from the seeds is used to treat migraine and also used as pain relief.

Flowering and Fruiting: March to July
IUCN status: Not Evaluated
Native: India

Scientific name: *Cullen corylifolium* (L.) Medik *Syn. Psoralea coryifolia* L.
Vernacular name: Bavanchi gida
Habit: Herb

Description: A small herb; found in agricultural field. Leaves are simple, alternate, and dentate. Flowers are bluish purple, cluster. Pod is closely pitted.

Flowering and Fruiting: July to November
IUCN status: Not Evaluated
Native: Southeast Asia

Scientific name: *Rhynchosia minima* (L.) DC.
Vernacular name: Ghattavare
Habit: Herb

Description: A twinning herb found in agricultural field and roadside. Stem is sparsely pubescent. Leaves are pinnately compound, 3 folioate, stipules small, terminal leaflet rhomboid–circular, glabrous, apex obtuse. Flowers are small, corolla yellow, pedicle short. Fruit is legume, oblanceolate.

Flowering and Fruiting: November to March
IUCN status: Least Concern
Native: Tropical Africa, Temperate Asia, tropical Asia, Australia, South America, and North America.

Scientific name: *Stylosanthes fruticosa* (Retz.) Alston
Vernacular name:
Habit: Herb

Description: An herb with hairy, up to 90 cm tall; distributed in all agricultural fields and waste lands. Leaves are alternate, compound; leaflets are 3, lanceolate–linear. Flower small, axillary, and terminal inflorescence. Fruit is compressed pod with hook at the apex.

Flowering and Fruiting: July to November
IUCN status: Not Evaluated
Native: India, Sri Lanka, and Old World Tropics

Scientific name: *Tephrosia purpurea* (L.) Pers.
Vernacular name: Koggi gida
Habit: Shrub

Description: A small deciduous undershrub, slightly hairy; found in throughout study area in all habitats. Leaves are pinnate, leaflets 7–10 pairs. Inflorescence is many, raceme, petals pink. Fruit is pod and slightly curved.

Flowering and Fruiting: January to December
IUCN status: Not Evaluated
Native: India, Sri Lanka, and Old World Tripics

Scientific name: *Tephrosia villosa* (L.) Pers
Vernacular name:
Habit: Shrub

Description: Erect under shrub; occasionally found in the study area. Leaves are pinnately compound, leaflets are 9–15. Flowers are many, pink or white in color. Pod is curved, covered with hairs, wolly.

Flowering and Fruiting: July to November
IUCN status: Least Concern
Native: Africa and India

Scientific name: *Teramnus labialis* (L. f.) Spreng
Vernacular name: Adavi uddu
Habit: Climber

Description: A perennial climber; rarely found in the field near agricultural field. Stem is slender, covered with hairs. Leaves are trifolioate, 3 leaflets, rounded, elliptic, ovate shapes, densely covered with hairs. Flowers are small, axillary, pink or purplish color, sepal tube is hairy. Fruit is pod, linear, densely covered with hairs.

Flowering and Fruiting: November to March
IUCN status: Not Evaluated
Native: Tropical Africa, tropical Asia, and South America

Scientific name: *Vigna aconitifolia* (Jacq.) Marechal
Vernacular name:
Habit: Herb

Description: A pubescent herb; recorded near agricultural field. Leaves are compound; leaflets are 3, lobed, dentate. Stipules are lanceolate. Flowers are yellow; stamens are 9 + 1, pod glabrous.

Flowering and Fruiting: August to November
IUCN status: Not Evaluated
Native: India

Scientific name: *Vigna trilobata* (L.) Verdc.
Vernacular name:
Habit: Climber

Description: Trailing herb; occasionally recorded in the study area. Branches are prostate, trailing, smooth, or velvety. Leaves are compound, 3-leaflets, lobed. Inflorescence is raceme, flowers are few, peduncle up to 20 cm long, corolla yellow. Pod is cylindrical up to 3–4 cm long.

Flowering and Fruiting: August to November
IUCN status: Not Evaluated
Native: India, Pakistan, Sri Lanka, Burma, and Afghanistan.

31. Family: Gentianaceae

Scientific name: *Canscora diffusa* (Vahl) R. Br. ex Roem. & Schult.
Vernacular name:
Habit: Herb

Description: Annual herb up to 30 cm tall, glabrous, stem is branched; recorded waste lands and near wet lands. Leaves are sessile, basal leaves ovate, apex acute. Flowers occur in lax cymes at the end of the branches, pale pink or rose pink flowers are 1–1.5 cm across 3 visible petals.

Flowering: August to November
IUCN status: Not Evaluated
Native: Tropical Africa, Asia, and Australia

Scientific name: *Enicostemma axillare* (Lam.) Raynal
Vernacular name:
Habit: Herb

Description: A small erect herb, up to 10 cm tall; occasionally recorded mainly near to rocky lands and waste places. Leaves are linear, oblong, entire, and smooth surface. Flowers are clusters on the stem, white.

Flowering: July to September
IUCN status: Not Evaluated
Native: Tropical Africa, India, Southeast Asia, and Malaysia

Scientific name: *Exacum pedunculatum* L.
Vernacular name:
Habit: Herb

Description: Annual erect herb up to 30 cm high, stem is angular, branches can observe in upper half. Leaves ovate, 5 nerved from the base, acute, petioles short. Flowers are terminal, corolla pale to bright blue, lobes 4 obliquely erect.

Flowering: July to September
IUCN status: Not Evaluated
Native: India and Sri Lanka

32. Family: Hernandiaceae

Scientific name: *Gyrocarpus americanus* Jacq.
Vernacular name:
Habit: Tree

Description: A deciduous tree; recorded in Mahammadapur Reserve Forest at Shahapur. Trunk and branches are smooth, bark gray bark. Leaves are simple, entire, broadly ovate, and often more or less 3 lobed, velvet on both the surfaces and 3 veined. Flowers are yellowish green. Fruit is woody nut with 2 thin wings; wings are up 10 cm long. This helps in seed dispersal.

Flowering: March to July
IUCN status: Not Evaluated
Native: Western Australia

33. Family: Hydrocharitaceae

Scientific name: *Hydrilla verticillata* (L. f.)
Vernacular name:
Habit: Aquatic Herb

Description: A complete aquatic herb; found in all the water bodies in the study area. Stem is slender, green. Leaves are whorl at the nodes.

IUCN status: Least Concern
Native: Tropical Asia

34. Family: Lamiaceae

Scientific name: *Anisochilus carnosus* (L. f.) Wall.
Vernacular name:
Habit: Herb

Description: An aromatic small herb; recorded in near rocky lands and near water ponds. Leaves are ovate–oblong, stalk 1–3 cm long, base heart shape, and margin crenate. Flowers are spike, long stalk, upper lip is ovate and lower lip is truncate, flowers are purplish.

Flowering: August to November
IUCN status: Not Evaluated
Native: Africa and Asia

Scientific name: *Anisomeles malabarica* (L.) R. Br. ex Sims
Vernacular name: Gandu beerana gida
Habit: Herb

Description: A medium-sized aromatic shrub up to 2 m tall. Stems are angular, densely covered with hairs. Leaves are aromatic, elliptic–oblong, soft hairs lower surface, dentate. Inflorescence is a terminal long spike, flowers up to 1.5 cm long, pale pink and white color.

Uses: Leaf extract is used to relive cold and cough.

Flowering: March to May
IUCN status: Not Evaluated
Native: Tropical and subtropical Australia, Malaysia, and Southeast Asia.

Scientific name: *Hyptis suaveolens* (L.) Poit.
Vernacular name:
Habit: Herb

Description: An aromatic, hairy herb; occasionally recorded. Leaves are crenate–dentate, hairy. Flowers are axillary, blue.

Flowering: March to May
IUCN status: Not Evaluated
Native: Tropical America

Scientific name: *Leonotis nepetaefolia* (L.) R. Br
Vernacular name: Kadu tumbe gida
Habit: Shrub

Description: A medium-sized shrub, recorded toward roadside, and agricultural fields. Stem is glabrous, angular. Leaves are ovate, dentate. Flowers are verticillaster, calyx end with sharp spine, corolla orange.

Flowering: August to January
IUCN status: Not Evaluated
Native: Tropical Africa

Scientific name: *Leucas aspera* (Willd.) Spreng.
Vernacular name: Thumbe
Habit: Herb

Description: A branched, erect herb annual herb; found in all habitats in the study region and abundantly recorded in cultivated fields. Stem up to 40 cm tall, quadrangular with hispid branches. Leaves are opposite, elliptic, lanceolate, crenate. Flowers are white, terminal heads.

Flowering: January to December
IUCN status: Not Evaluated
Native: Asia

Scientific name: *Ocimum americanum* L.
Vernacular name:
Habit: Herb

Description: An herb much branched; recorded entire study region. Leaves are strongly aromatic, elliptic, toothed margin, softly on both the surface and small, minute glands are present. Flowers are terminal inflorescence, bilipped, small, and white. Seeds are mucilaginous when wetted.

Uses: A common medicinal plant in every home. Leaves are used to cure cold, cough, and fever.

Flowering: July to November
IUCN status: Not Evaluated
Native: Tropical Africa and Asia.

Scientific name: *Orthosiphon glabratus* Benth.
Vernacular name:
Habit: Shrub

Description: A small shrub; recorded in waste lands. Leaves are elliptic, ovate, and dentate. Flowers are pale pink, stamins 5, long.

Flowering: July to August
IUCN status: Not Evaluated
Native: Southeast Asia

35. Family: Lauraceae

Scientific name: *Cassytha filiformis* L.
Vernacular name: Aakaasha balli
Habit: Climber

Description: A twining parasitic climber; recorded twice in the study area. Stem is pale yellow green, filiform, glabrous. Leaves alternate, small. Inflorescences spikes, flowers white.

Flowering: March to August
IUCN status: Not Evaluated
Native: Florida, South America, and Central America

36. Family: Lecythidaceae

Scientific name: *Couroupita guianensis* Aubl.
Vernacular name:
Habit: Tree

Description: A large semievergreen tree; recorded once at Bheemarayangudi. Leaves are simple, elliptic–oblong, clustered end of the branches. Flowers are attractive, directly forming from trunk, pink in color, large, six fleshy perianth lobes, pedicle long 5–7 cm, stamens many. Fruits are globose, woody, large size.

Flowering: March to May
IUCN status: Not Evaluated
Native: French Guiana

37. Family: Liliaceae

Scientific name: *Scilla hyacinthina* (Roth) J.F. Macbr.
Vernacular name: Adavi Ulla gadde
Habit: Herb

Description: An herb with under ground bulb; found in rocky lands and waste lands. Leaves are simple, entire, oblong, apex obtuse, originated from the base with purple spots. Flowers are terminal raceme, pinkish green.

Flowering: July to December
IUCN status: Not Evaluated
Native: India and Sri Lanka

38. Family: Lythraceae

Scientific name: *Ammannia baccifera* L.
Vernacular name: Agnivendrapaaku
Habit: Aquatic Herb

Description: An erect annual herb; found in near water bodies, wet lands and waste lands. Stem is erect, branching, slender, purplish, and angled. Leaves are narrow, oblong, linear, two leaves present oppositely at nodes. Flowers are small, purplish, dense cluster in leaf axils.

Flowering: July to December
IUCN status: Not Evaluated
Native: India

39. Family: Lythraceae

Scientific name: *Nesaea brevipes* Koehne
Vernacular name:
Habit: Aquatic Herb

Description: An aquatic erect herb up to 30 cm tall; recorded at near Shahapur mountain region. Stem is 4 angled. Leaves are elliptic or oblong, margin serrate. Flowers are cyme, axillary, stamins 4.

Flowering: July to December
IUCN status: Least Concern
Native: India, Sri Lanka, and Bangaladesh

Scientific name: *Lawsonia inermis* L.
Vernacular name: Mehandi gida
Habit: Shrub

Description: Dry deciduous shrub; recorded at agricultural fields and found in house holds. Leaves are simple, opposite, elliptic, lanceolate, and acute. Flowers are white; fruit is rounded end with small beak.

Uses: Leaf paste is used to control hair fall and dandruff and leaf paste and powder is used for mehandi.

Flowering and Fruiting: August to December
IUCN status: Not Evaluated
Native: India, Arabia, Persia, and Egypt

40. Family: Malvaceae

Scientific name: *Abutilon indicum* (L.) Sweet
Vernacular name: Mudre gida
Habit: Shrub

Description: A perennial shrub with soft hairs on branches; distributed throughout study region mainly roadside, waste lands, agricultural boundaries. Leaves are cordate, crenate, alternate, and long stalk, velvet, soft hairs on the surface. Flowers are solitary, axillary, yellow color with long stalk. Stamins are many. Fruit is circular, splits into 15–20 single seeded bits with hairy carpels, brown when dry.

Flowering and Fruiting: August to April
IUCN status: Not Evaluated
Native: Southeast Asia

Scientific name: *Bombax ceiba* L.
Vernacular name: Kempu buragada mara
Habit: Tree

Description: A tall tree, trunk is large covered with hard, sharp, conical prickles; recorded in near agricultural field and households. Leaves are compound, 3–7 leaflets, lanceolate to oval, pointed apex. Flowers are large, clustered at end of the branches, pink or crimson color. Fruits are ovoid with five angled. Seeds are covered with silky white hairs.

Flowering and Fruiting: February to June
IUCN status: Not Evaluated
Native: Southeast Asia and northern Australia

Scientific name: *Hibiscus hispidissimus* Griff.
Vernacular name: Kempu buragada mara
Habit: Shrub

Description: Shrub covered with hooked prickles. Leaves are alternate, 5–8 cm long, palmately lobed 3–5 lobed, hairy, cordate shape at base, margins toothed, lobes are pointed and leaf stalk 5–8 cm long. Flowers are solitary, axillary, yellow, and dark purple in center, stalk prickly. Fruit is capsule.

Flowering and Fruiting: August to February
IUCN status: Not Evaluated
Native: Tropical Africa

Scientific name: *Hibiscus micranthus* L. f.
Vernacular name:
Habit: Subshrub

Description: A subshrub, scabrid branches, stellate, pubescent; recorded near agricultural field and waste lands. Leaves are simple, ovate, elliptic, dentate, apex acute, soft hairy. Flowers are axillary, solitary, pedicle 2–3 cm long, curved, white and purple, stamins column, anthers yellow. Fruits are capsule, globose. Seeds are covered with densely hairy.

Flowering and Fruiting: August to February
IUCN status: Not Evaluated
Native: Tropical Africa, Arabia, India, and Pakistan.

Scientific name: *Hibiscus rosa-sinensis* L.
Vernacular name: Dasavala
Habit: Shrub

Description: An ornamental plant; found in gardens and agricultural fields. Leaves are simple, alternate, ovate, margin toothed, acuminate. Flowers are solitary, red or white, in the leaf axils, stamins tube bearing anthers. Fruits are capsules.

Uses: An ornamental plant and also used for medicinal purpose to control hair loss.

Flowering and Fruiting: January to December
IUCN status: Not Evaluated
Native: East Asia and China

Scientific name: *Diplopenta odorata* Alef. *Syn Pavonia odorata* Willd.
Vernacular name: Bala raakshasi
Habit: Herb

Description: An erect perennial herb covered with sticky hairs; recorded in waste lands, roadside and agricultural boundaries. Leaves are cordate, 3–5 angled, 4–6 cm long. Flowers are solitary, single in leaf axils, sepals 5; flowers are light pink or white pink. Fruit is spherical and mericarps smooth.

Flowering and Fruiting: September to November
IUCN status: Not Evaluated
Native: Tropical Asia and Myanmar

Scientific name: *Pavonia zeylanica* Cav.
Vernacular name: Chittaamutti gida
Habit: Herb

Description: A viscid herb; found throughout the study region. Leaves are cordate, 3–5 nerved, hairy. Flowers solitary, axillary, stalk long 2–3 cm, corolla white or pink. Fruits are mericarp, smooth.

Flowering and Fruiting: September to November
IUCN status: Not Evaluated
Native: India, Sri Lanka, and Southwest Asia

Scientific name: *Sida acuta* Burm. f.
Vernacular name: Bhimana kaddi
Habit: Herb

Description: An annual herb, growing up to 70–80 cm tall; distributed in throughout study area. Stem is slender, branched and covered with small hairs. Leaves simple, alternate, ovate, elliptic, pointed tip, margin toothed. Flowers are solitary, axillary from the leaf base, pale yellow. Fruits are mericarp, small. Seeds are dark brown.

Flowering and Fruiting: January to December
IUCN status: Not Evaluated
Native: North and South America, tropical Asia, and Temperate Asia

Scientific name: *Sida cordata* (Burm. f) Borss. Waalk.
Vernacular name:
Habit: Herb

Description: A prostrate ascending herb, up to 1 m tall, woody base, rooting at nodes; recorded in waste lands and roadsides. Leaves simple, alternate, ovate, and cordate at base and margin toothed. Flowers are solitary, stalk long up to 1–1.5 cm, pale yellow, about 1 cm across. Fruits are mericarp, ovate.

Flowering and Fruiting: July to February
IUCN status: Not Evaluated
Native: India, Sri Lanka, and tropical Africa

Scientific name: *Sida cordifolia* L.
Vernacular name: Bala
Habit: Herb

Description: An erect herb, branches covered with soft hairs; recorded in waster lands and roadside. Leaves are ovate, cordate at base, smooth surface, margin dentate. Flowers are axillary, small, yellow. Fruits are rounded and fruitlets having 2 short spines.

Flowering and Fruiting: August to February
IUCN status: Not Evaluated
Native: India

Scientific name: *Sida spinosa* L.
Vernacular name:
Habit: Herb

Description: An erect herb; distributed in throughout study region mainly in agricultural field and roadside. Leaves are simple, alternate, elliptic, dentate and a pair of spines at their base. Flowers are solitary, axillary, pale yellow.

Uses: Leaf paste is used to treat sexual problems in men.

Flowering and Fruiting: July to February
IUCN status: Not Evaluated
Native: Tropical America

Scientific name: *Thespesia populnea* (L.) Sol. ex Corr.
Vernacular name: Adavibende mara
Habit: Tree

Description: Tree, recorded roadside, agricultural field boundaries and conserved forests. Leaves are simple, alternate, long petioled, heart shaped, 5 nerves from the base, acuminate. Flowers are single or pairs in leaf axil, 8–10 cm long and 4–5 cm across, bell shaped, pale yellow and base purple. Fruits are capsule, brown when dry.

Flowering and Fruiting: May to November
IUCN status: Not Evaluated
Native: India, Africa, and Pacific Isles

Scientific name: *Waltheria indica* L.
Vernacular name:
Habit: Shrub

Description: A subshrub; distributed throughout study region. Leaves are simple, alternate, elliptic–oblanceolate, margin, crenate, velvet on both surfaces. Flowers are small, axillary and terminal fascicles, petals are yellow and spathulate. Fruit is capsule.

Flowering and Fruiting: August to December
IUCN status: Not Evaluated
Native: India

Scientific name: *Martynia annua* L.
Vernacular name: Huli uguru
Habit: Herb

Description: Annual herb covered with smooth hairs; occasionally recorded in the study area. Leaves are broad, slightly cordate at base, dentate, petiole long, hairy, leaves velvet, 5 nerves, smooth sticky hairy. Flowers are raceme, many, tubular, 3–5 cm long, white to pink, each flower coated with yellow line spots. Fruit with two hooked beak, hard, dark brown or black when ripe.

Flowering and Fruiting: August to November
IUCN status: Not Evaluated
Native: Mexico

41. Family: Meliaceae

Scientific name: *Azadirachta indica* A. Juss.
Vernacular name: Bevina mara
Habit: Tree

Description: A tall, much branched tree; found in entire study region. Bark is grayish black vertical ridges. Leaves are compound, leaflets are oppositely arranged along a central stalk, oblique, and leaflets toothed margins. Flowers are many, small, drooping from the branches, white. Fruit berries, oval to oblong shape, yellow when ripe.

Uses: A multipurpose useful plant. Whole plant is used as medicinal and woody is used as timber and fuel wood.

Flowering and Fruiting: February to July
IUCN status: Not Evaluated
Native: India

42. Family: Menispermaceae

Scientific name: *Cocculus hirsutus* (L.) Deils
Vernacular name: Dhagadi balli
Habit: Climber

Description: A creeper associated with tree species; recorded in entire study region. Leaves are densely velvety, ovate, ovate–oblong, 3–5 basal nerves. Flowers are axillary, fruit is a berry, dark purple.

Flowering and Fruiting: October to February
IUCN status: Not Evaluated
Native: North America, Asia, and Africa.

Scientific name: *Tinospora cordifolia* (Willd.) Hook. f. & Thoms.
Vernacular name: Amrutha balli
Habit: Climber

Description: A climber; recorded throughout study area. Stem shows aerial roots, stem green when young covered with loose, dry, papery bark. Leaves are simple, broadly ovate, base deeply cordate with 7 nerves. Flowers are small, axillary, yellow. Fruits are drupe, rounded, red when ripe.

Uses: Leaf paste is used to treat jaundice and fever.

Flowering and Fruiting: October to February
IUCN status: Not Evaluated
Native: India, Myanmar, and Sri Lanka

43. Family: Mimosaceae

Scientific name: *Acacia auriculiformis* Benth.
Vernacular name:
Habit: Tree

Description: A tree; recorded near roadside. Leaves are phyllodes, falcate shape and entire. Inflorescence elongated, flowers are densely arranged on axis. Fruit is pod and twisted.

Flowering and Fruiting: July to November
IUCN status: Not Evaluated
Native: Australia

Scientific name: *Acacia catechu* (Roxb.) Willd.
Vernacular name: Kachu gida
Habit: Tree

Description: A small deciduous tree; recorded in dry lands and barren land. Stem is dark brown with rough surface. Leaves are pinnate compound, leaflets are many, small, oblong. Glands are presented at leaf axils. Flowers are many on cylindrical spike, flowers white or pale yellow. Pod have short stalk, flat, curved, and oblong, 10–12 cm long, dark brown.

Flowering and Fruiting: March to August
IUCN status: Not Evaluated
Native: India

Scientific name: *Acacia farnesiana* (L.) Willd.
Vernacular name: Kasturi jail
Habit: Tree

Description: A small tree found in barren land and waste lands. Stem is dark brown, sharp thorns are occurred on branches. Flowers are small, fragrant, and yellow. Fruit is 5–8 cm long, cylindrical, glossy coat, and dark brown when dry.

Flowering and Fruiting: August to December
IUCN status: Not Evaluated
Native: Central and South America

Scientific name: *Acacia ferruginea* DC.
Vernacular name: Banni gida
Habit: Tree

Description: A deciduous tree; rarely recorded in the study area. Branches are slender, armed with conical prickles. Leaves are alternate, compound, prickles twin, infra stipular, slightly curved. Flowers are pale yellow, numerous, axillary spike. Pods are smooth, greenish to brown.

Flowering and Fruiting: August to February
IUCN status: Vulnerable
Native: Sri Lanka

Scientific name: *Acacia horrida* (L.) Willd.
Vernacular name: Dodda mullina jail
Habit: Tree

Description: A small deciduous tree or shrub, recorded waste lands and roadside. Stem greenish brown, large prickles. Leaves are compound, leaflets small and many. Flowers are yellowish white.

Flowering and Fruiting: August to December
IUCN status: Not Evaluated
Native: East Africa

Scientific name: *Acacia latronum* (L. f.) Willd.
Vernacular name:
Habit: Tree

Description: A deciduous tree, spreading like umbrella; occasionally recorded in the study area. Spines are two types large, white; hallow type, up to 10–12 cm long and other short, slender. In hallow spines ant's symbiotic association can see. Flowers are creamy white.

Flowering and Fruiting: August to February
IUCN status: Not Evaluated
Native: India and Myamnar

Scientific name: *Acacia leucophloea* (Roxb.) Willd.
Vernacular name: Bili jalli
Habit: Tree

Description: A medium-sized dry deciduous tree, occasionally recorded in the study area. Leaves are alternate, bipinnately compound; leaves are having small glands on the rachis. Flowers are small, cream to yellow colored. Pod is brown, flat, curved, and densely hairy.

Uses: Leaf extract is used to treat skin diseases and wood is used as timber and fuel wood.

Flowering and Fruiting: August to November
IUCN status: Not Evaluated
Native: Indomalayan region

Scientific name: *Acacia nilotica subsp. indica* (L.) Willd.
Vernacular name: Kari jail
Habit: Tree

Description: A deciduous tree; distributed throughout study region. Bark is black, ridged, branches with sharp long spines. Leaves are alternate, compound with pair of spines at their base. Leaflets are small up to 20 pairs. Flowers are clustered, yellow and globose. Pods are long, many, grayish white with many compartments.

Uses: Bark is used to cure wounds and common timber plant in the study area as well as fuel wood.

Flowering and Fruiting: August to December
IUCN status: Not Evaluated
Native: India

Scientific name: *Albizia amara* (Roxb.) Boivin
Vernacular name: Tugli mara
Habit: Tree

Description: A small tree, umbrella-shaped canopy; occasionally recorded in the study area. Bark grayish; Leaves are compound 15–20 cm long with 25–30 pairs of leaflets. Leaflets are oblong, small, and hairy. Flowers are cream color, globose. Pods are oblong, 20–25 cm long, 2–4 cm wide and brown when mature.

Uses: Fuel wood.

Flowering and Fruiting: April to October
IUCN status: Not Evaluated
Native: India, Sri Lanka, and tropical Africa

Scientific name: *Albizia lebbeck* (L.) Benth.
Vernacular name: Baage mara
Habit: Tree

Description: A large deciduous tree; recorded roadside and agricultural boundaries. Leaves compound; leaflets oblong, 10–15 pairs. Inflorescence terminal, flowers are greenish white. Fruits are pod, yellowish brown, flat, seeds 8–10.

Uses: Fuel wood.

Flowering and Fruiting: March to December
IUCN status: Not Evaluated
Native: Asia

Scientific name: *Dichrostachys cinerea* (L.) Wight & Arn.
Vernacular name: Vadavarada gida
Habit: Tree

Description: A small deciduous tree, up to 10 m tall; recorded in waste lands. Leaves are compound, bipinnate, leaflets small, 20–30 pairs. Flowers are attractive, brush like; heads are half pink and half yellow. Pod narrowly oblong, curved and some time coiled.

Flowering and Fruiting: March to October
IUCN status: Not Evaluated
Native: India, SE Asia, Africa, and northern Australia

Scientific name: *Leucaena leucocephala* (Lam.) de Wit
Vernacular name: Sanna sogachu
Habit: Tree

Description: A medium-sized tall tree; maximum number of species were recorded along with agricultural field boundaries. Bark is gray brown with vertical lines and older stems are having rough surface. Flowers are globose, cream color. Fruit is pod, linear, 15–20 cm long. Seeds are many.

Uses: A common plant is used for a variety of purposes by the local people such as fuel wood, preparing sticks, making traditional house, and medicinal purpose. Leaves are used as green manure to animals.

Flowering and Fruiting: March to December
IUCN status: Not Evaluated
Native: Central America

Scientific name: *Mimosa intisia* L.
Vernacular name:
Habit: Tree

Description: A small prickly deciduous tree; recorded in waste lands and dry deciduous forests. Leaves are compound, leaflets are small; thorns are presented at leaf base. Sharp hooked thorns are occurred all branches. Pod flat, curved with hooked spines. Seeds 6–8.

Flowering and Fruiting: April to December
IUCN status: Not Evaluated

Scientific name: *Mimosa pudica* L.
Vernacular name: Lajja
Habit: Herb

Description: Prostrate herb; distributed in wild, waste lands. Stems are branching with prickly. Leaves are alternate, bipinnate, leaflets are 25–35, small; leaflets are highly sensitive to touch and heat. Flowers are small, pink, many in globose heads. Fruits are pods, slightly curved, joint, covered with bristles.

Flowering and Fruiting: January to December
IUCN status: Not Evaluated
Native: South America

Scientific name: *Pithecellobium dulce* (Roxb.) Benth.
Vernacular name: Seeme hunase
Habit: Tree

Description: A medium-sized deciduous tree. Branches are many with spines. Leaves are alternate, compound with small leaflets. Flowers are many and minute on raceme inflorescence and aggregated into globose. Pod is twisted. Seeds are black. The white spongy part is edible.

Uses: Fruits are edible.

Flowering and Fruiting: March to June
IUCN status: Not Evaluated
Native: Mexico, South America, and West Indies

Scientific name: *Prosopis cineraria* (L.) Druce
Vernacular name: Shastra mara
Habit: Tree

Description: A medium-sized thorny tree, occasionally recorded in the study area. Stem is gray in color; branches are slender with scattered spines. Leaves are double compound; leaflets are dark green, end with small point. Flowers are small, yellow to cream white color. Pod is straight or curved, 10–15 cm long and smooth surface.

Flowering and Fruiting: March to June
IUCN status: Not Evaluated
Native: Indian subcontinent

Scientific name: *Prosopis juliflora* (Sw.) DC.
Vernacular name: Sarkari jail
Habit: Tree

Description: A medium-sized thorny tree; abundantly distributed in the study area in all kinds of habitats. Stem is greenish brown, much branched with spines. Leaves are small, compound, leaflets 20–25 pairs. Flowers are cream colored, small, many are occurred. Pod is pale yellow, straight or curved up to 25 cm long.

Uses: A common plant is abundantly is growing in the study area. It is a primary source for the local people for fuel wood.

Flowering and Fruiting: March to November
IUCN status: Not Evaluated
Native: Mexico, South America, and Caribbean

Scientific name: *Samanea saman* (Jacq.) Merr.
Vernacular name:
Habit: Tree

Description: A rain tree with umbrella-like canopy; occasionally found in the study region. Leaves are alternate, bipinnate, leaflets have glands. Flowers are globose, clustered arranged in raceme. Flowers are pinkish white. Pod is with thick mesocarp.

Flowering and Fruiting: March to August
IUCN status: Not Evaluated
Native: Tropical America and West Indies

44. Family: Moraceae

Scientific name: *Ficus arnottiana* (Miq.) Miq.
Vernacular name: Bettada arali
Habit: Tree

Description: A fig tree, single species were recorded in the study area. Leaves are like peepal tree veins are thick, pale yellow green color, leaves caudate. Fruit is fig.

Flowering and Fruiting: March to August
IUCN status: Not Evaluated
Native: India, Sri Lanka

Scientific name: *Ficus benghalensis* L.
Vernacular name: Alada mara
Habit: Tree

Description: A large evergreen tree much branched with latex, branches supported by prop roots that grow into the soil. Leaves are big 20–25 cm long, 10–13 cm wide, leathery, simple, alternate, entire, and ovate shapes with strong veins. Fruit is fig, axillary, sessile, and fruits are yellow orange.

Flowering and Fruiting: March to August
IUCN status: Not Evaluated
Native: India

Scientific name: *Ficus racemosa* L.
Vernacular name: Attimara
Habit: Tree

Description: A large tree with latex; recorded in Bheema River basin. Leaves are alternate, simple, ovate, glossy green, 15–20 cm long, and 5–7 cm wide. Fruits are fig, many, clustered, and directly formed on the branches and trunk.

Uses: Fruits are edible.

Flowering and Fruiting: March to August
IUCN status: Not Evaluated
Native: India

Scientific name: *Ficus religiosa* L.
Vernacular name: Aralimara
Habit: Tree

Description: A tall deciduous tree; recorded in throughout the study region. Bark gray brown. Leaves are heart shaped with long petiole up to 10 cm and apex long tail-like structure. Fruits are figs, globose, green, and purplish when ripe.

Uses: Crushed leaves are used to treat block skin spots.

Flowering and Fruiting: March to August
IUCN status: Not Evaluated
Native: India and Burma

Scientific name: *Streblus asper* Lour.
Vernacular name:
Habit: Tree

Description: A small deciduous tree; only few individuals recorded in the study area. Leaves are oblong–ovate, 4–10 cm long, rough on both the surface and pointed tip. Flowers are greenish yellow. Fruit is pericarp, pale yellow.

Flowering and Fruiting: March to August
IUCN status: Not Evaluated
Native: India and Sri Lanka

45. Family: Moringaceae

Scientific name: *Moringa oleifera* Lam.
Vernacular name: Nugge mara
Habit: Tree

Description: A medium-sized tree; cultivated and also growing wild in roadside. Leaves are multipinnately compound; leaflets are small. Flowers are many from axillary branches, white in color. Fruit is a cylindrical, long up to 40–50 cm, with ridges. Seeds are 10–15, winged.

Uses: Leaves and fruits are edible as vegetable.

Flowering and Fruiting: February to November
IUCN status: Not Evaluated
Native: India

46. Family: Myrtaceae

Scientific name: *Eucalyptus globulus* Labill.
Vernacular name:
Habit: Tree

Description: A tall evergreen tree cultivated and planted in forest lands. Trunk and branches are smooth, gray green color. Leaves are long, curved, end with tapering apex and strong odor. Flowers are many, clustered, stamins are many. Food is woody.

Uses: A highly timber yielding plant and used medicinal purpose also.

Flowering and Fruiting: October to February
IUCN status: Not Evaluated
Native: Australia

Scientific name: *Psidium guajava* L.
Vernacular name: Seebe gida
Habit: Tree

Description: A small-sized woody tree, bark is light brown. Leaves are opposite decussate, elliptic, oblong, acute. Flowers are solitary and often pairs at the axils, white petals; stamins are many. Fruits are berry.

Uses: Fruits are edible.

Flowering and Fruiting: March to November
IUCN status: Not Evaluated
Native: Tropical America

Scientific name: *Syzygium cumini* (L.) Skeels.
Vernacular name: Nerale
Habit: Tree

Description: A tall tree with a dense crown; occasionally recorded in the study region. Leaves are simple, opposite, elliptic, oblong, acuminate, and leathery. Flowers are fragrant, numerous, cluster, stamens white. Fruits are oblong, rounded, berries, and deep purple to black when ripe. Fruits are sweet, sour, and cooling.

Uses: Fruits are edible and medicinal.

Flowering and Fruiting: March to September
IUCN status: Not Evaluated
Native: South and Southeast Asia

47. Family: Nelumbonaceae

Scientific name: *Nelumbo nucifera* Gaertn.
Vernacular name: Tavere
Habit: Aquatic herb

Description: An aquatic, free-floating, rhizomatous herb. Rhizomes are whitish, a bed-like structure under the soil. Leaves are rounded, plate like, peltate, upper surface coated with wax and stalk is very long with small prickles. Flowers are solitary, large, long stalk, beautiful, pinkish white. Bud looks like oval shape. Fruits are large, spongy structure. Seeds are many with individual cavities.

Flowering and Fruiting: January to December
IUCN status: Not Evaluated
Native: Tropical Asia, Queensland, and Australia.

48. Family: Nyctaginaceae

Scientific name: *Boerhavia diffusa* L.
Vernacular name: Punarnava
Habit: Herb

Description: A creeping annual herb; distributed throughout the study region, mainly waste lands and agricultural fields. Stem and branches are slightly hairy, light brown to light pink color. Leaves are opposite, deltoid to cordate, entire. Flowers are axillary, perianth campanulate, sub-sessile. Fruits are small 5 ribbed.

Flowering and Fruiting: August to December
IUCN status: Not Evaluated
Native: India, Sri Lanka, and Pantropics

Scientific name: *Boerhavia erecta* L.
Vernacular name:
Habit: Herb

Description: An annual erect herb up to 1 m tall; occasionally recorded. Stems are sticky, woody at base, green or purple color. Leaves are pair, opposite, slightly hairy on surface. Flowers are axillary, cluster at the end of the branches; flowers are whit to pink, divide with 5 lobes. Fruits are small and ribbed.

Flowering and Fruiting: August to December
IUCN status: Not Evaluated
Native: USA, Mexico, Central America, and South America.

Scientific name: *Bougainvillea spectabilis* Willd
Vernacular name:
Habit: Woody climber

Description: A woody perennial climber, cultivated as ornamental and growing in wild also. Leaves are alternate, simple, and ovate to elliptic. Flowers are axillary, clusters, variations in colors purple, red, or pink. Fruit is achene, elongated.

Flowering and Fruiting: January to December
IUCN status: Not Evaluated
Native: South America

49. Family: Nymphaeaceae

Scientific name: *Nymphoides cristata* (Roxb.) Kuntze
Vernacular name: Neela thavarae
Habit: Aquatic herb

Description: An aquatic, free-floating herb with rhizome. Rhizomes are short and rounded. Leaves are heart-shaped, long stalk, and margin toothed, purple in lower surface. Flowers are large, attractive, petals many, rose in color.

Flowering and Fruiting: January to December
IUCN status: Least Concern
Native: Asia

50. Family: Olacaceae

Scientific name: *Ximenia americana* L.
Vernacular name:
Habit: Shrub

Description: A small shrub with spines branches. Leaves are oblong, occur in the axils of straight spines, 2–3 cm long, gray green. Flowers are clusters, small, greenish white.

Flowering and Fruiting: March to July
IUCN status: Not Evaluated
Native: Central and South America

51. Family: Orchidaceae

Scientific name: *Eulophia epidendraea* (J. Koenig ex Retz.) C.E.C. Fischer
Vernacular name:
Habit: Herb

Description: An orchid species with rhizome; very rare recorded in the study area. Stem is smooth surface, unbranched. Flowers are many raceme, yellow purple and leathery.

Flowering and Fruiting: March to April
IUCN status: Not Evaluated
Native: India

Scientific name: *Eulophia ramentacea* Lindl.
Vernacular name:
Habit: Herb

Description: An herb with terrestrial rhizome; recorded near water bodies. Leaves appear after flowers. Flowers many, raceme, yellowish. Fruit is capsule.

Flowering and Fruiting: September to December
IUCN status: Not Evaluated
Native: India

52. Family: Oxalidaceae

Scientific name: *Biophytum sensitivum* (L.) DC
Vernacular name: Ghaati horamuni
Habit: Herb

Description: A small herb up to 30 cm tall; found in waste lands and agricultural fields. Leaves are pinnately compound, numerous, crowded at the apex of the stem, leaflets are many 10–15 pairs, oblong–ovate. Flowers are many, crowded, yellow with red markings. Fruit is capsule.

Flowering and Fruiting: September to March
IUCN status: Not Evaluated
Native: Tropical Africa and Asia

Scientific name: *Oxalis corniculata* L.
Vernacular name: Pullampurachi soppu
Habit: Herb

Description: A creeping herb; distributed in waste lands and moisture places. Root nodes can observe from the branches, branches are slightly hairy. Leaves are alternate, long leaf stalk, palmately trifoliate. Flowers are axillary, simple, yellow. Fruit is capsule, cylindrical, and ridged.

Flowering and Fruiting: September to December
IUCN status: Not Evaluated
Native: Europe

53. Family: Papaveraceae

Scientific name: *Argemone mexicana* L.
Vernacular name: Daturi gida
Habit: Herb

Description: A perennial prickly herb, distributed throughout the study region. Branches are with yellow latex. Leaves are with sharp spines, lobed, alternate. Flowers are solitary, present at terminal end, 3–5 cm across, yellow. Fruit is capsule, prickly with many black seeds.

Uses: Root powder is used to treat bleeding and paste is used to cure skin diseases.

Flowering and Fruiting: January to December
IUCN status: Not Evaluated
Native: West Indies

54. Family: Passifloraceae

Scientific name: *Passiflora foetida* L.
Vernacular name: Kukki balli
Habit: Climber

Description: A climber with dense hairy and associated with tree and shrub species. Leaves are simple, alternate, and hairy on both the surface. Flowers are axillary, solitary, 3–5 cm across, white color. Fruit is flesh and globose, covered with dense hairy.

Flowering and Fruiting: July to December
IUCN status: Not Evaluated
Native: South America

55. Family: Plumbaginaceae

Scientific name: *Plumbago zeylanica* L.
Vernacular name: Chitra moola
Habit: Shrub

Description: A small low shrub found in waste lands and agricultural fields. Leaves are simple, alternate, and elliptic–cordate. Flowers are 1–2 cm across with a tube. Sepals are sticky. Fruit is elongated capsule.

Flowering and Fruiting: November to April
IUCN status: Not Evaluated
Native: India, Sri Lanka, and Pantropics

56. Family: Poaceae

Scientific name: *Aristida setacea* Retz.
Vernacular name: Dodda hanchi hullu
Habit: Grass

Description: A grass species distributed throughout the study region. Culms stout and erect. Leaves are linear, convolute, sheaths long. Ligule is a fringe of short hairs.

Flowering and Fruiting: May to December
IUCN status: Not Evaluated
Native: India, Sri Lanka, and Myanmar

Scientific name: *Arundinella setosa* (L.) Link
Vernacular name:
Habit: Grass

Description: Annual grass up to 40 cm long, nodes glabrous. Leaves are 7–10 cm long, lanceolate, densely hairy. Panicle is long, effuse. Branches are scabrid. Spikes are glabrous.

Flowering and Fruiting: June to December
IUCN status: Not Evaluated
Native: Europe and Asia

Scientific name: *Arundo donax* L.
Vernacular name:
Habit: Grass

Description: Annual or perennial grass, grows up 5–6 m. Stem is hallow, leaves are alternate, long 30–40 cm with short tip, gray green and hairy. Flowers are long.

Flowering: May to June
IUCN status: Least Concern
Native: Eastern Asia

Scientific name: *Bambusa arundinacea* (Retz.) Willd.
Vernacular name: Bedru
Habit: Grass

Description: A tall, perennial, erect grass species. Stem is divided nodes and internodes with thorny. Internodes are hallow. Leaves are sheathing, alternate, entire, linear, and acute.

Uses: Bamboo sticks are used to make agricultural tools.

Flowering: Once in life time
IUCN status: Not Evaluated
Native: Southeast Asia

Scientific name: *Chloris barbata* Sw.
Vernacular name: Manchada kalu hullu
Habit: Grass

Description: A grass up to 80–100 cm tall; abundantly found in the study area. Nodes are often purplish. Peduncles are below the inflorescence, glabrous.

Flowering: July to December
IUCN status: Not Evaluated
Native: Tropical Africa

Scientific name: *Cymbopogon martini* (Roxb.) S. Watson
Vernacular name:
Habit: Grass

Description: A grass, panicles are erect.

Flowering: August to November
IUCN status: Not Evaluated
Native: Southeast Asia

Scientific name: *Cynodon dactylon* (L.) Pers.
Vernacular name: Karike
Habit: Grass

Description: A common weed grass distributed throughout the study region. Leaves are flat, lanceolate, acuminate margin, sheath glabrous, hairy.

Flowering: August to November
IUCN status: Not Evaluated
Native: Africa

Scientific name: *Dactyloctenium aegyptium* (L.) Willd.
Vernacular name:
Habit: Grass

Description: Annual grass, culms erect or prostrate, rooting at nodes. Leaves are linear, acuminate, and glaucous. Ligule is short. Spike digitately reading 1–4 cm long.

Flowering: August to November
IUCN status: Not Evaluated
Native: Tropical Africa, tropical Asia, and temperate Asia

Scientific name: *Digitaria bicornis* (Lam.) Roem. & Schult
Vernacular name:
Habit: Grass

Description: An annual culms decumbent grass, rooting at nodes up to 60 cm tall. Leaves are sheath, linear. Spikelets are paired.

Flowering: August to November
IUCN status: Not Evaluated
Native: Indian subcontinent, China, and Australia

Scientific name: *Dinebra retroflexa* (Vahl) Panz.
Vernacular name: Naribalada hullu
Habit: Grass

Description: Annual erect herb with tufted at base. Branches are many, rooting at lower parts. Leaves are sheaths glabrous, linear up to 30 cm long. Inflorescence raceme, spikelets are sessile.

Flowering: November to December
IUCN status: Not Evaluated
Native: India, Africa, and Madagascar

Scientific name: *Echinochloa colona* (L.)
Vernacular name: Kadu haraka hullu
Habit: Grass

Description: An Annual grass, up to 70 cm tall. Leaves are hairless on surface. Ligule is absent. Leaves are 10–30 cm long. Flowers are raceme, arise along central axis.

Flowering: September to December
IUCN status: Least Concern
Native: North Africa, India, and Arabian Peninsulas

Scientific name: *Eragrostis japonica* (Thunb.) Trin.
Vernacular name:
Habit: Grass

Description: An annual grass. Leaves are flat, branches of panicles whorled.

Flowering: September to December
IUCN status: Least Concern

Scientific name: *Ischaemum rugosum* Salisb.
Vernacular name:
Habit: Grass

Description: An annual herb up to 100 cm tall. Leaf blade 10–30 cm long, 3–5 cm wide. Inflorescence terminal, axillary, and paired raceme, each 7–12 cm long.

Flowering: September to December
IUCN status: Not Evaluated
Native: Tropical Africa, India, Indochina, China, Malay Peninsula, Australia, Mexico, Central and south America, and West Indies.

Scientific name: *Iseilema hackelii* Shrestha & Gandhi
Vernacular name:
Habit: Grass

Description: An annual grass. Culms compressed at base. Spikelets are broad at apex as long.

Flowering: September to December
IUCN status: Not Evaluated
Native: Europe, western Asia, and North America

Scientific name: *Saccharum spontaneum* L.
Vernacular name: Kadu kabbu
Habit: Grass

Description: An annual or perennial grass, recorded in near river bed and along with streams and sandy places. Clums are erect, slightly woody, glumes are reddish brown. Ligule is fringe of hairs.

Flowering: September to December
IUCN status: Least Concern
Native: Asia

Scientific name: *Setaria pumila* (Poir.) Roem. & Schult.
Vernacular name:
Habit: Grass

Description: A tufted erect annual herb. Leaves are linear, flat, acuminate, sheaths glabrous. Ligules ridge of hairs. Panicle cylindrical, spikelets are 2–3 mm long, ellipsoid.

Flowering and Fruiting: September to December
IUCN status: Not Evaluated
Native: India, Sri Lanka, and Paleotropics

57. Family: Polygonaceae

Scientific name: *Antigonon leptopus* Hook. & Arn.
Vernacular name:
Habit: Climber

Description: A fast growing vine, recorded near Bheemarayangudi. Leaves are heart shaped to arrowhead shape up to 10–13 cm long. Flowers are axillary, pink color.

Flowering and Fruiting: September to December
IUCN status: Not Evaluated
Native: Mexico and Central America

58. Family: Polygonaceae

Scientific name: *Persicaria glabra* (Willd.) M. Gómez
Vernacular name: Nerru kanagilu
Habit: Aquatic herb

Description: An aquatic herb; recorded near water bodies. Stem is smooth and branches are red, swollen, often rooting at nodes. Leaves are oblong–lanceolate, elliptic–lanceolate 10–20 cm long and 2–4 cm wide. Inflorescence is terminal raceme, each flower is jointed with pedicle, flowers are white and pink.

Flowering and Fruiting: June to September
IUCN status: Not Evaluated
Native: Madagascar, China, Indian subcontinent, and South America

Scientific name: *Polygala chinensis* L.
Vernacular name:
Habit: Herb

Description: A small erect herb with branches up to 25 cm tall; recorded near paddy fields. Lower roots are stout. Stem woody at base, pubescent. Leaves are obovate, elliptic to lanceolate, papery. Inflorescence is axillary, densely crowded; flowers are yellowish white with pink.

Flowering and Fruiting: April to September
IUCN status: Not Evaluated
Native: India, China, Thailand, Indochina, Sumatra, and Peninsular Malaysia

59. Family: Pontederiaceae

Scientific name: *Eichhornia crassipes* (Mart.) Solms
Vernacular name:
Habit: Aquatic herb

Description: A free-floating perennial aquatic herb. Leaves are broad up to 15–20 cm across thick, glossy and ovate. Leaves have a long, spongy stalk. Flowers are attractive, pink in color.

Flowering and Fruiting: March to November
IUCN status: Not Evaluated
Native: Brazil

Scientific name: *Monochoria vaginalis* (Burm. f.) C. Presl ex Kunth
Vernacular name: Neelothpala
Habit: Aquatic herb

Description: An aquatic herb. Leaves are broad, up to 15–20 cm long, base heart shape, thick. Flowers are blue colored.

Flowering: July to November
IUCN status: Not Evaluated
Native: South Asia

60. Family: Portulacaceae

Scientific name: *Portulaca oleraceae* L.
Vernacular name:
Habit: Herb

Description: A small prostrate succulent herb. Stem is reddish green color. Leaves are thick, small, obovate, and fleshy. Flowers are small, 1 cm across. Fruit is globose, small.

Flowering and Fruiting: August to February
IUCN status: Not Evaluated
Native: Europe

61. Family: Punicaceae

Scientific name: *Punica granatum* L.
Vernacular name: Dalimbre
Habit: Tree

Description: A small tree, up to 8 m tall; cultivated and recorded wild also. Stem and branches are smooth, dark gray bark. Leaves are oblong–lanceolate, ovate with short stalk. Flowers are terminal clusters and solitary, orange red color. Fruits are spherical and rounded, crowned with persistent calyx and covered with thick leathery wall. Seeds are many, small.

Uses: Fruits are edible and also medicinal.

Flowering and Fruiting: August to February
IUCN status: Least Concern
Native: Central and Middle Asia

62. Family: Rhamnaceae

Scientific name: *Scutia myrtina* (Burm. f.) Kurz
Vernacular name:
Habit: Shrub

Description: A large prickly shrub; recorded in waste lands and roadsides. Leaves are small, simple, opposite. Flowers are axillary clusters. Fruit is small, globose, and smooth.

Flowering and Fruiting: August to February
IUCN status: Not Evaluated
Native: India, Sri Lanka, Myanmar, and Africa

Scientific name: *Ziziphus mauritiana* Lam.
Vernacular name: Borae hannu
Habit: Tree

Description: A medium-sized tree; occasionally recorded in the study area. Branches are many with spines, spines are small, hard, and hook shaped. Leaves are simple, elliptic to ovate with white tomentose. Stipules are modified into hooked spines. Flowers are axillary, small, clusters. Fruits are globose, berry, orange red when ripe.

Uses: Fruits are edible. Leaf paste is used to treat scorpion bite.

Flowering and Fruiting: February to August
IUCN status: Not Evaluated
Native: India

Scientific name: *Ziziphus oenoplia* Mill.
Vernacular name:
Habit: Shrub

Description: A straggling shrub; found in waste lands and roadside. Young branches are slightly velvety with paired thorns. Leaves are simple, alternate, ovate, ovate to lanceolate, oblique, and 3-nerved. Flowers are born from the leaf axils.

Uses: Fruits are edible.

Flowering and Fruiting: July to October
IUCN status: Not Evaluated
Native: India, Sri Lanka, and Pantropics

Scientific name: *Ziziphus xylopyrus* (Retz.) Willd
Vernacular name: Chuttimullu
Habit: Tree

Description: A medium-sized tree; recorded in waste lands and conserve forests. Branches are many with hooked spines. Leaves are dark green, 3-nerved, ovate to oblong. Flowers are small, many. Fruits are berry, globose, ovate. Seeds are hard.

Uses: Fruits are edible.

Flowering and Fruiting: August to January
IUCN status: Not Evaluated
Native: India and Sri Lanka

63. Family: Rubiaceae

Scientific name: *Canthium coromandelicum* (Burm. f.) Alston
Vernacular name: Mullukhare
Habit: Shrub

Description: A thorny shrub recorded throughout the study region; thorns are strong and opposite. Leaves are ovate, oblong, entire, and smooth surface. Flowers are axillary raceme, small, pale yellow to white. Fruits are obovate, separate like two parts and yellow and dark purple when ripe.

Uses: Fruits are edible.

Flowering and Fruiting: July to November
IUCN status: Not Evaluated
Native: India, Sri Lanka, and tropical east Africa.

Scientific name: *Catunaregam spinosa* (Thunb.) Tiruv.
Vernacular name: Kaarigida
Habit: Shrub

Description: Shrub, abundantly found in near mountain regions. Leaves are simple, entire, opposite, obovate, obtuse, upper surface smooth and lower surface tomatoes and stipules are ovate. Flowers are solitary on lateral and terminal branches, attractive white color. Fruits are berry, woody and glabrous. Seeds are many.

Uses: Fruits are edible.

Flowering and Fruiting: July to November
IUCN status: Not Evaluated
Native: India, South tropical Africa, and Southern Africa

Scientific name: *Hedyotis puberula* (G. Don) Arn.
Vernacular name: Chaaya vaeru
Habit: Herb

Description: A small herb up to 60 cm tall, rarely recorded in the study area. Leaves are linear, apex acute. Flowers are small, white color, pedicelled. Fruit is capsule.

Flowering and Fruiting: September to November
IUCN status: Not Evaluated
Native: Old World tropics

Scientific name: *Ixora arborea* Roxb. ex J.E. Sm.
Vernacular name:
Habit: Shrub

Description: A perennial shrub, ornamental, and growing wild also. Leaves are opposite, elliptic, oblong, and shortly acuminate. Flowers are clusters, numerous, white or orange red color. Fruits are drupe, red color.

Flowering and Fruiting: January to December
IUCN status: Not Evaluated
Native: India and Sri Lanka

Scientific name: *Meyna laxiflora* Robyns
Vernacular name:
Habit: Shrub

Description: A shrub, rarely recorded in the study area. Thorns are strong, curved structure and oppositely presented. Leaves are ovate to oblong, opposite. Flowers are axil cymes, greenish white. Fruits are berry, smooth, yellow when ripe.

Flowering and Fruiting: March to August
IUCN status: Not Evaluated
Native: India

Scientific name: *Morinda pubescens* J.E. Sm.
Vernacular name: Maligana mara
Habit: Tree

Description: A medium-sized tree, occasionally recorded in the study area. Leaves are opposite, elliptic to lanceolate, apex acuminate and glabrous on both the surfaces. Flowers are axillary, small, greenish white. Fruits are drupe, 2–3 cm across, globose.

Flowering and Fruiting: March to August
IUCN status: Not Evaluated
Native: Southern Asia

Scientific name: *Pavetta indica* L.
Vernacular name:
Habit: Shrub

Description: An erect shrub, occasionally recorded in the study area. Leaves are elliptic–oblong to elliptic–lanceolate up 15 cm long and pointed tip. Flower are cluster, many, white color, slender tube up to 5–7 cm long.

Flowering and Fruiting: March to August
IUCN status: Not Evaluated
Native: India

Scientific name: *Spermacoce hispida* L.
Vernacular name:
Habit: Herb

Description: A small hispid herb, up to 50 cm high. Stem is bluish; leaves oblong to elliptic, apex acute, hairy on both the surfaces. Stipules are sheathing. Flowers are verticillate, axillary, small, pinkish white, tube slender.

Flowering and Fruiting: June to December
IUCN status: Not Evaluated
Native: Indian subcontinent

Scientific name: *Spermacoce stricta* L. f.
Vernacular name:
Habit: Herb

Description: Annular procumbent hispid herb; branches are quadrangular. Leaves are elliptic to lanceolate or ovate–oblong, scabrid on both the surfaces, apex obtuse. Flowers are axillary, violet. Fruit capsule, oblong, hispid above.

Flowering and Fruiting: June to December
IUCN status: Not Evaluated
Native: Tropical Africa and Asia.

64. Family: Rutaceae

Scientific name: *Aegle marmelos* (L.) Corr.
Vernacular name: Bilvapatre gida
Habit: Tree

Description: A tree, occasionally recorded in the study area. Branches are many; bark grayish rough and axillary spines. Leaves are 3–5 foliolate, terminal leaflets is larger than other leaflets. Flowers are many, axillary panicles, white color with fragrant. Fruits are berries, variable shape, and woody. Seeds are many.

Uses: Leaf powder and extract is used to treat stomach pain and fever, and fruit extract is used to digestion.

Flowering and Fruiting: March to December
IUCN status: Not Evaluated
Native: India

Scientific name: *Chloroxylon swietenia* DC.
Vernacular name: Masivala
Habit: Tree

Description: Tree up to 10 m tall, recorded throughout the study region, especially in forest lands recorded abundantly. Stem bark thick, pale yellow, deeply fissured. Leaves are pinnate, leaflets 10–13 pairs, oblong, lanceolate, entire and gland dotted. Flowers are terminal panicle and white color. Fruits are capsule, oblong, or ovate. Seeds are compressed, many.

Flowering and Fruiting: March to August
IUCN status: Vulnerable
Native: India and Sri Lanka

Scientific name: *Limonia acidissima* L.
Vernacular name: Belada gida
Habit: Tree

Description: A deciduous tree up to 20 m tall; recorded near agricultural fields. Branches are many with axillary spines; bark is brown, rough, and deep vertical fissures. Leaves are pinnate, 5–7 leaflets, ovate, margins slightly toothed with presence of glands and strong aroma. Flowers are axillary panicles, brown color. Fruits are berries, rounded and covered with woody layer. Seeds are many, embedded in thick pulp.

Uses: Fruits are edible.

Flowering and Fruiting: March to September
IUCN status: Not Evaluated
Native: India

Scientific name: *Murraya koenigii* (L.) Spreng.
Vernacular name: Karebevu
Habit: Tree

Description: A shrub or small tree; found in near agricultural field and cultivated. Stem is smooth, bark is brownish black. Leaves are compound, leaflets 20–25, leaflets with minute glands and strongly aromatic; compound leaves are crowded at the end of branchlets. Flowers are small, many in cluster, white color, fragrant. Fruits are berries, smooth.

Uses: Leaves are used as vegetable. Leaves are used to cure piles diseases.

Flowering and Fruiting: March to August
IUCN status: Not Evaluated
Native: India, Sri Lanka, China, Myanmar, and Laos

65. Family: Salvadoraceae

Scientific name: *Salvadora persica* L.
Vernacular name:
Habit: Tree

Description: A small tree up to 10 m tall; rarely recorded in the study area. Branchlets are drooping, pale yellowish. Leaves are elliptic, oblong, glabrous, thick, entire, and obtuse. Flowers are axillary and terminal panicles, yellowish green. Fruits are drupe, red when ripe and smooth.

Flowering and Fruiting: December to April
IUCN status: Not Evaluated
Native: Tropical Africa, India, and Arabia Peninsula

66. Family: Santalaceae

Scientific name: *Santalum album* L.
Vernacular name: Gandadamara
Habit: Tree

Description: A medium-sized semiparasitic tree, distributed throughout the study region. Stems are slender, drooping branches, bark gray and rough. Leaves are opposite, elliptic–ovate. Flowers are many, small, terminal axillary and dark brown color. Fruits are drupes, purplish black when ripe.

Uses: A valuable medicinal plant in the study area. Leaf paste and stem, wood parts are used to cure skin diseases, fever, and headache.

Flowering and Fruiting: August to February
IUCN status: Vulnerable
Native: India

67. Family: Sapindaceae

Scientific name: *Cardiospermum halicacabum* L.
Vernacular name: Agni balli
Habit: Climber

Description: Pubescent climber; distributed throughout the study region and associated with trees and shrub species. Leaves are bipinnate, leaflets ovate to lanceolate, dark green, dentate, acuminate. Flowers are axillary, umbellate. Fruits are capsule, winged an angles and pubescent. Seeds are black.

Uses: Leaf paste is used to treat Joint pains.

Flowering and Fruiting: August to December
IUCN status: Not Evaluated
Native: India

Scientific name: *Dodonaea viscosa* (L.) Jacq.
Vernacular name: Bandarike
Habit: Shrub

Description: A dry deciduous shrub up to 2–3 m tall, distributed throughout the study area. Leaves are oblong to lanceolate, viscid, and entire. Inflorescence axillary cymes, flowers greenish yellow, shorter than leaves. Fruit is capsule and winged. Seeds are black.

Uses: Leaf extract is used to treat wounds and swelling burns. Used for fuel wood.

Flowering and Fruiting: September to December
IUCN status: Not Evaluated
Native: New Zealand

Scientific name: *Sapindus laurifolia* Vahl
Vernacular name: Antuvala
Habit: Tree

Description: A medium-sized tree up to 15 m tall. Branches are woody, bark ash gray, rough. Leaves are compound, 5–7 leaflets, leaflets large 5–10 × 3–5 cm, ovate, oblong and leathery, pubescent below and terminal leaflets is larger than others. Flowers are small, many, terminal panicle, and fragrant. Fruits are drupes, rounded and ovate, 2–3 lobed, soft hairs on surface when young. Seeds are black.

Uses: Fruits are used as soap nut and crushed fruits are used for vomiting.

Flowering and Fruiting: October to March
IUCN status: Not Evaluated
Native: India

68. Family: Scrophulariaceae

Scientific name: *Bacopa monnieri* (L.) Wett.
Vernacular name: Niru brahmi
Habit: Herb

Description: A sub-succulent creping herb, densely recorded near water bodies and wet lands. Roots arise from nodes and stem is dull pink to white color, cylindrical. Leaves are small, opposite, smooth, fleshy, and rounded apex. Flowers are axillary and bluish white with purple ting. Fruits are capsule.

Uses: The whole plant is used to cure kidney problems.

Flowering and Fruiting: January to December
IUCN status: Least Concern
Native: India

Scientific name: *Stemodia viscosa* Roxb.
Vernacular name: Antikamini
Habit: Herb

Description: A small herb up to 50–60 cm tall; occasionally recorded in the study area. Stem covered with hairy. Leaves are stalkless, 2–5 cm long, oblong, acute and heart shaped at the base. Flowers are axillary and presence at terminal branches, violet with two lipped.

Flowering and Fruiting: September to December
IUCN status: Not Evaluated
Native: India, Sri Lanka, Pakistan, and Afghanistan

69. Family: Simaroubaceae

Scientific name: *Ailanthus excelsa* Roxb.
Vernacular name: Hebbevu
Habit: Tree

Description: A large deciduous tree. Bark light gray and smooth and aromatic. Leaves are large pinnate compound up to 70–90 cm, alternate, leaflets 10–15 pairs, long stalked, leaflets ovate to lanceolate and base unequal. Flowers are many, clusters, droop at leaf base and greenish yellow color. Fruits are samara with single seeded.

Flowering and Fruiting: January to May
IUCN status: Not Evaluated
Native: India, Sri Lanka, and Australia

70. Family: Solanaceae

Scientific name: *Datura metel* L.
Vernacular name: Ummatti gida
Habit: Herb

Description: An herb or subshrub; recorded throughout the study area. Leaves are large, 10–15 cm long, 5–8 cm wide, simple, alternate, dentate and petiole long 5–6 cm long. Flowers are solitary, axillary, trumpet shaped, up to 10 cm across and corolla creamy white. Stamens 5, attached toward the base of corolla tube. Fruits are capsule, many, spherical, and covered with short spines.

Uses: Leaf extract is used to treat scorpion bite.

Flowering and Fruiting: January to December
IUCN status: Not Evaluated
Native: India, Sri Lanka, and tropical Asia

Scientific name: *Nicandra physalodes* (L.) Gaertn.
Vernacular name:
Habit: Shrub

Description: A subshrub. Stem is angular; leaves are large, alternate, ovate, apex acuminate and dentate. Flowers are solitary, axillary, large, white-centered funnel shaped blue color. Fruits are berries enclosed with calyx.

Flowering and Fruiting: July to December
IUCN status: Not Evaluated
Native: South America

Scientific name: *Physalis minima* L.
Vernacular name: Sanna budda gida
Habit: Herb

Description: A small herb up to 3 ft. Leaves are alternate, up to 10 cm long, dentate, and hairy on both the surfaces. Flowers are small, 5 angular, pedicle hairy and greenish yellow color. Fruits are berry, enclosed within papery shell.

Uses: Fruits are edible.

Flowering and Fruiting: August to December
IUCN status: Not Evaluated
Native: Southeast Asia and northern Australia

Scientific name: *Solanum nigrum* L.
Vernacular name: Kakki hannu
Habit: Herb

An annual herb; up to 1 m tall. Stem angled, dichotomously branched, smooth. Leaves are ovate to lanceolate, papery and apex acute. Flowers are small, drooping, white color, umbellate. Fruits are berries, spherical and purplish black when ripe. Seeds are many, disk shaped and yellow.

Flowering and Fruiting: September to March
IUCN status: Not Evaluated
Native: India and Sri Lanka

Scientific name: *Solanum trilobatum* L.
Vernacular name: Mullu mustha
Habit: Shrub

Armed trailing subshrub. Leaves are ovate, angular, 3–5 nerved, prickly on both the side, base cordate and obtuse. Flowers are axillary, violet color. Fruits are berries, globose.

Flowering and Fruiting: November to March
IUCN status: Not Evaluated
Native: Southeast Asia

Scientific name: *Solanum xanthocarpum* Schrad. & Wendl.
Vernacular name: Kantakari
Habit: Herb

A prickly diffuse herb, recorded in waste lands, roadside and near agricultural fields. Leaves are lanceolate, dissected, prickly on the surfaces, prickly yellow. Flowers are axillary, purple color. Fruits are berries, rounded, smooth, and yellow when ripe.

Uses: Boiled fruits are used to treat jaw pains.

Flowering and Fruiting: January to December
IUCN status: Not Evaluated
Native: India, Bangladesh, and Nepal

Scientific name: *Withania somnifera* (L.) Dunal
Vernacular name: Ottatti gida
Habit: Shrub

An erect shrub; densely covered with soft gray hairs. Leaves ovate, entire, pointed at apex and covered with soft dense hairs. Flowers are small, axillary, and pale yellow. Fruits are berries, orange red when ripe, spherical shape.

Flowering and Fruiting: January to December
IUCN status: Not Evaluated
Native: India, Sri Lanka, China, and Japan

71. Family: Sterculiaceae

Scientific name: *Melhania incana* Heyne ex Wight
Vernacular name:
Habit: Herb

An erect woody herb, occasionally recorded in the study area. Branches are densely tomentose. Leaves are elliptic–lanceolate, glabrescent above, and gray pilose below, crenate–serrate margin. Flowers are axillary and yellow color. Fruit capsule, globose–ovoid and hairy.

Flowering and Fruiting: August to January
IUCN status: Not Evaluated
Native: India

72. Family: Tiliaceae

Scientific name: *Corchorus aestuans* L.
Vernacular name:
Habit: Herb

An annual branched erect herb and hairy; common weed recorded toward roadside, waste lands and agricultural fields. Leaves are alternate, elliptic, ovate–oblong, sparsely hairy, apex acute, and margin serrate. Flowers are axillary, solitary and often pairs, small, and yellow color. Fruits are capsules, angled, septate. Seeds are many and dark brown.

Flowering and Fruiting: May to December
IUCN status: Not Evaluated
Native: West Indies, Mexico, and Central and South America

Scientific name: *Corchorus trilocularis* L.
Vernacular name: Yenne pundit gida
Habit: Herb

Annual herb, many branched; recorded in roadside, cultivated fields and waste lands. Leaves are oblong–lanceolate, 3–5 cm long, serrate margin, slightly hairy on both the surfaces. Flowers are axillary cyme and yellow in color. Fruit capsule, 3-angled and beak short.

Flowering and Fruiting: July to December
IUCN status: Not Evaluated
Native: Africa, India, and Arabian Peninsula

Scientific name: *Grewia flavescens* Juss.
Vernacular name: Kireegara kela
Habit: Shrub

A shrub or small tree, often recorded roadside and waste lands. Branches are many, rough surface. Leaves are simple, alternate, oblong–ovate, 3-nerved, and hairy on both the surface, but more hairy lower surface, margins toothed, stalk is short, velvety, and pointed tip. Flowers are small and yellow. Fruits are berry, hard, and densely covered with hairy.

Flowering and Fruiting: October to March
IUCN status: Not Evaluated
Native: Africa, Asia, and Australia

Scientific name: *Grewia tenax* (Forssk.) Fiori
Vernacular name:
Habit: Shrub

A shrub up to 4–5 m tall; recorded in roadside and waste lands. Branchlets are tomentose and leaves are various in shapes like, ovate, rounded and elliptic, serrate margin, scabrid upper surface, and sparsely hairy lower surface. Flowers are axillary cyme and white color. Fruits are 4-lobed, glabrescent, and yellow when ripe.

Flowering and Fruiting: June to December
IUCN status: Not Evaluated
Native: Tropical Africa, tropical Asia, and Arabian Peninsula

Scientific name: *Grewia villosa* Willd.
Vernacular name:
Habit: Shrub

A shrub, 5–6 m tall; branches are velvety. Leaves are orbicular, cordate, velvet on both the surfaces, base cordate, and margin serrate. Flowers are axillary and yellow. Fruits are drupes, 4-lobed and densely covered with hairy.

Flowering and Fruiting: June to December
IUCN status: Not Evaluated
Native: Africa

Scientific name: *Triumfetta rotundifolia* Lam.
Vernacular name:
Habit: Shrub

A small much branched perennial shrub; found in throughout the study region. Branches are slender and hairy. Leaves are alternate, slightly 3-lobed, dentate margin, and apex acute. Flowers are small, terminal or leaf opposed clusters. Fruits are small, globose, and bristly.

Flowering and Fruiting: June to December
IUCN status: Not Evaluated
Native: India and Nepal

73. Family: Turneraceae

Scientific name: *Turnera ulmifolia* L.
Vernacular name:
Habit: Shrub

A shrub; leaves are dark green and clustered end of the branches, lanceolate to oblong or elliptic, margins serrate. Flowers are solitary, terminal, and bright yellow.

Flowering: March–April
IUCN status: Not Evaluated
Native: West Indies and Mexico

74. Family: Typhaceae

Scientific name: *Typha angustifolia* L.
Vernacular name: Aapu
Habit: Aquatic herb

An aquatic or marshy rhizomatous erect herb up to 2 m tall. Leaves arise from the base, leaf sheath white and green. Inflorescence is terminal spike and densely flowered. Spike 20–30 cm, flowers tiny. Fruit is small, dry, and indehiscent.

Uses: Dried root powder is used to treat cure urinary problems and leaves are used to make traditional house.

Flowering: March to April
IUCN status: Not Evaluated
Native: North America, Europe, and South America

75. Family: Ulmaceae

Scientific name: *Holoptelea integrifolia* (Roxb.) Planch.
Vernacular name: Tapsi mara
Habit: Tree

A deciduous tree; bark is gray and covered with blisters, peeling corky scale on old stems and trees. Leaves are alternate, elliptic–ovate, smooth, and pointed tip. Leaf base rounded or heart shaped. Flowers are small, greenish yellow to brownish, sepals are velvety. Fruits are samara, circular. Seeds are flat.

Flowering and Fruiting: November to March
IUCN status: Not Evaluated
Native: India, Nepal, Sri Lanka, and Indochina

76. Family: Verbenaceae

Scientific name: *Clerodendrum inerme* (L.) Gaertn.
Vernacular name: Vishamadari
Habit: Shrub

A climbing shrub; branches are many. Leaves are opposite, simple, small, entire and ovate. Flowers are axillary, 3-flowers, fragrant and corolla tube slender. Fruits are small enclosed with persistent calyx.

Flowering and Fruiting: January to December
IUCN status: Not Evaluated
Native: India and Sri Lanka

Scientific name: *Clerodendrum phlomidis* L. f.
Vernacular name:
Habit: Shrub

A large pubescent woody shrub, growing up to 4–5 m tall. Leaves are deltoid, ovate, margin crenate, subacute–obtuse. Flowers are creamy white color, c. 1–5 cm across and corolla tube 1.5–2.5 cm long, calyx pubescent. Fruits are drupes, obovoid, black, and wrinkled.

Flowering and Fruiting: March to July
IUCN status: Not Evaluated
Native: Indian subcontinent and Myanmar

Scientific name: *Duranta repens* L.
Vernacular name: Duranta kanti
Habit: Shrub

A shrub with thorny branches. Leaves are simple, opposite, ovate apex acute. Flowers are drooping clusters, 1–2 cm across, pale violet color. Fruits are globose and orange colored when ripe.

Flowering and Fruiting: January to December
IUCN status: Not Evaluated
Native: Central America

Scientific name: *Lantana camara* L.
Vernacular name: Hunni gida
Habit: Shrub

A prickly bushy shrub. Branches are slender, angled, prickly, and umbel shaped. Leaves are opposite, ovate, cordate, serrate margin and rough surface on both the side. Inflorescence terminal, umbellate, many, small, pink and yellow color. Fruits are aggregate of small globose berries.

Uses: Stem parts are used for fuel wood.

Flowering and Fruiting: January to December
IUCN status: Not Evaluated
Native: Tropical America

Scientific name: *Phyla nodiflora* (L.) Greene
Vernacular name:
Habit: Herb

A small aquatic herb found in moisture places and near water bodies. Rooting formation at nodes; leaves small 1 cm across, ovate, thick, and serrate margin. Flowers occur on spike, spike solitary, long pedicle, and flowers are white tinged.

Flowering: December to March
IUCN status: Least Concern
Native: India, Sri Lanka, and Nepal

Scientific name: *Premna latifolia* Roxb.
Vernacular name: Nerugala
Habit: Shrub

A shrub or small tree. Branches are dark brown, sub-glabrescent. Leaves are broad ovate–oblong, ovate, cordate, papery, and apex acuminate. Inflorescence is cluster, pale yellow and 2-lipped. Fruits are black and tuberculate.

Flowering and Fruiting: September to April
IUCN status: Not Evaluated
Native: India

Scientific name: *Priva cordifolia* (L.) Druce
Vernacular name:
Habit: Herb

An erect herb; growing as weed in many places in the study area. Stem is quadrangular and rough. Leaves are broadly ovate; petiole is long up to 4 cm, apex acute, crenate, and membranous. Inflorescence terminal up to 20 cm long, flowers are small, white, 5-ribbed, pubescent, and corolla tube 2 lipped, curved. Fruits are enclosed in calyx tube, sticky, and pubescent.

Flowering and Fruiting: August to December
IUCN status: Not Evaluated
Native: India, Pakistan, and Nepal

Scientific name: *Stachytarpheta indica* (L.) Vahl
Vernacular name:
Habit: Herb

An erect herb, annual or perennial, branched at the base up to 2 m tall. Stem is quadrangular, solid, and glabrous. Leaves are simple, opposite, elliptic to ovate, 3–5 cm long 1–2 cm wide, margin dentate and apex acute. Flowers are terminal spike, blue color with white throat. Fruit is a nut.

Flowering and Fruiting: August to February
IUCN status: Not Evaluated
Native: Africa

Scientific name: *Tectona grandis* L. f.
Vernacular name: Tega mara
Habit: Tree

A deciduous, erect tree, not much branched; bark yellowish brown and rough. Leaves are simple, opposite, large in size 4–60 cm long and 20–30 cm wide, ovate and obovate, petiole stout, apex acute or obtuse, margin entire, and rough surface above and pubescent below. Flowers are many, terminal cyme, and white. Fruits are drupe, globose, brown, densely hairy, and covered by the calyx.

Uses: Wood is used as timber.

Flowering and Fruiting: August to December
IUCN status: Not Evaluated
Native: Southeast Asia

Scientific name: *Vitex negundo* L.
Vernacular name: Lakki gida
Habit: Shrub

A subshrub or small tree up to 7 m tall; occasionally recorded in the study area. Branches are many and bark is thin, gray and smooth surface. Leaves are compound, leaflets 3–5 various in size. Flowers are terminal cymes. Pale blue to bluish purple color. Fruits are drupes, ovoid and black when ripe.

Uses: Leaf paste is used to reduce bone pains and skin diseases.

Flowering and Fruiting: March to December
IUCN status: Not Evaluated
Native: India, Nepal, Bhutan, Pakistan, and Sri Lanka

77. Family: Violaceae

Scientific name: *Hybanthus enneaspermus* (L.) F. Muell.
Vernacular name: Ratna purusha
Habit: Herb

A small herb up to 20 cm high. Leaves are linear–lanceolate, margin crenate. Flowers are small, axillary, long petiole, bilipped, lowers lip is larger, attractive pink dark color. Fruit is capsule.

Flowering and Fruiting: July to December
IUCN status: Not Evaluated
Native: South Asia

78. Family: Vitaceae

Scientific name: *Cissus glauca* Roxb.
Vernacular name:
Habit: Climber

A climber, stem is cylindrical. Leaves are alternate, long petiole up to 4 cm, cordate, serrate margin apex, glabrous and acuminate. Inflorescence umbellate, axillary, terminal, pale yellow or white color. Fruits are berry and globose.

Flowering and Fruiting: July to December
IUCN status: Not Evaluated
Native: Austrilia

Scientific name: *Cissus quadrangularis* L.
Vernacular name: Mangaravalli
Habit: Shrub

A succulent rambling shrub, perennial. Stem is quadrangular, glabrous, fleshy, and stout tendrils at nodes. Leaves are small, rounded, kidney shaped, sub-fleshy, and entire. Flowers are many, small, greenish yellow with red tipped. Fruits are berries, globose and reddish when ripe. Single seeded.

Uses: Crushed stem part is used to treat bone fractures.

Flowering and Fruiting: January to December
IUCN status: Not Evaluated
Native: India

Scientific name: *Cyphostemma setosum* (Roxb.) Alston
Vernacular name:
Habit: Herb

Prostrate herb; roots are tuberous and stem is reddish, covered with bristly hairs. Leaves are foliate, sessile, leaflets are elliptic. Fruits are berries, ovate and red when ripe.

Flowering and Fruiting: February to July
IUCN status: Not Evaluated
Native: Temperate western Asia

79. Family: Zygophyllaceae

Scientific name: *Fagonia arabica* L.
Vernacular name:
Habit: Herb

A small spiny bushy herb with many branches. Leaves are trifoliate, terminal leaflet is larger than others, linear to lanceolate, semifleshy, and pubescent. Flowers are solitary and violet color. Fruit is 5 lobed and densely hairy.

Flowering and Fruiting: December to March
IUCN status: Not Evaluated
Native: Africa

Scientific name: *Tribulus terrestris* L.
Vernacular name: Neggele mullu
Habit: Herb

A prostrate to decumbent herb, annual and perennial; distributed throughout the study region. Branches are slender covered with hairs. Leaves compound, pairs unequal, leaflets small with oblique base. Flowers are solitary, axillary, small, c. 1 cm across and yellow. Fruits are schizocarps, hairy, divide into 4–5 units and each part consist 2–4 sharp spines.

Flowering and Fruiting: January to December
IUCN status: Not Evaluated
Native: Old World and India

Phytoplankton:

1. Family: Bacillariaceae

Scientific name: *Hantschia amphioxys* (Ehr.) Grun.
Class: Basillariophyceae

Hantschia amphioxys is a rectangular shape and slightly asymmetrical having one face convex and the opposite concave. The average size is varies from 60 to 80 micra length, and shows two transverse and radiate chromoplasts. The raphe system runs along the ventral edge of the valve face and is subtended by large rectangular fibulae. In a frustules, the raphe systems of both valves lie on the same side.

IUCN status: Not Evaluated

Scientific name: *Nitzschia closterium* (Ehr.) Wm. Smith
Class: Bacillariophyceae

Nitzschia closterium cells are cylindrical, fusiform, not twisted the apical axis. Valve face weekly silicified, traversed by more or less transpical silicified thickening. Raphe traversed by a serious of fibulae, joined directly to the valve face.
IUCN status: Not Evaluated

Scientific name: *Nitzschia paradoxa* (J.F. Gmelin) Grunow
Class: Bacillariophyceae

Nitzschia paradoxa forms motile-like colonies. Frustules isopolar, cells seen in valve or girdle view. Valves shallow, bilaterally symmetrical, linear, linear to lanceolate with rounded, cuneate poles. The raphe is invisible but its presence can be detected because of the fibulae.

IUCN status: Not Evaluated

2. Family: Characeae

Scientific name: *Chara sp.*
Class: Charophyceae

Chara is green algae, plant-like appearance that can grow in all water ponds. The branches are whorled, garlicky odor with rough surface hence it is called stonewort. The branches consists nodes and internodes and branches arise from nodes; multicellular rhizoids and needle-shaped stipulodes are presented.

IUCN status: Least Concern

3. Family: Chlorophyceae

Scientific name: *Cladophora glomerata* (L.) Kützing
Class: Cladophoraceae

Cladophora glomerata is uniseriate branched filaments with apical intercalary growth. Branches sparse to profuse, branching rhizoids arise from basal cell and other cells in basal region. Thick-walled rhizoids and basal portions of main axes also perennating devices. Chloroplasts parietal, densely packed discoid and united in a reticulum.

IUCN status: Not Evaluated

4. Family: Desmidiaceae

Scientific name: *Staurastrum gracile forma* Iyengar et Vimala bai
Class: Cladophoraceae

Staurastrum gracile is very small desmid; cells variable, usually of small or medium size. They spin and tumble, constriction slight, usually an acute notch, semicells variable in form, usually more or less cup shaped, lower angles broadly rounded, lateral margins nearly vertical or slightly diverging. Semicells usually broadening slightly toward the apex, which is very slightly convex.

IUCN status: Not Evaluated

Scientific name: *Cosmarium hammeri* Var. africanum F.E. Fritsch
Class: Chlorophyceae

Cosmarium hammeri is a nonmotile unicell, characterized by a constriction in the middle of the cell which divides it into two symmetrical halves. Semicells are sub-quadrate with sides straight or convex. Each semicells is partly filled by a green chloroplast containing two pyrenoids.

IUCN status: Not Evaluated

5. Family: Fragilariaceae

Scientific name: *Synedra ulna* (Nitsch) Ehr.
Class: Bacillariophyceae

Synedra ulna is a linear or some times linear–lanceolate shape, valves narrowing to blunt sub-rostrate or rostrate apices. Central area is distinct, roughly squire in outline and usually reaching the valve margin.

IUCN status: Not Evaluated

Scientific name: *Synedra ulna (Nitz.) Ehr. Var. oxyrynchus (Kutz.) Van Heurck*
Class: Bacillariophyceae

Valves are lanceolate; suddenly attenuated toward the very acute extremities.

IUCN status: Not Evaluated

Scientific name: *Fragilaria fonticola* var. chandolensis H.P. Gandhi.
Class: Bacillariophyceae

The plankton cells coined at valve face to form ribbon-like colonies. Valves are linear to linear lanceolate with rounded, rostrate ends. The central margin is unilaterally expanded. Valve flat, slightly undulate to costae. The colonies are attached to the substratum at one end.

IUCN status: Not Evaluated

6. Family: Melosiraceae

Scientific name: *Melosira granulata (Kutz.) Ralfs*
Class: Bacillariophyceae

Frustules are cylindrical, join face to face, and form filamentous colonies. Valves are 4–17 μm in diameter. The mantle has straight sides and the valve face is flat. Colonies are usually straight, but sometimes spiral, with curved cells.

IUCN status: Not Evaluated

7. Family: Microcystaceae

Scientific name: *Gloeocapsa punctata* Nägeli
Class: Cyanophyceae

Gleocapsa punctata is spherical cells is irregularly arranged and enclosed with a thick sheath of mucilage. Colony 12–60 diameter, composed number of subfamilies.

IUCN status: Not Evaluated

8. Family: Nostocaceae

Scientific name: *Nostoc piscinale* Kutz. ex Born. & Flah.
Class: Cyanophyceae

This is commonly occurred cyanobacteria in the field. The colonies are covered with thick mucilage threads. The cells are arranged bed-like chains that are grouped together in a gelatinous mass.

IUCN status: Not Evaluated

9. Family: Oscillatoriaceae

Scientific name: *Lyngbya majuscula* (Dillwyn) Harvey
Class: Cyanophyceae

A filamentous cyanobacteria. Filamentous consists of a sheathed group of trichomes.

IUCN status: Not Evaluated

Scientific name: *Oscillatoria tenuis* Ag. Ex Gomont
Class: Cyanophyceae

Fresh water filamentous algae; trichome straight, blue green, slightly curved at terminal. Septum slightly constricted, granular present along septum, apical cells semispherical and wall sometime thickened.

IUCN status: Not Evaluated

10. Family: Phormidiaceae

Scientific name: *Phormidium corium* var. captatum Gardner
Class: Cyanophyceae

A fresh water plankton; occurred up to 3 m deep. Thallus is mucilaginous, soft puffs; trichomes 3–6 μm, unbranched, fragile, densely tangled, apices slightly tapering, straight. Apical cells rounded or obtuse cone shaped at tip with distinct thickening of outer membrane.

IUCN status: Not Evaluated

11. Family: Scenedesmaceae

Scientific name: *Coelastrum microporum Naegeli*
Class: Cladophoraceae

Colonies spherical with inner empty space; cells attached to each other by a protrusion of cell wall arranged in single layer. Cell body mostly spherical with globular bulges. Chloroplasts cup shaped with pyrenoid.

IUCN status: Not Evaluated

12. Family: Zygnemataceae

Scientific name: *Zygnema khanne* Skuja
Class: Cladophoraceae

Zygnema khanne is very common algae found along side *Spirogyra*. It is unbranched, mucilaginous filaments, each cell with two star-shaped embedded chloroplast with central pyrenoids.

IUCN status: Not Evaluated

Scientific name: *Spirogyra hyalina* Cleve
Class: Cladophoraceae

A filamentous green algae commonly found in fresh water as skeins of fine green threads; unbranched mucilaginous filaments, varying 10–100 μm diameter. The threads consist of individual cells attached end-to-end. Inside each cell is one or more ribbon-like, spirally arranged chloroplast.

IUCN status: Not Evaluated

Scientific name: *Spirogyra condensata* (Vaucher) Kutzing
Class: Chlorophyceae

Fresh water filamentous green algae; the cells are arranged ladder-like and lateral conjugation.

IUCN status: Not Evaluated

Photographs of plant species are given in photoplates 3.2–3.26 (tree species photoplates 3.2, 3.3, 3.4, 3.5, 3.6; shrub species 3.7, 3.8, 3.9, 3.10, 3.11; herbs 3.12, 3.13, 3.14, 3.15, 3.16, 3.17, 3.18, 3.19, 3.20, 3.21, 3.22; climbers 3.23, 3.24, 3.25, 3.26) and phytoplankton in photoplate 3.27

Acacia auriculiformis

Acassia catechu

Acacia farnesiana

Acacia latronum

Acacia leucophloea

Acacia horrida

Acacia nillotica sub sp. indica

Aegle marmelos

Ailanthus excelsa

Albizia amara

Albizia lebbeck

Annona squamosa

Photoplate 3.2 Tree species of the study region

Azadirachta indica

Bauhinia purpurea

Borassus flabellifer

Cassia fistula

Chloroxylon swietenia

Cordia dichotoma

Dalbergia sissoo

Delonix elata

Delonix regia

Dichrostachys cinerea

Dolichandrone falcata

Dolichandrone atrovirens

Photoplate 3.3 Tree species continued

Photoplate 3.4 Tree species continued

Moringa oleifera

Parkinsonia aculeata

Phoenix sylvestris

Phyllanthus emblica

Plumeria rubra

Pithecellobium dulce

Pongamia pinnata

Prosopis cineraria

Prosopis juliflora

Samanea saman

Santalum album

Sapindus laurifolia

Photoplate 3.5 Tree species continued

Senna siamea

Syzygium cumini

Tamarindus indica

Tectona grandis

Terminalia bellirica

Tecoma stans

Terminalia catappa

Thespesia populnea

Wrightia tinctoria

Ziziphus mauritiana

Ziziphus oenoplia

Ziziphus xylopyrus

Photoplate 3.6 Tree species continued

Photoplate 3.7 Shrub species of the study region

Capparis divaricata — *Carissa spinarum* — *Dodonaea viscosa*

Euphorbia sp. — *Euphorbia tirucalli* — *Grewia flavescens*

Grewia tenax — *Grewia villosa* — *Guilandina bonduc*

Gyrocarpus americanus — *Hibiscus hispidissimus* — *Indigofera tinctoria*

Photoplate 3.8 Shrub species continued

Ipomoea carnea

Jatropha glandulifera

Kirganelia reticulata

Lantana camara

Lawsonia inermis

Leonotis nepetaefolia

Maytenus emarginata

Nerium indicum

Opuntia dillenii

Orthosiphon glabratus

Pavetta indica

Pavonia zeylanica

Photoplate 3.9 Shrub species continued

Plumbago zeylanica

Premna latifolia

Randia brandisii

Ricinus communis

Salvadora persica

Securinega leucopyrus

Senna auriculata

Sida acuta

Sida cordifolia

Streblus asper

Tephrosia purpurea

Thevetia neriifolia

Photoplate 3.10 Shrub species continued

Photoplate 3.11 Shrub species continued

Acalypha indica

Achyranthes aspera

Aerva lanata

Ageratum conyzoides

Aloe vera

Alternanthera pungens

Alysicarpus vaginalis

Alternanthera sessilis

Amaranthus spinosus

Amaranthus viridis

Anisochilus carnosus

Ammannia baccifera

Photoplate 3.12 Herbaceous plant species of the study region

Andrographis paniculata

Argemone mexicana

Aristida setacea

Aristolochia bracteolata

Arundinella setosa

Bacopa monnieri

Bambusa arundinacea

Barleria prionitis

Biophytum sensitivum

Blainvillea acmella

Blumea eriantha

Boerhavia erecta

Photoplate 3.13 Herbaceous plant species continued

Boerhavia diffusa

Bulbostylis densa

Canscora diffusa

Caralluma attenuata

Celosia argentea

Chloris barbata

Chrozophora plicata

Cleome viscosa

Coldenia procumbens

Commelina benghalensis

Corchorus aestuans

Corchorus trilocularis

Photoplate 3.14 Herbaceous plant species continued

Crotalaria pallida *Crotalaria retusa* *Crotalaria ramosissima*
Croton bonplandianum *Cynodon dactylon* *Cyanotis tuberosa*
Cyperus compressus *Cyperus pangorei* *Cyperus rotundus*
Cyperus squarrosus *Dactyloctenium aegyptium* *Datura metel*

Photoplate 3.15 Herbaceous plant species continued

Photoplate 3.16 Herbaceous plant species continued

Flaveria trinervia | *Gomphrena celosioides* | *Grangea maderaspatana*

Hedyotis puberula | *Heliotropium indicum* | *Heliotropium ovalifolium*

Hibiscus micranthus | *Hybanthus enneaspermus* | *Hymenotherum tenniflоum*

Indigofera linnaei | *Indigofera cordifolia* | *Lagascea mollis*

Photoplate 3.17 Herbaceous plant species continued

Photoplate 3.18 Herbaceous plant species continued

Parthenium hysterophorus

Pavonia odorata

Peristrophe paniculata

Persicaria glabra

Phyla nodiflora

Phyllanthus amarus

Phyllanthus maderaspatensis

Physalis minima

Phyllanthus virgatus

Polycarpaea aurea

Portulaca pilosa

Pulicaria wightiana

Photoplate 3.19 Herbaceous plant species continued

Photoplate 3.20 Herbaceous plant species continued

Spermacoce hispida

Sphaeranthus amaranthoides

Stachytarpheta indica

Stemodia viscosa

Stylosanthes fruticosa

Trianthema portulacastrum

Tribulus lanuginosus

Trichodesma indicum

Tridax procumbens

Typha angustifolia

Vigna trilobata

Vinca rosea

Photoplate 3.21 Herbaceous plant species continued

Actiniopteris radiata

Adiantum sp.

Marchantia sp.

Marsilea quadrifolia

Selaginella sp.

Photoplate 3.22 Herbaceous plant species continued

Abrus precatorius

Antigonon leptopus

Aristolochia indica

Basella alba

Bougainvillea spectabilis

Capparis zeylanica

Cardiospermum halicacabum

Cassytha filiformis

Cissus glauca

Cissus quadrangularis

Clitoria ternatea

Coccinia grandis

Photoplate 3.23 Climbing plant species of the study region

Cocculus hirsutus

Combretum ovalifolium

Corallocarpus epigaeus

Cryptolepis buchanani

Cryptostegia grandiflora

Cucumis callosus

Ctenolepis cerasiformis

Cyphostemma setosum

Diplocyclos palmatus

Gymnema sylvestre

Hemidesmus indicus

Ipomoea aquatica

Photoplate 3.24 Climbing plant species continued

Ipomoea cairica

Ipomoea obscura

Ipomoea pescaprae

Ipomoea hederifolia

Ipomoea quomoclit

Jacquemontia pentantha

Maerua oblongifolia

Merremia tridentata

Mukia maderaspatana

Oxystelma esculentum

Passiflora foetida

Photoplate 3.25 Climbing plant species continued

Pentatropis capensis

Pergularia daemia

Rhynchosia minima

Rivea hypocrateriformis

Telosma pallida

Teramnus labialis

Tinospora cordifolia

Tylophora indica

Wattakaka volubilis

Photoplate 3.26 Climbing plant species continued

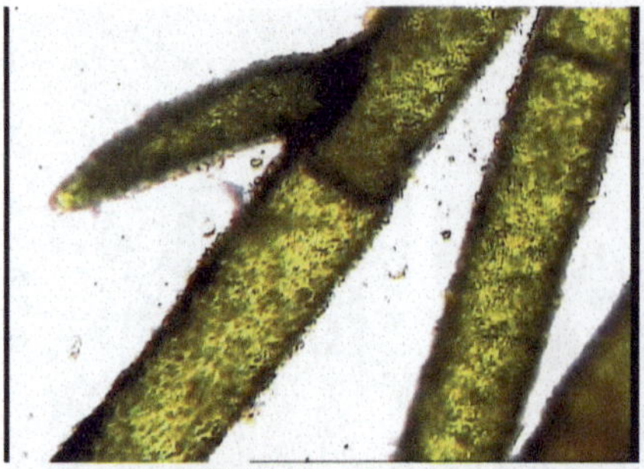

Cladophora glomerata

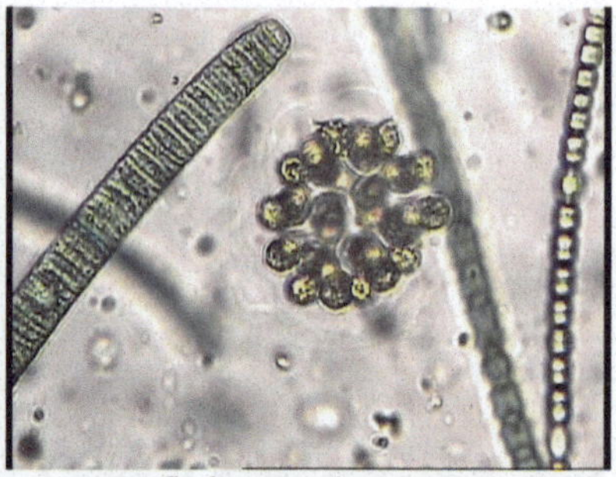

Coelastrum microporum

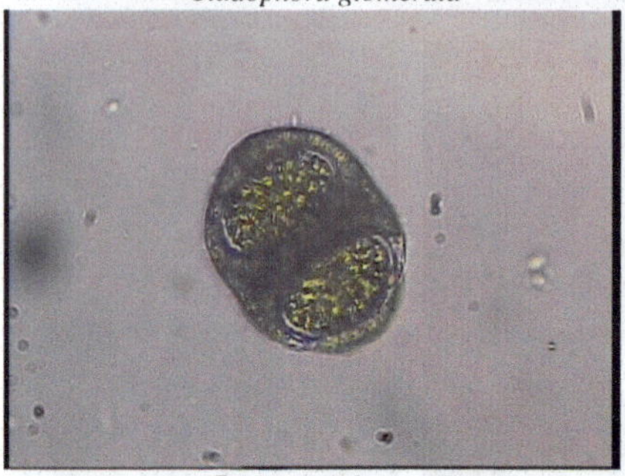

Cosmarium hammeri

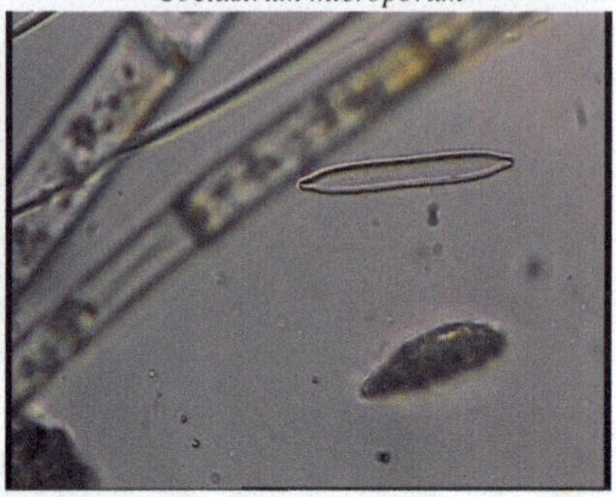

Fragilaria fonticola

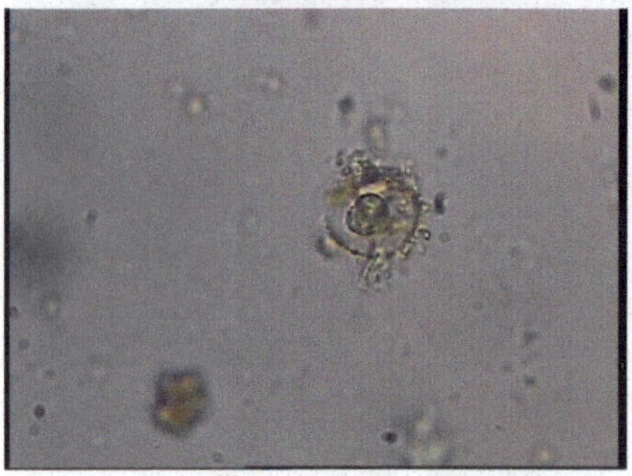

Gleocapsa punctata

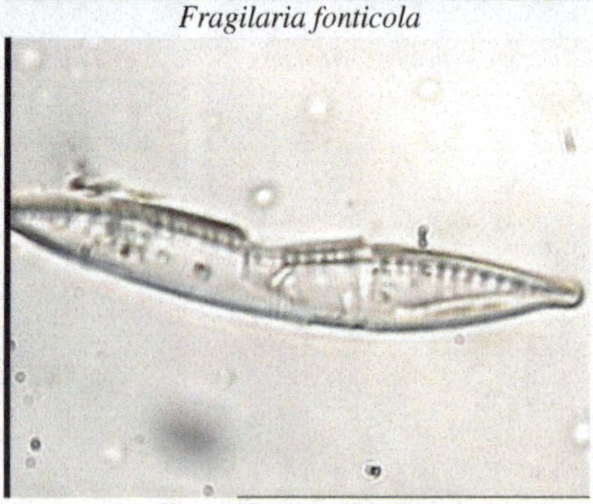

Hantzschia amphioxys

Photoplate 3.27 Phytoplankton of the study region

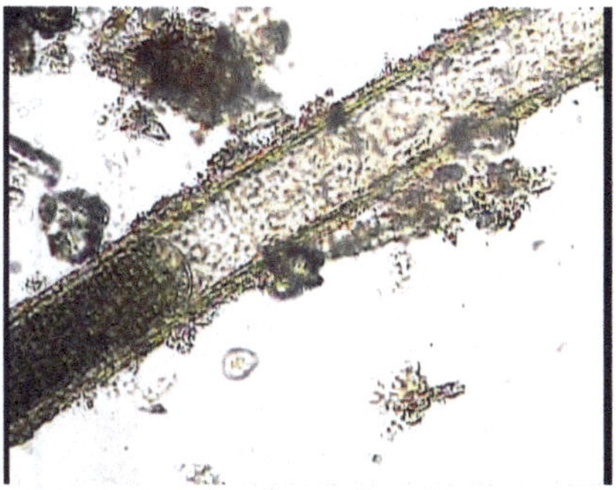

Lyngbya majuscula

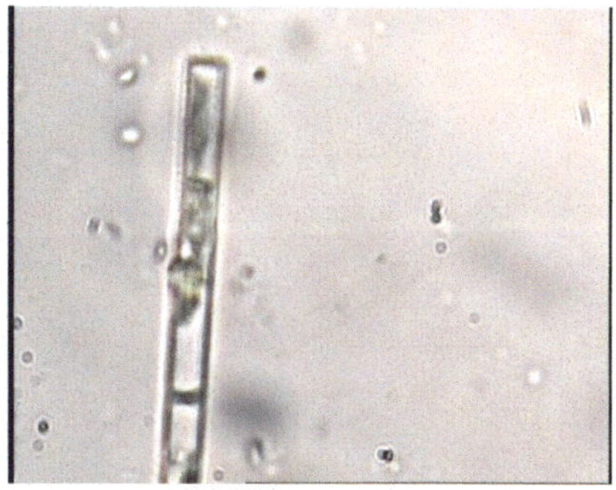

Melosira granulata

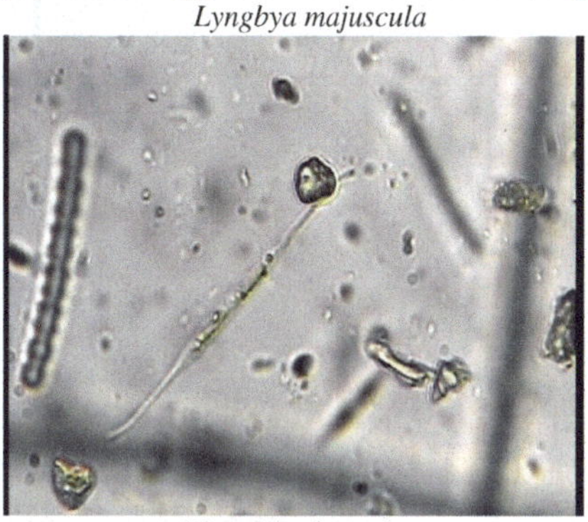

Nitzschia closterium

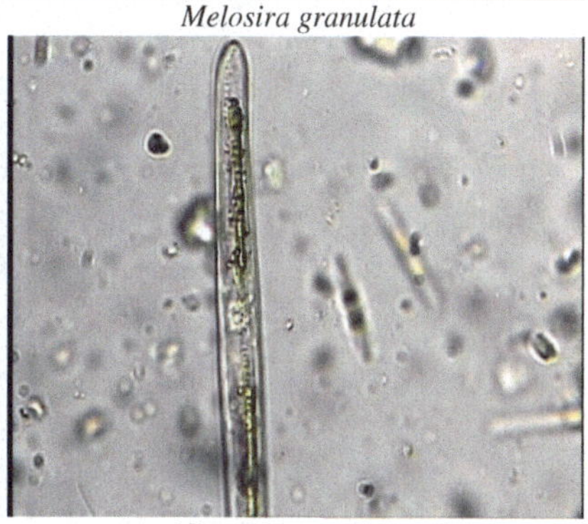

Nitzschia paradoxa

Photoplate 3.27 (continued)

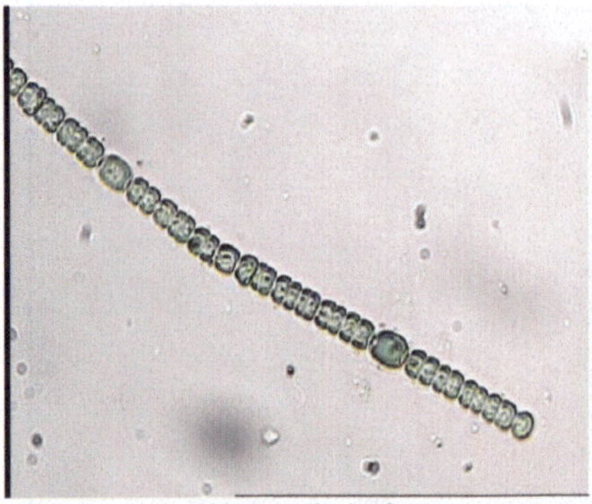

Nostoc piscinale

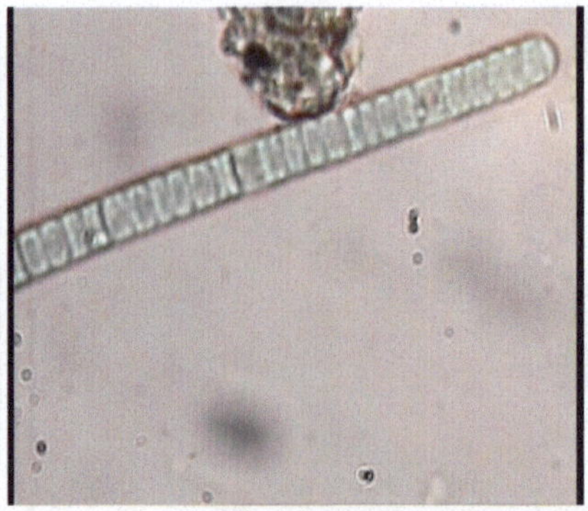

Oscillatoria tenuis

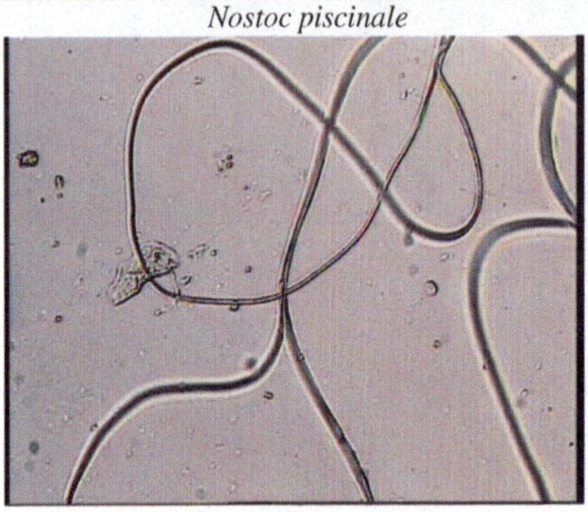

Phormidium corium

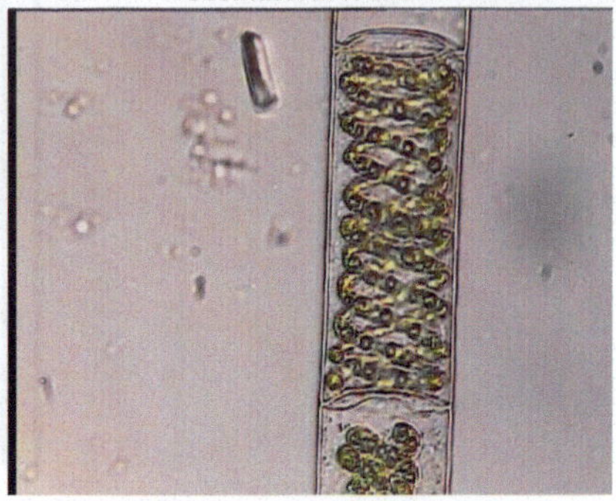

Spirogyra condensata

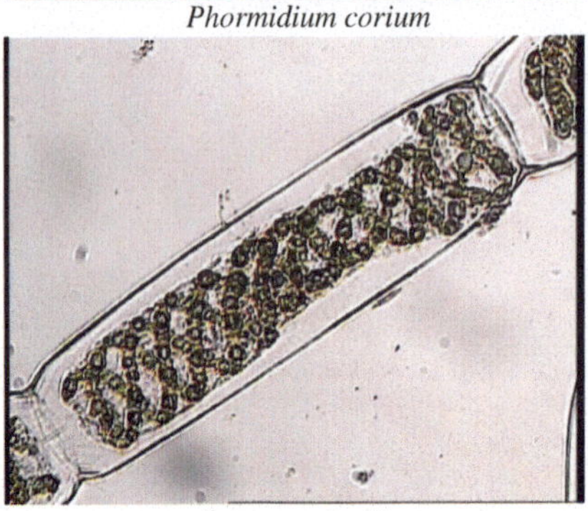

Spirogyra hyalina

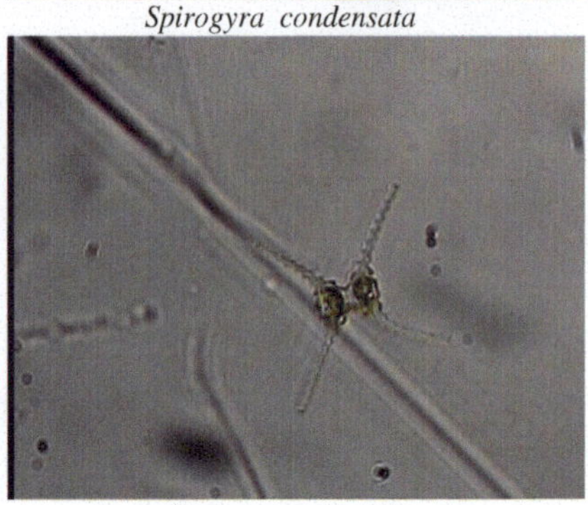

Staurastrum gracile

Photoplate 3.27 (continued)

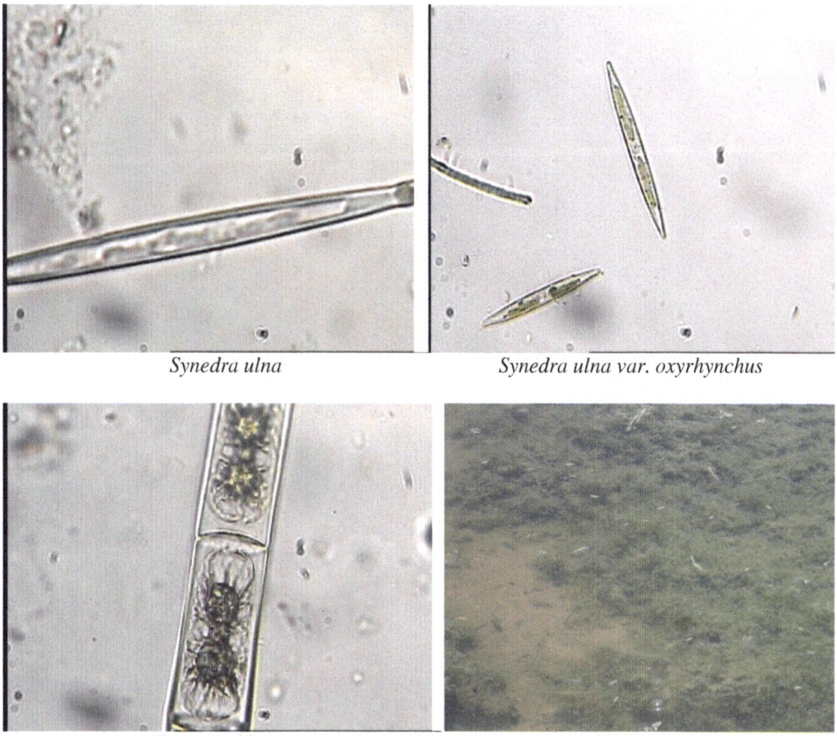

Synedra ulna *Synedra ulna var. oxyrhynchus*

Zygnema khanne *Chara sp.*

Photoplate 3.27 (continued)

References

Arulmurugan P, Nagaraj S, Anand N (2011) Biodiversity of fresh water algae from Guindy campus of Chennai, India. J Ecobiotechnology 3:19–29

Mahendra Perumal G, Anand N (2009) Biodiversity of freshwater Algae in Tamil Nadu II. In: Anand N (ed) Biology and Biodiversity of Microalgae. Centre for Advanced Studies in Botany, University of Madras, Chennai, pp 302–308

Nautiyal S, Rao KS, Maikhuri RK, Negi KS, Kala CP (2002) Status of medicinal plants on way to Vashukital in Mnadakini valley, Garhwal Himalaya, Uttaranchal. J Non Timber Forest Prod 9:124–131

Rathore M (2009) Nutrient content of important fruit trees from arid zone of Rajasthan. J Hort Forest 1:103–108

Suseela MR (2009) Conservation and diversity of fresh water algae. In: Anand N (ed) Biology and Biodiversity of Microalgae. Centre for Advanced Studies in Botany, University of Madras, Chennai, India, pp 41

Visser ME, Caro SP, Kees van Oers KV, Schaper SV, Helm B (2010) Phenology, seasonal timing and circannual rhythms: towards a unified framework. Philos Trans Biol Sci 365:3113–3127

Chapter 4
Animal Biodiversity

4.1 Introduction

Fauna are defined as the availability, diversity, and distribution of animals of a particular location during particular time period. The animals are subdivided into invertebrates and vertebrates. The biological classification of the species as per major taxonomic ranks is given in Fig. 4.1.

Invertebrates and Vertebrates

The animal kingdom was classified into two main groups: (1) invertebrates (without backbone) and (2) vertebrates (with notochord or backbone). The classification procedure for faunal diversity in general is given in Fig. 4.2. The invertebrates are the animals without backbone. They represent enormous number of species on this earth which are classified into different phyla. Arthropoda is the largest phylum in invertebrates.

- Protozoa
- Porifera (sponges)
- Cnidaria (sea anemone and jellyfishes)
- Platyhelminthes (flatworms)
- Nematoda (roundworms)
- Annelida (segmented worms)
- Arthropoda (insects, crustaceans)
- Mollusca (squids and snails)
- Echinodermata (sea urchins and starfishes)

These invertebrates represent the evolutionary history of the vertebrates. These invertebrates are both single-celled animals and multicellular animals and available in each and every part of the earth such as soil, air, water, and polar regions.

S. Nautiyal et al., *Biodiversity of Semiarid Landscape*,
Environmental Science, DOI 10.1007/978-3-319-15464-0_4

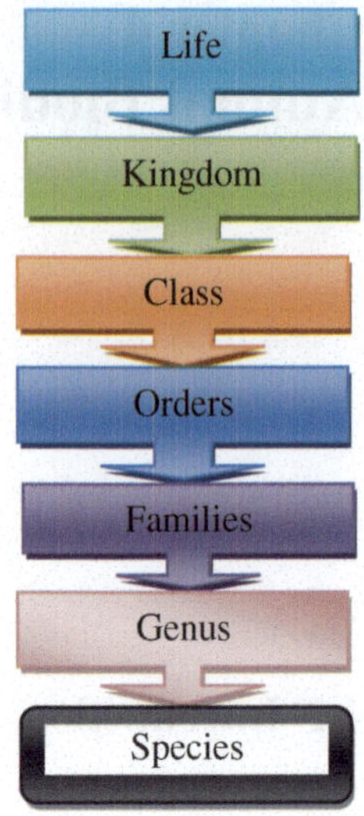

Fig. 4.1 Hierarchy of biological classification of major taxonomic ranks

Vertebrates

Vertebrates are the animals which are having notochord (a cartilaginous skeletal rod supporting the body in some stages of life cycle), and in majority of animals, it is replaced by vertebral column or backbone. And in these vertebrates, the body is divided into head, neck, trunk, and tail. All vertebrates have common diagnostic character, that is, the presence of skull or cranium which will protect the brain and other sense organs. They were found in each and every part of the world from the high mountains to deep sea and from the polar regions to tropics. And they will show much diversification in size ranging from small to bigger. Vertebrates originated about 500 million years ago during the early Paleozoic period. The vertebrate's evolution started with the evolution of fishes, followed by amphibians, then reptiles, and then birds and mammals. The earliest known vertebrates had the basic body structure with a notochord, rudimentary vertebrae, and a well-defined head and tail and they lacked jaws. Vertebrates play a very important role in ecosystems and they also play a vital role in food web. The classification procedure is further narrowed down for the faunal diversity of the study region located in semiarid landscape (Fig. 4.3).

Based on anatomy, genetics, and evolutionary relationship, vertebrates are classified into the following classes: (1) Fishes, (2) Amphibians, (3) Reptiles, (4) Aves, and (5) Mammals.

Fishes are the first vertebrates which are present on the earth. The first jawed vertebrates appeared in the Devonian period which is known as the "age of fishes."

Amphibians are the first vertebrates which are adapted life on land. They can lead their life both in water (only freshwater) and in land. They are cold-blooded animals. For respiratory purpose, they use both lungs and skin as respiratory organs.

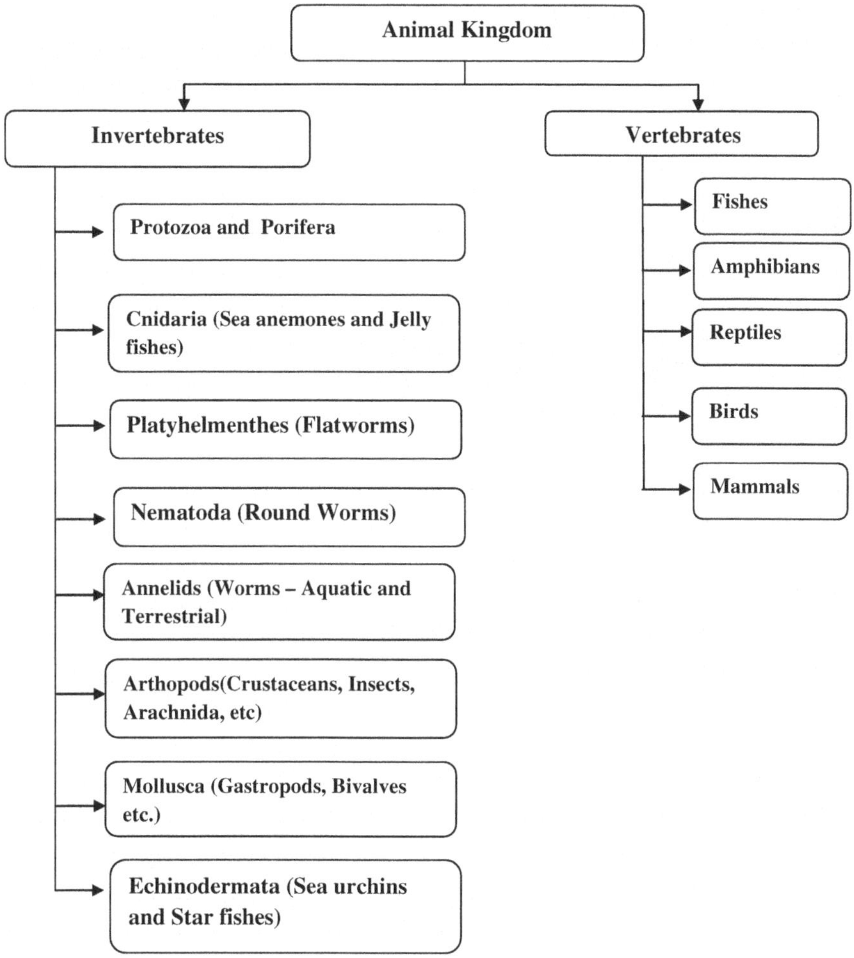

Fig. 4.2 General classification of animal kingdom

They have three-chambered heart with two atria and one ventricle. They play a very important role in ecosystem as secondary consumers in food chains. They prey on both invertebrates and vertebrates. They are considered as biological pest controllers. Several studies indicated that the amphibians evolved during the Devonian period around 370 million years ago.

Reptiles are the cold-blooded animals; generally, they have scales on their body. These are the first successful animals started their life on terrestrial systems. Reptiles are evolved from reptile-like amphibians during the Carboniferous period (around 320 years ago). Reptiles play important roles in ecosystem as the important element of food web. Reptiles play both the roles, as a predator and as a prey. Some reptiles help in controlling the harmful agricultural pests (rodent and insect pests).

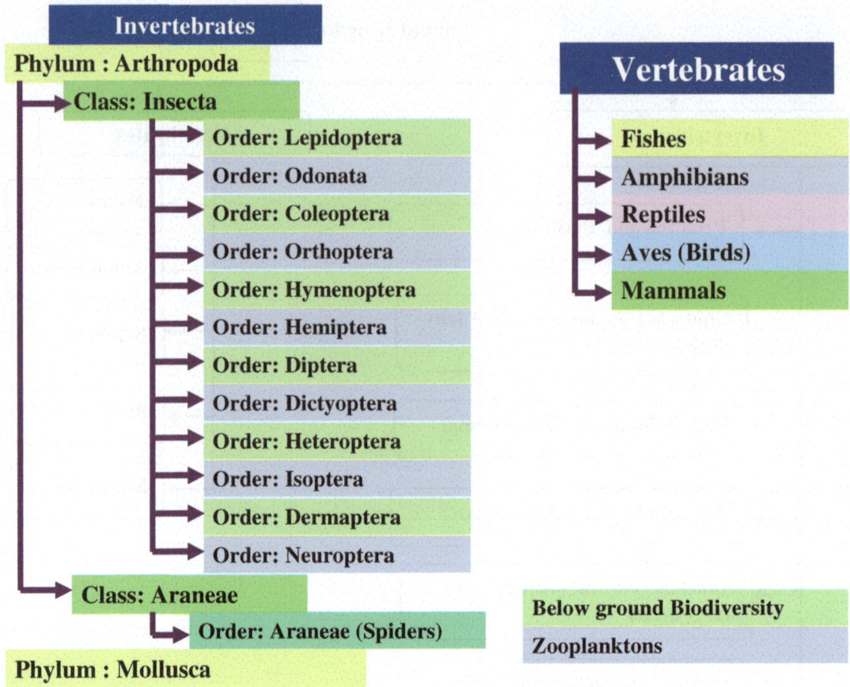

Fig. 4.3 Classification of faunal diversity in the study region

Birds are one of the most beautiful creatures on the earth. They are warm-blooded vertebrates. They have symbiotic relationships with other organisms, so that they become an integral part of the ecosystem. Evolution of birds took place during the Mesozoic era, around 150–200 million years ago. Most of the birds make the regular and seasonal journeys from one geographic location to another and back again called migrations. Birds do migration two times a year, during nesting time and in winter due to scarcity of food supply and also escape from cold weather.

Mammals are the animals which have mammary glands for nourishing their young ones. Generally, body of the mammals is covered by hair or fur. Evolution of mammals took place in Triassic period, but the maximum number of different types of mammals has arisen in Tertiary period. For this reason, Tertiary period is considered as age of mammals. Mammals are the highly developed animals in animal kingdom.

Some of the species are studied in more detail in order to obtain quantified information on their availability in the semiarid region of India. The data collected from the empirical field studies were calculated to obtain density, frequency, and abundance of some of the important indicator species of the semiarid region. Among the invertebrates, butterflies were chosen for detailed investigation and quantified information is being given in the following section. However, all the classes (fishes, amphibians, reptiles, birds, and mammals) of vertebrate group of animal kingdom were chosen for the in-depth study to obtain the data on density, frequency, and abundance in their habitats of the semiarid region.

4.2 Invertebrates

A total of 164 insect species were recorded and identified. The insects collected from the study area belong to 13 orders, 67 families, and 146 genera (Table 4.1). The order Coleoptera with the maximum number of families (13) showed the highest diversity with 40 genera and 45 species, followed by the orders—Lepidoptera (34 species), Orthoptera (25 species), Hemiptera (14 species), Odonata (13 species), Hymenoptera and Diptera (11 species each), Dictyoptera (3 species), Heteroptera and Isoptera (2 species each), and Dermaptera, Mantophasmatidae, and Neuroptera (1 species each). An inventory of insects identified from the semiarid region during the study period is given in Table 4.2.

A total of 164 insect species were recorded from the study region, but assessment and examination of all the recorded species for the density, frequency, and abundance are difficult in one study. Therefore, among the invertebrates, only butterflies are taken under consideration for obtaining the data on various aspects, for example, density, frequency, and abundance. The butterflies are chosen for detailed examination because among the insects, they are most preferred as indicator species in most of the studies related to environmental impact assessment. The butterflies are highly sensitive to several driving forces such as weather and climate, habitat degradation, and impact of anthropogenic activities on ecosystems.

Table 4.1 Insect diversity based on order, family, genus, and species in the studied semiarid region

S. no.	Order	Family	Genus	Species
1	Coleoptera	13	40	45
2	Hymenoptera	6	10	12
3	Hemiptera	11	14	14
4	Diptera	9	10	11
5	Lepidoptera	12	27	34
6	Orthoptera	4	22	25
7	Dictyoptera	2	3	3
8	Heteroptera	2	2	2
9	Isoptera	1	2	2
10	Dermaptera	1	1	1
11	Mantophasmatidae	1	1	1
12	Neuroptera	1	1	1
13	Odonata	4	13	13
	Total	**67**	**146**	**164**

Table 4.2 Inventory of insect species in the studied semiarid region, Karnataka, India

S. no.	Scientific name	Common name	Family
Order: *Coleoptera*			
1	*Adoretus* sp.		Scarabaeidae
2	*Agestrata* sp.	Cetoniid beetles	Scarabaeidae
3	*Catharsius molossus* Linnaeus, 1758		Scarabaeidae
4	*Cetonia aurata* Linnaeus, 1758	Rose chafer	Scarabaeidae
5	*Copris* sp.		Scarabaeidae
6	*Gymnopleurus* sp.		Scarabaeidae
7	*Holotrichia serrata* Hope, 1837	Root grub	Scarabaeidae
8	*Liatongus radamistus*		Scarabaeidae
9	*Odontria xanthosticta* White, 1846	Chafer beetle	Scarabaeidae
10	*Onitis philemon* Fabricius, 1801		Scarabaeidae
11	*Onthophagus gazella* Fabricius	Dung beetle	Scarabaeidae
12	*Oryctes rhinoceros* Linnaeus, 1758	Rhinoceros beetle	Scarabaeidae
13	*Oxycetonia versicolor* Fabricius, 1775	Citrus flower beetle	Scarabaeidae
14	*Scarabaeus* sp.		Scarabaeidae
15	*Xylotrupes ulysses* Guérin-Méneville, 1830	Rhinoceros beetle	Scarabaeidae
16	*Spilopyra sumptuosa* Baly, 1860	Chrysomelid beetle	Chrysomelidae
17	*Zygogramma bicolorata* Pallister, 1953	Mexican beetle	Chrysomelidae
18	*Zygogramma tortuosa* Rogers, 1856	Parthenium beetle	Chrysomelidae
19	*Epilachna vigintioctopunctata* Fabricius, 1775	28-spotted ladybird beetle	Coccinellidae
20	*Harmonia octomaculata* Fabricius, 1781	Eight-spotted ladybird	Coccinellidae
22	*Acanthophorus serraticornis* Olivier, 1795		Cerambycidae
27	*Apomecyna pertigera*		Cerambycidae
28	*Batocera rubus* Linnaeus, 1758	Longhorn beetle	Cerambycidae
21	*Neocerambyx paris* Wiedemann, 1821		Cerambycidae
25	*Sthenias grisator* Fabricius, 1787		Cerambycidae
26	*Sthenias* sp.		Cerambycidae
24	*Stromatium barbatum* Fabricius, 1775	Kulsi teak borer	Cerambycidae
23	*Xystrocera globosa* Olivier, 1795	Monkey pod/samanea round-headed borer	Cerambycidae
29	*Cardiophorus quadrimaculatus*	Click beetles	Elateridae
30	*Chalcophora virginiensis* Drury, 1770	Sandalwood borer	Buprestidae

(continued)

Table 4.2 (continued)

S. no.	Scientific name	Common name	Family
31	*Psiloptera fastuosa* Fabricius, 1775	Jewel beetle or metallic wood borers	Buprestidae
32	*Sphenoptera gossypii* Cotes, 1891	Large brown cotton borer	Buprestidae
33	*Sphenoptera* sp.		Buprestidae
34	*Alcidodes collaris* Pascoe	Stem weevils	Curculionidae
35	*Alcidodes* sp.		Curculionidae
36	*Hyopolixus truncatulus*	Amaranthus stem weevil	Curculionidae
37	*Epicauta* sp.	Blister beetles or meloid beetles	Meloidae
38	*Cicindela sexpunctata*	Tiger beetle	Cicindelidae
39	*Cicindela* sp.	Tiger beetle	Cicindelidae
40	*Amblystomus quttatus* Bates, 1873		Carabidae
41	*Chlaenius nepalensis*	Ground beetles	Carabidae
42	*Omphra pilosa* Klug, 1834	Carabid beetle or ground beetle	Carabidae
43	*Chiloloba acuta* Wiedmann, 1823	Chafer	Cetoniidae
44	*Hydrophilus triangularis*	Water scavengers	Hydrophilidae
45	*Tribolium castaneum* Herbst, 1797	Rust red flour beetle	Tenebrionidae
Order: *Hymenoptera*			
46	*Camponotus compressus* Fabricius, 1787	Black common ant	Formicidae
47	*Monomorium indicum* Forel, 1902	Small black ant	Formicidae
48	*Solenopsis geminata* Fabricius, 1804	Small red ant	Formicidae
49	*Apis cerana* Fabricius, 1793	Indian bee	Apidae
50	*Apis dorsata* Fabricius, 1793	Rock bee	Apidae
51	*Apis florea* Fabricius, 1787	Little bee	Apidae
52	*Xylocopa* sp.	Carpenter bee	Apidae
53	*Polistes dominulus* Christ, 1791	Paper wasp	Vespidae
54	*Vespa cincta* Degeer, 1773	Yellow band wasp	Vespidae
55	*Evania appendigaster* Linnaeus,	Ensign wasp	Evaniidae
56	*Megachile* sp.	Leaf-cutter bee	Megachilidae
57	*Podium luctuosum* Smith, 1856	Sphecid wasp	Sphecidae
Order: *Hemiptera*			
58	*Anoplocnemis phasiana* Fabricius,	Squash bug	Coreidae
59	*Euthochtha galeator* Fabricius 1767	Coreid bug	Coreidae
60	*Leptoglossus phyllopus* Linnaeus, 1767	Leaf-footed bug	Coreidae
61	*Eurybrachis tomentosus* Blanchard, 1840	Eury brachid bug	Eurybrachidae
62	*Halyomorpha halys* Stal, 1855	Stinkbug	Pentatomidae

(continued)

Table 4.2 (continued)

S. no.	Scientific name	Common name	Family
63	*Nezara viridula* Linnaeus, 1758	Green stinkbug	Pentatomidae
64	*Chrysocoris stolli* Wolf	Green jewel bug	Scutelleridae
65	*Zelus renardii* Kolenati, 1856	Assassin bugs	Reduviidae
66	*Dysdercus cingulatus* Fabricius, 1775	Red cotton bug	Pyrrhocoridae
67	*Lygaeus equestris* Linnaeus, 1758	Black and red bug	Lygaeidae
68	*Gerris* sp. (cf. lacustris)	Pond skaters	Gerridae
69	*Lethocerus americanus* Leidy, 1847	Giant water bug	Belostomatidae
70	*Notonecta glauca* Linnaeus, 1758	Backswimmer	Notonectidae
71	*Cimex lectularius* Linnaeus	Bedbug	Cimicidae
Order: *Diptera*			
72	*Tabanus striatus* Fabricius, 1787	Horsefly	Tabanidae
73	*Promachus rufipes* Fabricius, 1775	Robber fly	Asilidae
74	*Sarcophaga* sp.	Flesh fly	Sarcophagidae
75	*Sarcophaga* sp.	Sarcophagid fly	Sarcophagidae
76	*Cochliomyia hominivorax* Coqurel, 1858	Blue bot fly	Calliphoridae
77	*Culex* sp.	Mosquito	Culicidae
78	*Eristalis* spp.	Syrphid fly	Syrphidae
79	*Ischiodon scutellaris* Fabricius	Common hoverfly	Syrphidae
80	*Musca domestica* Linnaeus, 1758	Housefly	Muscidae
81	*Stomoxys calcitrans* Linnaeus, 1758	Stable fly	Hippoboscidae
82	*Simulium* sp.	Neirida fly	Simuliidae
Order: *Lepidoptera*			
83	*Amata* sp.	Ammatid moth	Ammatidae
84	*Creatonotos gangis* Linnaeus, 1763		Arctiidae
85	*Sesamia inferens* Walker, 1556		Noctuidae
86	*Endoxyla leucomochla* Turner, 1915	Witjuti grub	Cossidae
87	*Melittia cucurbitae* Harris, 1828	Pumpkin vine borer	Sesiidae
88	*Nisaga simplex* Walker, 1855		Eupractidae
89	*Acraea terpsicore* Linnaeus, 1758	Tawny coster	Nymphalidae
90	*Ariadne ariadne* Linnaeus, 1763	Angled castor	Nymphalidae
91	*Byblia ilithyia* Drury, 1773	Joker butterfly	Nymphalidae
92	*Charaxes solon* Fabricius, 1793	Black rajah	Nymphalidae
93	*Danaus genutia* Cramer, 1779	Striped tiger	Nymphalidae
94	*Danaus chrysippus* Linnaeus, 1758	Plain tiger	Nymphalidae
95	*Ergolis merione* Cramer	Castor butterfly	Nymphalidae
96	*Euploea core* Cramer, 1780	Common Indian crow	Nymphalidae
97	*Hypolimnas misippus* Linnaeus, 1764	Danaid egg fly	Nymphalidae

(continued)

Table 4.2 (continued)

S. no.	Scientific name	Common name	Family
98	*Junonia almana* Linnaeus, 1758	Peacock pansy	Nymphalidae
99	*Junonia lemonias* Linnaeus, 1758	Lemon pansy	Nymphalidae
100	*Junonia orithya* Linnaeus, 1758	Blue pansy	Nymphalidae
101	*Tirumala limniace* Caramer, 1775	Blue tiger	Nymphalidae
102	*Graphium Agamemnon* Linnaeus, 1758	Tailed jay	Papilionidae
103	*Graphium doson* C.R. Felder, 1864	Common jay	Papilionidae
104	*Pachliopta aristolochiae* Fabricius, 1775	Common rose	Papilionidae
105	*Pachliopta hector* Linnaeus, 1758	Crimson rose	Papilionidae
106	*Papilio demoleus* Linnaeus, 1758	Lime or citrus butterfly	Papilionidae
107	*Papilio polytes* Linnaeus, 1758	Common Mormon	Papilionidae
108	*Catopsilia pomona* Fabricius, 1775	Common emigrant	Pieridae
109	*Catopsilia pyranthe* Latrelle, 1758	Mottled emigrant	Pieridae
110	*Cepora nerissa* Fabricius, 1775	Common gull	Pieridae
111	*Eurema hecabe* Linnaeus, 1758	Common grass yellow	Pieridae
112	*Pieris rapae* Linnaeus, 1768	Cabbage butterfly	Pieridae
113	*Melanitis leda* Linnaeus	Common evening brown	Satyridae
114	*Freyeria trochilus* Freyer, 1845	Grass jewel	Lycaenidae
115	*Jamides celeno* Cramer, 1775	Common cerulean	Lycaenidae
116	*Caprona ransonnetti* Felder, 1868	Golden angle	Hesperiidae
Order: Orthoptera			
117	*Gryllus domesticus* Linnaeus, 1758	House crickets	Gryllidae
118	*Gryllus* sp.	Crickets	Gryllidae
119	*Acrida exaltata* Walker, 1859	Short-horned grasshoppers	Acrididae
120	*Acrida* sp.	Short-horned grasshoppers	Acrididae
121	*Anacridium flavescens* Fabricius, 1793	Short-horned grasshoppers	Acrididae
122	*Atractomorpha crenulata* Fabricius, 1793	Short-horned grasshoppers	Acrididae
123	*Brachycrotaphus longiceps*	Short-horned grasshoppers	Acrididae
124	*Chrotogonus oxypterus* Blanchard, 1836	Short-horned grasshoppers	Acrididae
125	*Dioabolocatantops pinguis* Stal, 1860	Short-horned grasshoppers	Acrididae
126	*Epistaurus sinetyi* Bolivar, 1902	Short-horned grasshoppers	Acrididae
127	*Eyprepocnemis alacris* Serville, 1838	Short-horned grasshoppers	Acrididae
128	*Gastrimargus africanus* Saussure, 1888	Short-horned grasshoppers	Acrididae
129	*Oedaleus abruptus* Thunberg, 1815	Short-horned grasshoppers	Acrididae

(continued)

Table 4.2 (continued)

S. no.	Scientific name	Common name	Family
130	*Oxya fuscovittata* Marschall, 1836	Short-horned grasshoppers	Acrididae
131	*Phlaeoba panteli* Bolivar, 1902	Short-horned grasshoppers	Acrididae
132	*Trilophidia annulata* Thunberg, 1815	Short-horned grasshoppers	Acrididae
133	*Tristria pulvinata* Uvarov, 1921	Short-horned grasshoppers	Acrididae
134	*Truxalis indica* Bolivar, 1902	Short-horned grasshoppers	Acrididae
135	*Tylotropidius varicornis* Walker, 1870	Short-horned grasshoppers	Acrididae
136	*Chrotogonus trachypterus* Blanchard, 1836	Short-horned grasshoppers	Pyrgomorphidae
137	*Conocephalus fasciatus* De Geer, 1773	Long-horned grasshoppers	Tettigoniidae
138	*Conocephalus indicus* Redtenbacher	Long-horned grasshoppers	Tettigoniidae
139	*Mecopoda elongata* Linnaeus, 1758	Long-horned grasshoppers	Tettigoniidae
140	*Neoconocephalus ensiger* Harris, 1841	Long-horned grasshoppers	Tettigoniidae
141	*Scudderia furcata* Brunner, 1878	Fork-tailed bush katydid	Tettigoniidae
Order: *Dictyoptera*			
142	*Blattella germanica* Linnaeus, 1767	German cockroach	Blattellidae
143	*Periplaneta americana* Linnaeus, 1758	American cockroach	Blattidae
144	*Prosoplecta* sp.	Cockroach	Blattidae
Order: *Heteroptera*			
145	*Cimex lectularius* Linnaeus, 1802	Bedbug	Cimicidae
146	*Acanthaspis* sp.	Assassin bug	Reduviidae
Order: *Isoptera*			
147	*Microtermis* sp.	Termites	Termitidae
148	*Odontotermes obesus* Rambur, 1842	Termites	Termitidae
Order: *Dermaptera*			
149	*Labia minora*	Ear wig	Labiidae
Order: *Mantophasmatidae*			
150	*Mantis* sp.	Praying mantid	Mantidae
Order: *Neuroptera*			
151	*Chrysoperla* sp.	Green lacewing	Chrysopidae
Order: *Odonata*			
152	*Anaciaeschna jaspidea* Burmeister, 1839	Rusty darner	Aeshnidae
153	*Anax guttatus* Burmeister, 1839	Blue-tailed green darner	Aeshnidae
154	*Ictinogomphus rapax* Rambur, 1842	Common clubtail	Gomphidae
155	*Paragomphus lineatus* Selys, 1850	Lined hooktail	Gomphidae

(continued)

Table 4.2 (continued)

S. no.	Scientific name	Common name	Family
156	*Brachythemis contaminata* Fabricius, 1793	Ditch jewel	Libellulidae
157	*Crocothemis servilia* Drury, 1773	Ruddy marsh skimmer	Libellulidae
158	*Diplacodes trivialis* Rambur, 1842	Ground skimmer	Libellulidae
159	*Neurothemis tullia* Drury, 1773	Pied paddy skimmer	Libellulidae
160	*Orthetrum sabina* Drury, 1770	Green marsh hawk	Libellulidae
161	*Pantala flavescens* Fabricius, 1798	Wandering glider	Libellulidae
162	*Rhyothemis variegata* Linnaeus, 1763	Common picture wing	Libellulidae
163	*Epophthalmia vittata* Burmeister, 1839	Blue-eyed cruiser	Macromiidae
164	*Macromia cingulata* Rambur, 1842	Macromia cingulata	Macromiidae

4.2.1 Butterflies

A total of 28 butterfly species were recorded, which belong to six families (Hesperiidae, Lycaenidae, Nymphalidae, Papilionidae, Pieridae, and Satyridae) of the order Lepidoptera. Among the 28 species (Table 4.3), plain tiger (*Danaus chrysippus*) was observed at the highest density (23.37), frequency (20.21), and abundance (21.28), followed by cabbage butterfly (*Pieris rapae*), with density (17.17), frequency (19.15), and abundance (15.09), and common grass yellow (*Eurema hecabe*) with density (11.29). The least values for density (0.94) and frequency (0.81) were calculated for common evening brown (*Melanitis leda*) from the studied semiarid region. *Pachliopta aristolochiae* is found to be the third highest in abundance (12.39), followed by *Danaus chrysippus* and *Pieris rapae*. Among the species of butterfly, the plain tiger is the dominant species of the studied semiarid region. The codominant species are cabbage butterfly and common rose (Table 4.3). The semiarid ecosystems are very sensitive to many factors such as climate change and other anthropogenic activities. Therefore, collecting quantitative information and obtaining the final results of the indicator species are highly important. This information on indicator species would help researchers in future to understand the implications and impacts of environmental and socioeconomic changes. From the studied semiarid region, four species of butterfly, *viz*. plain tiger, cabbage butterfly, common rose, and common grass yellow, are showing high values of density, frequency, and abundance. These species can be considered for regular monitoring as indicator species, so the impact of the various driving factors on biodiversity can be assessed at the micro-level.

A total of 82 spider species that are male, female, and juveniles were collected from different types of habitats of the studied semiarid region (Table 4.4). The spiders identified during the study period come under 19 families of the order

Table 4.3 Density (D), frequency (F), and abundance (A) of butterfly in the studied semiarid region

S. no.	Scientific name	D	F	A
1	*Caprona ransonnetti*	2.7	2.8	2.85
2	*Jamides celeno*	3.03	3.01	3.33
3	*Freyeria trochilus*	2.81	2.73	3.08
4	*Ariadne ariadne*	3.26	3.31	2.86
5	*Charaxes solon*	1.8	2.16	2.4
6	*Junonia orithya*	2.25	2.64	2.23
7	*Tirumala limniace*	3.5	3.35	3.3
8	*Ergolis merione*	3.08	3.21	3.04
9	*Euploea core*	2.07	2.93	7.03
10	*Hypolimnas misippus*	2.23	3.19	2.38
11	*Byblia ilithyia*	1.8	1.92	2.46
12	*Junonia lemonias*	4.02	4.12	3.9
13	*Junonia almana*	4.05	3.59	3.87
14	*Danaus chrysippus*	23.37	20.21	21.28
15	*Danus genutia*	1.59	1.86	2.91
16	*Acraea violae*	1.55	1.77	2.5
17	*Graphium doson*	3.03	2.97	4.23
18	*Papilio polytes*	2.43	2.59	3.29
19	*Pachliopta aristolochiae*	10.17	11.17	12.39
20	*Pachliopta hector*	4.01	4.12	4.39
21	*Papilio demolias*	4.55	4.7	4.57
22	*Graphium Agamemnon*	2.92	2.83	3.49
23	*Pieris rapae*	17.17	19.15	15.09
24	*Catopsilia pomona*	3.29	3.74	2.85
25	*Eurema hecabe*	11.29	10.64	9.95
26	*Cepora nerisa*	3.31	2.97	4.46
27	*Catosilia pyranthe*	1.96	2.44	2.22
28	*Melanitis leda*	0.94	0.81	2.39

Araneae in the class Arachnida (Table 4.5). The family Lycosidae represented by 17 species was the most dominant one, followed by Tetragnathidae (13), Araneidae (12), Theridiidae (8), Salticidae (6), Thomisidae (5), Oxyopidae (4), Philodromidae (3), Miturgidae (3), Uloboridae (2), and Hersiliidae, Corinnidae, Pisauridae, Filistatidae, Eresidae, Pholcidae, Scytodidae, Selenopidae, and Sparassidae (1 each) (Table 4.5).

Table 4.4 Spiders (order Araneae) from the studied semiarid region

S. no.	Scientific name	U	F	M	Fj	Mj	Fsa	Msa	Family
1	*Arctosa* sp.1					Mj			Lycosidae
2	*Hippasa pisaurina*		F						Lycosidae
3	*Lycosa bistriata*		F						Lycosidae
4	*Lycosa madani*		F						Lycosidae
5	*Lycosa pictula*		F						Lycosidae
6	*Lycosa prolifica*		F						Lycosidae
7	*Lycosa* sp.		F	M					Lycosidae
8	*Pardosa birmanica*		F	M		Mj	Fsa		Lycosidae
9	*Pardosa altitudes*		F						Lycosidae
10	*Pardosa mysorensis*		F	M			Fsa		Lycosidae
11	*Pardosa oakleti*			M					Lycosidae
12	*Pardosa pseudoannulata*		F	M					Lycosidae
13	*Pardosa pusiola*						Fsa		Lycosidae
14	*Pardosa sangosa*		F						Lycosidae
15	*Pardosa shyamae*		F						Lycosidae
16	*Pardosa* sp.1			M	Fj	Mj	Fsa		Lycosidae
17	*Pardosa sumatrana*		F	M					Lycosidae
18	*Larinia* sp.1				Fj				Tetragnathidae
19	*Leucauge celebesiana*		F				Fsa		Tetragnathidae
20	*Tetragnatha fletcheri*		F	M					Tetragnathidae
21	*Tetragnatha javana*		F						Tetragnathidae
22	*Tetragnatha mandibulata*		F	M					Tetragnathidae
23	*Tetragnatha maxillosa*		F						Tetragnathidae
24	*Tetragnatha moulmeinensis*		F						Tetragnathidae
25	*Tetragnatha okumae*			M					Tetragnathidae
26	*Tetragnatha* sp.1		F	M	Fj	Mj			Tetragnathidae
27	*Tetragnatha sutherlendi*			M					Tetragnathidae
28	*Zygiella indica*		F	M	Fj				Tetragnathidae
29	*Zygiella malanocrania*		F						Tetragnathidae
30	*Zygiella* sp.1						Fsa		Tetragnathidae
31	*Araneus* sp.1				Fj				Araneidae
32	*Argiope anasuja*		F						Araneidae
33	*Argiope minute*						Fsa		Araneidae
34	*Argiope* sp.1		F		J		Fsa		Araneidae
35	*Cyclosa hexatuberculata*		F						Araneidae
36	*Cyclosa mulmeinensis*		F	M	Fj				Araneidae
37	*Cyrtophora cicatrosa*		F		Fj				Araneidae
38	*Gibbaranea bituberculata*	U							Araneidae
39	*Neoscona excelsus*		F						Araneidae
40	*Neoscona mukerjei*		F	M					Araneidae
41	*Neoscona* sp.1		F		Fj	Mj			Araneidae

(continued)

Table 4.4 (continued)

S. no.	Scientific name	U	F	M	Fj	Mj	Fsa	Msa	Family
42	*Neoscona theisi*		F	M	Fj				Araneidae
43	*Argyrodes cyrtophore*		F						Theridiidae
44	*Argyrodes* sp.1		F						Theridiidae
45	*Argyrodes* sp.2	U							Theridiidae
46	*Theridula* sp.1	U							Theridiidae
47	*Theridula* sp.2	U		M					Theridiidae
48	*Theridion manjithar*	U				J			Theridiidae
49	*Theridion* sp.1		F				Fsa		Theridiidae
50	*Theridion* sp.2		F						Theridiidae
51	*Marapissa dhakuriensis*		F						Salticidae
52	*Marapissa* sp.1			M	Fj	Mj		Msa	Salticidae
53	*Phidippus* sp.1		F	M	Fj		Fsa		Salticidae
54	*Plexyppus payakullii*			M	Fj				Salticidae
55	*Rhene* sp.		F	M					Salticidae
56	*Plexyppus* sp.1					Mj			Salticidae
57	*Mesumenoides* sp.1				Fj				Thomisidae
58	*Thanatus stripatus*		F						Thomisidae
59	*Thomisus andamanensis*		F				Fsa		Thomisidae
60	*Thomisus pugilis*			M			Fsa		Thomisidae
61	*Xysticus* sp.1		F						Thomisidae
62	*Oxyopes ratnae*		F	M	Fj	Mj			Oxyopidae
63	*Oxyopes* sp.1		F		Fj				Oxyopidae
64	*Peucetia* sp.1		F	M	Fj				Oxyopidae
65	*Peucetia viridans*		F						Oxyopidae
66	*Philodromus shillongensis*		F						Philodromidae
67	*Philodromus* sp.1				Fj				Philodromidae
68	*Tibellus* sp.1		F		Fj				Philodromidae
69	*Cheiracanthium* sp.		F		Fj				Miturgidae
70	*Cheiracanthium saraswati*			M					Miturgidae
71	*Cheiracanthium danieli*		F						Miturgidae
72	*Uloborus krishnae*		F						Uloboridae
73	*Uloborus danolius*		F						Uloboridae
74	*Castianeria* sp.1		F	M					Corinnidae
75	*Stegodyphus sarasinorum*		F						Eresidae
76	*Pritha* sp.1		F						Filistatidae
77	*Hersilia savignyi*		F	M	Fj		Fsa		Hersiliidae
78	*Crossopriza lyoni*		F						Pholcidae
79	*Tinus* sp.1		F	M					Pisauridae
80	*Scytodes* sp.						Fsa		Scytodidae
81	*Selenops* sp.				Fj				Selenopidae
82	*Olios* sp.					Mj			Sparassidae

U Unknown, *F* Female, *M* Male, *Fj* Female juvenile, *Mj* Male juvenile, *Fsa* Female subadults, *Msa* Male subadults

Table 4.5 Family-wise distribution of spider species in the study region

No.	Family	Species
1	Lycosidae	17
2	Tetragnathidae	13
3	Araneidae	12
4	Theridiidae	8
5	Salticidae	6
6	Thomisidae	5
7	Oxyopidae	4
8	Miturgidae	3
9	Philodromidae	3
10	Uloboridae	2
11	Corinnidae	1
12	Eresidae	1
13	Filistatidae	1
14	Hersiliidae	1
15	Pholcidae	1
16	Pisauridae	1
17	Scytodidae	1
18	Selenopidae	1
19	Sparassidae	1
	Total	**82**

4.2.2 Vertebrates

A total of 110 species of vertebrates were recorded from the study region under various phyla (Table 4.5). Among the total identified vertebrates, 11 were fishes, 5 amphibians, 13 reptiles and 71 birds and 10 mammals. Among the vertebrates, aves were the most dominant in terms of numbers, followed by reptiles, and amphibians (Table 4.6).

4.2.2.1 Fishes

A total of 11 species of fishes were recorded which belong to four families. The fish species belong to four families, viz. Channidae, Cichlidae, Cyprinidae, and Siluridae (Table 4.7). The maximum number of species belongs to the family Cyprinidae. Six species were found in river only, and five were found in both the ecosystems—rivers and lakes. Density and abundance of species were very low when compared to the other vertebrates. Among the 11 species recorded, mrigal (*Cirrhinus cirrhosus*) showed high density, abundance, and frequency, followed by catla (*Catla catla*). Indian butter catfish (*Ompok bimaculatus*) showed the least density, abundance, and frequency. The region is scarce in water and hence the low density, abundance, and frequency of fishes as compared to the other vertebrates.

Table 4.6 Vertebrates from the study region

S. no.	Phylum	Number of species
1	Fishes	11
2	Amphibians	5
3	Reptiles	13
4	Aves	71
5	Mammals	10

Table 4.7 Fishes of the study region and their density (D), frequency (F), and abundance (A) in the habitats

S. no.	Scientific name	Common name	Family	Habitat	D	F	A
1	*Channa striata* (Bloch, 1793)	Snakehead murrel	Channidae	River	0.04	4.16	0.12
2	*Tilapia mossambica* (Peters, 1852)	Tilapia	Cichlidae	Lake and river	0.06	6.6	0.2
3	*Puntius denisonii* (FDay, 1865)	Denison barb	Cyprinidae	River	0.2	15	0.3
4	*Catla catla* (F. Hamilton, 1822)	Catla	Cyprinidae	River	0.4	25	0.8
5	*Labeo rohita* (F. Hamilton, 1822)	Rohu	Cyprinidae	Lake and river	0.1	10	0.2
6	*Cirrhinus cirrhosus* (Bloch, 1795)	Mrigal	Cyprinidae	River	0.4	27	0.9
7	*Cyprinus carpio carpio* (Linnaeus, 1758)	Common carp	Cyprinidae	Lake and river	0.1	6.3	0.1
8	*Puntius sarana* (F. Hamilton, 1822)	Olive barb	Cyprinidae	Lake and river	0.16	13.3	0.25
9	*Rasbora daniconius* (F. Hamilton, 1822)	Rasbora	Cyprinidae	River	0.1	6.3	0.2
10	*Wallago attu* (Bloch and Schneider, 1801)	Wallago/boal	Siluridae	Lake and river	0.16	13.3	0.25
11	*Ompok bimaculatus* (Bloch, 1794)	Indian butter catfish	Siluridae	River	0.1	4.2	0.1

4.2.2.2 Amphibians

A total of five amphibian species which belong to 3 families were recorded from the study area (Table 4.8). Interestingly, all the five species were recorded to be of more or less similar density. Of these five species, common Indian tree frog (*Polypedates maculatus*) shows high density, abundance, and frequency, while Indian bull frog (*Hoplobatrachus tigerinus*) shows low density, abundance, and frequency. These five species are from three families and are more or less equally distributed in the study area.

Table 4.8 Amphibians and their density (D), frequency (F), and abundance (A) in the studied semiarid region

S. no.	Scientific name	Common name	Family	D	F	A
1	*Duttaphrynus melanostictus* (Schneider, 1799)	Common Indian toad	Bufonidae	0.26	16	0.16
2	*Hylarana erythraea* (Schlegel, 1837)	Green paddy frog	Ranidae	0.25	18	1.34
3	*Euphlyctis cyanophlyctis* (Schneider, 1799)	Skittering frog	Ranidae	0.2	16.7	1.29
4	*Polypedates maculatus* (J.E. Gray, 1834)	Common Indian tree frog	Rhacophoridae	0.3	25.3	1.49
5	*Hoplobatrachus tigerinus* (Daudin, 1803)	Indian bull frog	Ranidae	0.1	8.66	0.93

4.2.2.3 Reptiles

During the study period a total of 13 species of reptiles were recorded from the study region. All of them come under order squamata and belong to six different families (Table 4.9). These reptiles were found in different habitats such as agricultural lands, aquatic lands, and rocky hill ranges and in mixed vegetation. The

Table 4.9 Reptiles of the study region and their density (D), frequency (F), and abundance (A) in the studied semiarid region

S. no.	Scientific name	Common name	Family	D	F	A
1	*Sitana ponticeriana* (Cuvier, 1829)	Fan throat lizard	Agamidae	0.36	12	1.21
2	*Calotes versicolor* (Daudin, 1812)	Indian garden lizard	Agamidae	0.13	9.33	0.29
3	*Psammophilus dorsalis* (Griffith and Pidgeon, 1837)	Peninsular rock agama	Agamidae	0.48	58.8	1.17
4	*Ptyas mucosus* (Linnaeus, 1758)	Rat snake	Colubridae	0.24	23.3	1.01
5	*Ahaetulla nasuta* (Lacepede, 1789)	Green snake	Colubridae	0.06	6	0.8
6	*Naja naja* (Linnaeus, 1758)	Indian cobra	Elapidae	0.03	2.67	0.4
7	*Bungarus caeruleus* (Schneider, 1801)	Krait	Elapidae	0.53	6.66	0.75
8	*Hemidactylus brookii* (Gray, 1845)	Brook's gecko	Gekkonidae	1.86	15.8	0.55
9	*Hemidactylus frenatus* (Dumeril and Bibron, 1836)	Common house gecko	Gekkonidae	0.9	61.3	2.61
10	*Mabuya carinata* (Schineider, 1801)	Common skink	Scincidae	0.9	10.5	1.21
11	*Eutropis macularia* (Blyth, 1853)	Bronze grass skink	Scincidae	1.63	12.1	0.42
12	*Eutropis carinata* (Schneider, 1801)	Keeled Indian mabuya	Scincidae	0.88	78.6	1.78
13	*Varanus bengalensis* (Daudin, 1802)	Indian monitor lizard	Varanidae	6.76	30.8	1.17

habitat-wise description is not given here. Among them, Indian monitor lizard (*Varanus bengalensis*), Brook's gecko (*Hemidactylus brookii*), and Bronze grass skink (*Eutropis macularia*) were recorded in abundance. The common house geckos were the most abundantly found amphibian species in the study region. Keeled Indian mabuya (*Eutropis carinata*) showed high frequency, followed by common house gecko (*Hemidactylus frenatus*) and Peninsular rock agama (*Psammophilus dorsalis*). These reptiles were found in different habitats such as agricultural lands, aquatic lands, and rocky hill ranges and in mixed vegetation. The semiarid regions are favorable for reptile population, and these reptiles are showing very good adaptation to the changing climatic conditions of the region.

4.2.2.4 Birds

The study area had a total of 71 bird species belonging to 37 families. Of the 71 bird species, 37 species were terrestrial, 18 species were aquatic, and 16 species were both aquatic and terrestrial (Table 4.10). The scientific names with author citation, common name, and family along with the information on density, frequency, and abundance for the bird species are given in Table 4.11. Among the bird species, the density value for *Lonchura punctulata* was calculated to be 3.33, followed by *Pycnonotus cafer* (0.93) and *Ceryle rudis* (0.85). *P. cafer, C. rudis*, *Acridotheres tristis*, and *Cinnyris asiaticus* spotted frequently, and the values for these species are calculated to be 48, 44.2, 40, and 38.8, respectively. Species, namely *Hirundo concolor*, *Mirafra erythroptera*, *Streptopelia senegalensis*, *Anthus cinnamomeus*, and *Lonchura punctulata*, found less frequently, and the values are calculated to be 1.43, 1.19, 0.95, 0.95, and 0.29, respectively. The values for abundance were recorded high for the species *Plegadis falcinellus* (2.04), *Ceryle rudis* (1.92), and *Pycnonotus cafer* (1.85) and least for the species *Hirundo concolor* (0.07) *and Lonchura punctulata* (0.03). A few bird species such as dusky crag martin (*Hirundo concolor*), African pipit (*Anthus cinnamomeus*), and laughing dove (*Streptopelia senegalensis*) showed least values for density, frequency, and abundance (Table 4.11).

The migratory behavior of the birds was studied during the study period. Seven bird species, viz. purple heron (April–July), painted stork (April–July), common coot (July–October), purple moorhen (November–January), yellow-wattled lapwing, white-necked stork (April–August), and black-winged stilt (April–June), are the migratory birds that visit the semiarid region during the different seasons of a year (Table 4.12). The timing of migratory birds with the weather data collected from the

Table 4.10 Birds of the study region and their preferred habitat

Habitat	Number of birds
Aquatic	18
Terrestrial	37
Both (aquatic and terrestrial)	16
Total	**71**

Table 4.11 Density (D), frequency (F), and abundance (A) of the bird species of the studied semiarid region, Karnataka, India

S. no.	Scientific name	Common name	Family	D	F	A
1	*Accipiter badius* (Gmelin, 1788)	Shikra	Accipitridae	0.62	32.1	1.41
2	*Aquila nipalensis* (Hodgson, 1833)	Stepped eagle	Accipitridae	0.06	4.67	0.26
3	*Elanus axillaris* (Latham, 1802)	Black-shouldered kite	Accipitridae	0.65	33.8	1.72
4	*Haliastur indus* (Boddaert, 1783)	Brahminy kite	Accipitridae	0.05	4.05	0.56
5	*Aegithina tiphia* (Linnaeus, 1758)	Common iora	Aegithinidae	0.06	4.29	0.46
6	*Eremopterix griseus* (Scopoli, 1786)	Ashy-crowned sparrow lark	Alaudidae	0.6	35.3	1.32
7	*Mirafra erythroptera* (Blyth, 1845)	Indian bush lark	Alaudidae	0.01	1.19	0.09
8	*Anas poecilorhyncha* (Forster, 1781)	Indian spot bill duck	Anatidae	0.43	23.3	1.06
9	*Ardea cinerea* (Linnaeus, 1758)	Gray heron	Ardeidae	0.27	10	0.55
10	*Ardea purpurea* (Linnaeus, 1766)	Purple heron	Ardeidae	0.06	4.67	0.26
11	*Ardeola grayii* (Sykes, 1832)	Indian pond heron	Ardeidae	0.4	15.8	1.06
12	*Bubulcus ibis* (Linnaeus, 1758)	Cattle egret	Ardeidae	0.3	17.3	0.95
13	*Casmerodius albus* (Linnaeus, 1758)	Great egret	Ardeidae	0.34	13.3	1.11
14	*Egretta garzetta* (Linnaeus, 1766)	Little egret	Ardeidae	0.53	24.7	1.26
15	*Caprimulgus indicus* (Latham, 1790)	Gray nightjar	Caprimulgidae	0.04	3.33	0.53
16	*Ceryle rudis* (Linnaeus, 1758)	Lesser pied kingfisher	Cerylidae	0.85	44.2	1.92
17	*Vanellus malabaricus* (Boddaert, 1783)	Yellow-wattled lapwing	Charadriidae	0.47	30.8	1.04
18	*Vanellus indicus* (Boddaert, 1783)	Red-wattled lapwing	Charadriidae	0.61	31.3	1.15
19	*Mycteria leucocephala* (Pennatit, 1769)	Painted stork	Ciconiidae	0.4	19.3	0.8
20	*Ciconia episcopus* (Boddaert, 1783)	White-necked stork	Ciconiidae	0.29	21.3	0.55
21	*Prinia socialis* (Sykes, 1832)	Ashy prinia	Cisticolidae	0.06	5.48	0.42
22	*Columba livia* (Gmelin, 1789)	Blue rock pigeon	Columbidae	0.48	29.2	1.39
23	*Streptopelia senegalensis* (Linnaeus, 1766)	Laughing dove	Columbidae	0.01	0.95	0.11
24	*Streptopelia chinensis* (Scopoli, 1786)	Spotted dove	Columbidae	0.62	38	1.58

(continued)

Table 4.11 (continued)

S. no.	Scientific name	Common name	Family	D	F	A
25	*Corvus splendens* (Vieillot, 1817)	House crow	Corvidae	0.17	11.33	0.61
26	*Centropus sinensis* (Stephens, 1815)	Greater coucal	Cuculidae	0.72	35.3	1.73
27	*Eudynamys scolopaceus* (Linnaeus, 1758)	Asian koel	Cuculidae	0.07	6.43	0.49
28	*Phaenicophaeus pyrrhocephalus* (Pennant, 1769)	Blue-faced malkoha	Cuculidae	0.07	5.24	0.65
29	*Dicrurus macrocercus* (Vierllot, 1817)	Black drongo	Dicruridae	0.17	9.33	0.37
30	*Lonchura malabarica* (Linnaeus, 1766)	Indian silverbill	Estrildidae	0.56	32	1.38
31	*Lonchura Malacca* (Linnaeus, 1766)	Black-headed munia	Estrildidae	0.06	4.52	0.4
32	*Lonchura punctulata* (Linnaeus, 1758)	Spotted munia	Estrildidae	3.33	0.29	0.03
33	*Halcyon smyrnensis* (Linnaeus, 1758)	White-breasted king fisher	Halcyonidae	0.53	27.3	1.12
34	*Hirundo daurica* (Laxmann, 1769)	Red-rumped swallow	Hirundinidae	0.03	2.38	0.32
35	*Hirundo smithii* (Leach, 1818)	Wire-tailed swallow	Hirundinidae	0.05	4.52	0.45
36	*Hirundo concolor* (Sykes, 1832)	Dusky crag martin	Hirundinidae	0.01	1.43	0.07
37	*Megalaima haemacephala* (Statius muller, 1776)	Coppersmith barbet	Megalaimidae	0.05	4.05	0.4
38	*Merops leschenaulti* (Vieillot, 1817)	Chestnut-headed bee-eater	Meropidae	0.52	28.3	1.21
39	*Merops orientalis* (LInnaeus, 1766)	Small bee-eater	Meropidae	0.43	18	0.94
40	*Anthus cinnamomeus* (Ruppell, 1840)	African pipit	Motacillidae	0.01	0.95	0.09
41	*Anthus rufulus* (Vieillot, 1818)	Paddy field pipit	Motacillidae	0.08	6.67	0.52
42	*Copsychus saularis* (Linnaeus, 1758)	Oriental magpie robin	Muscicapidae	0.05	3.57	0.44
43	*Saxicoloides fulicatus* (Linnaeus, 1766)	Indian robin	Muscicapidae	0.65	35.4	1.57
44	*Saxicola caprata* (Linnaeus, 1766)	Pied bushchat	Muscicapidae	0.05	4.52	0.47
45	*Nectarinia asiatica* (Latham, 1790)	Purple sunbird (eclipse male)	Nectariniidae	0.06	4.52	0.58
46	*Cinnyris asiaticus* (Latham, 1790)	Purple sunbird	Nectariniidae	0.73	38.8	1.64
47	*Nectarinia minima* (Sykes, 1832)	Purple-rumped sunbird	Nectariniidae	0.04	3.33	0.41

(continued)

Table 4.11 (continued)

S. no.	Scientific name	Common name	Family	D	F	A
48	*Oriolus kundoo* (Sykes, 1832)	Indian golden oriole	Oriolidae	0.07	5.71	0.61
49	*Oriolus oriolus* (Linnaeus, 1758)	Eurasian golden oriole	Oriolidae	0.09	5.46	1.23
50	*Passer domesticus* (Linnaeus, 1758)	House sparrow	Passeridae	0.45	30.7	1.18
51	*Phalacrocorax fuscicollis* (Stephens, 1826)	Indian cormorant	Phalacrocoracidae	0.58	35.3	1.28
52	*Francolinus pondicerianus* (Gmelin, 1789)	Gray francolin	Phasianidae	0.07	5.24	0.54
53	*Pavo cristatus* (Linnaeus, 1758)	Indian peafowl	Phasianidae	0.47	31.7	1.55
54	*Dinopium benghalense* (Linnaeus, 1758)	Lesser golden-backed woodpecker	Picidae	0.06	5.71	0.45
55	*Ploceus philippinus* (Linnaeus, 1766)	Baya weaver	Ploceidae	0.53	31.7	1.25
56	*Psittacula krameri* (Scopoli, 1769)	Rose-ringed parakeet	Psittaculidae	0.06	4.05	0.41
57	*Pycnonotus cafer* (Linnaeus, 1766)	Red-vented bulbul	Pycnonotidae	0.93	48	1.85
58	*Amaurornis phoenicurus* (Pennant, 1769)	White-breasted water hen	Rallidae	0.64	37.3	1.37
59	*Fulica atra* (Linnaeus, 1758)	Common coot	Rallidae	0.39	20	0.78
60	*Porphyrio porphyrio* (Linnaeus, 1758)	Purple moorhen	Rallidae	0.45	22	1.2
61	*Himantopus himantopus* (Linnaeus, 1758)	Black-winged stilt	Recurvirostridae	0.17	11.3	0.61
62	*Sterna aurantia* (Gray, JE, 1831)	River tern	Sternidae	0.17	10	0.33
63	*Acridotheres tristis* (Linnaeus, 1766)	Common myna	Sturnidae	0.68	40	1.55
64	*Sturnus malabaricus* (Gmelin, 1789)	Chestnut-tailed starling	Sturnidae	0.08	5.71	0.59
65	*Sturnus pagodarum* (Gmelin, 1789)	Brahminy starling	Sturnidae	0.05	4.76	0.42
66	*Sturnus roseus* (Linnaeus, 1758)	Rosy starling	Sturnidae	0.06	5.71	0.61
67	*Plegadis falcinellus* (Linnaeus, 1766)	Glossy ibis	Threskiornithidae	0.67	25.3	2.04
68	*Pseudibis papillosa* (Terminck, 1824)	Black ibis	Threskiornithidae	0.45	23.3	1.11
69	*Turdoides striata* (Dumont, 1823)	Jungle babbler	Timaliidae	0.52	30	1.04
70	*Turdoides affinis* (Jerdon, 1845)	White-headed babbler	Timaliidae	0.55	32	1.38
71	*Upupa epops* (Linnaeus, 1758)	Common hoopoe	Upupidae	0.41	24	1.01

Table 4.12 Season for migratory birds and their appearance with relation to weather conditions in the studied semiarid region

			Jan	Feb	Mar	Apr	May	Jun	July	Aug	Sep	Oct	Nov	Dec
Temperature (°C)			26.2	29.2	32.3	34.6	41.1	40.09	38.9	38.1	30.5	35.7	28.5	27.3
Rain fall (mm)			0	4	0	12	15	18	82	142	54	53	17	0
Humidity (%)			42.25	39.5	28.3	29	38	56.6	71	72	74.1	63	68.9	52.8
S. No.	Common name	Scientific name												
1	Purple heron	*Ardea purpurea*												
2	Painted stork	*Mycteria leucocephala*												
3	Common coot	*Fulica arta*												
4	Purple moorhen	*Porphyrio porphyrio*												
5	Yellow wattled lapwing	*Vanellus malabaricus*												
6	White necked stork	*Ciconia episcopus*												
7	Black winged stilt	*Himantopus himantopus*												

study region was correlated to understand the seasonal appearance of the birds in the region. It is that most of the migratory birds were found to visit the study region between April and September when the average rainfall ranges between 29 and 142 mm. During these months, their staple diet of fishes is available in plenty in the water bodies of the region. Purple moorhen was the only migratory bird, which was found to be available during the winter months between October and January. During this time, the rainfall is almost negligible, but this is reported as the season for the breeding of purple moorhen and the area provides favorable conditions for the breeding. The information compiled on the birds shows that 7 species are migratory species and 64 species belong to the semiarid regions.

4.2.2.5 Mammals

A total of 10 species of mammals were recorded belonging to 9 families from the studied semiarid region (Table 4.13). The density value is calculated to be less than 1.0 for all the mammals. Among all mammals, species, namely *Mus booduga*, *Acerodon jubatus*, *Funambulus palmarum*, and *Mus musculus*, shows high value for density as compared to others. The high value in terms of frequency was obtained for *M. booduga* (14), followed by *F. palmarum* and *M. musculus* (12.7 each), and however least for *Lepus nigricollis* (2.67). The value for abundance was found highest for *Acerodon jubatus* (2.28), followed by *M. booduga* (1.49) and *Funambulus palmarum* (1.28), and least for *Vulpes bengalensis* (0.38). Wild boar (*Sus Scrofa*) and Indian fox (*V. bengalensis*) were not found very frequently from the studied semiarid region (Table 4.13).

Table 4.13 Mammals with their density (D), frequency (F), and abundance (A) in the studied semiarid region

S. no.	Scientific name	Common name	Family	D	F	A
1	*Vulpes bengalensis* (Sharo, 1800)	Indian fox	Canidae	0.04	3.75	0.38
2	*Macaca radiata* (E. Geoffroy, 1812)	Common monkey	Cercopithecidae	0.11	5.33	1.2
3	*Axis axis* (Erxleben, 1777)	Chital	Cervidae	0.09	5.42	0.42
4	*Herpestes edwardsii* (E. Geoffroy saint, Hilaire, 1818)	Indian gray mongoose	Herpestidae	0.13	11.7	0.84
5	*Lepus nigricollis* (F. Cuvier, 1823)	Indian hare	Leporidae	0.03	2.67	0.47
6	*Mus booduga* (Gray, 1837)	Little Indian field mouse	Muridae	0.21	14	1.49
7	*Mus musculus* (Linnaeus, 1758)	House mouse	Muridae	0.2	12.7	1.28
8	*Acerodon jubatus* (Eschscholtz, 1831)	Giant golden-crowned flying fox	Pteropodidae	0.21	9.33	2.28
9	*Funambulus palmarum* (Linnaeus, 1766)	Three-striped palm squirrel	Sciuridae	0.21	12.7	1.28
10	*Sus scrofa* (Linnaeus, 1758)	Wild boar	Suidae	0.07	6.67	0.65

4.2.2.6 Zooplankton

A total of 24 species of zooplankton were identified from the collected samples of the water. Out of which, 14 species belong to phylum Rotifera and 7 belong to the arthropoda (Table 4.14). In Rotifera, the dominant species are *Keratella tropica, Asplanchna priodonta*, and *Habrotrocha bidens*. In Copepoda (subclass), *Mesocyclops leuckarti* and *Heliodiaptomus viduus* were dominant. In the Cladocera, group *Ceriodaphnia cornuta* is the dominant species in the present study.

4.3 Species Description

Species description is arranged as per orders for invertebrates and groups for vertebrates based on field records and surveying various field guides (Daniel 2002; Daniels 2002; Gay et al. 2009; Manakadan et al. 2012; Prater 2005; Sebastian and Peter 2008; Singh 2011; Subramanian 2005 and Encyclopedia of Life online).

Table 4.14 Zooplankton of the study region

S. no.	Scientific name	Family	Phylum
1	*Asplanchna priodonta* (Gosse, 1850)	Asplanchnidae	Rotifera
2	*Brachionus bidentata* (Anderson, 1889)	Brachionidae	Rotifera
3	*Brachionus calyciflorus* (Pallas, 1766)	Brachionidae	Rotifera
4	*Brachionus falcatus* (Zacharias, 1898)	Brachionidae	Rotifera
5	*Brachionus urceolaris* (Muller, 1773)	Brachionidae	Rotifera
6	*Cephalodella gibba* (Ehrenberg, 1832)	Notommatidae	Rotifera
7	*Ceriodaphnia cornuta* (Sars, 1885)	Daphniidae	Arthropoda
8	*Chydorus sphaericus* (O.F. Muller, 1785)	Chydoridae	Arthropoda
9	*Habrotrocha bidens* (Gosse, 1851)	Habrotrochidae	Rotifera
10	*Heliodiaptomus viduus* (Gurney, 1916)	Diaptomidae	Arthropoda
11	*Hemicypris fossiculata*	Cyprididae	Arthropoda
12	*Keratella tropica* (Apstein, 1907)	Brachionidae	Rotifera
13	*Lecane luna* (Muller, 1776)	Lecanidae	Rotifera
14	*Mesocyclops leuckarti* (Claus, 1857)	Cyclopidae	Arthropoda
15	*Moina brachiata* (Jurine, 1820)	Moinidae	Arthropoda
16	*Monostyla bulla* (Gosse, 1851)	Lecanidae	Rotifera
17	*Mytilina ventralis* (Gosses, 1886)	Mytilinidae	Rotifera
18	*Mytilina acanthophora* (Hauer)	Mytilinidae	Rotifera
19	*Platyias quadricornis* (Ehrenberg, 1832)	Brachionidae	Rotifera
20	*Rotifer tardus* (Ehrenberg, 1838)	Philodinidae	Rotifera
21	*Tropocyclops prasinus* (Fischer, 1860)	Cyclopidae	Arthropoda
22	*Horlla brehmi*		Rotifera
23	*Filinia longiseta*	Trochosphaeridae	Rotifera
24	*Paracy final*		Rotifera

Invertebrates

a. **Insects**

Order: Coleoptera
Family: Buprestidae

1. **Scientific name**: ***Chalcophora virginiensis*** **Drury, 1770**

   Common Name: Sandalwood borer
   Description: Elongated in shape and large and black in color with sculptured appearance on the back. Bright coloration under the wings.

2. **Scientific name**: ***Psiloptera fastuosa*** **Fabricius, 1775**

   Common Name: Jewel beetle or metallic wood borers
   Description: Elongated in shape and large and black in color with sculptured appearance on the back. Bright cataration under the wings.

3. **Scientific name**: ***Psiloptera fastuosa*** **Fabricius, 1775**

   Common Name: Jewel beetle or metallic wood borers
   Description: 20–25 mm in length. Bright metallic bluish green in color. Rugulose head and convex elytra, with rows of different punctuations.

4. **Scientific name**: ***Sphenoptera gossypii*** **Cotes, 1891**

   Common Name: Large brown cotton borer
   Description: This is longhorn beetle with a black-and-white coloration with black and large antenna.

5. **Scientific name**: ***Sphenoptera*** **sp**

   Common Name:
   Description: Not descried—specimen lost.

Family: Carabidae

6. **Scientific name**: ***Amblystomus quttatus*** **Bates, 1873**

   Common Name:
   Description: The average length of this beetle is 25 mm. Dark shiny in color with flat body. With striated elytra and straight front tibia.

7. **Scientific name**: ***Chlaenius nepalensis***

   Common Name: Ground beetles
   Description: 16–17 mm in length and 5–7 mm in width. Pronotum and head is black with green or coppery metallic in color. Elytra are black in color. Black ventral surface and palpi and tarsi nearly black to dark brown in color.

8. **Scientific name**: ***Omphra pilosa*** **Klug, 1834**

   Common Name: Carabid beetle or ground beetle
   Description: 5–8 mm in length. This species has concealed scutellum. Oval and convex in shape.

Family: Cerambycidae

9. **Scientific name**: ***Acanthophorus serraticornis*** **Olivier, 1795**

   Common Name:
   Description: 7–8.5 cm in length. Reddish tawny in color with shining. Conspicuous mandibles on the head. Sternum is pubescent on the ventral side.

10. **Scientific name**: ***Apomecyna pertigera***

    Common Name:
    Description: 7–8.5 cm in length. Reddish tawny in color with shining. Conspicuous mandibles on the head. Sternum is pubescent on the ventral side.

11. **Scientific name**: ***Apomecyna pertigera***

    Common Name: Longhorn beetle/timber beetle
    Description: Elongated body with blackish brown in color with white spots. Long antennae. This is a destructive pest of cucumbers, gourds, pumpkins, squashes, and watermelons.

12. **Scientific name**: ***Batocera rubus* Linnaeus, 1758**

    Common Name: Longhorn beetle
    Description: Moderate to large size with elongated body. Grow up to 40 mm in length. Body is light brown in color with white spots. Filiform antennae. Eyes are complete and reniform. Pronotum is transverse. Legs with smooth tarsal claws.

13. **Scientific name**: ***Neocerambyx paris* Wiedemann, 1821**

    Common Name: Longhorn beetles
    Description: This looks dark brown and head with a furrow between the eyes. The antennae are long.

14. **Scientific name**: ***Sthenias grisator* Fabricius, 1787**

    Common Name:
    Description: Medium sized and stout beetle. Gray in color with white spots in the center of elytra. Head is dark brown in color. Strong mandibles.

15. **Scientific name**: ***Sthenias* sp.**

    Common Name: Longhorn beetle
    Description: Body is small and moderate in size and elongated. The average length is 9.5–39 mm. Reniform and complete eyes. Filiform antennae. Tarsal claws are smooth.

16. **Scientific name**: ***Stromatium barbatum* Fabricius, 1775**

    Common Name: Kulsi teak borer
    Description: 11–30 mm in length. Reddish brown in color. Head with antennae. Antennae are long in females than in males. Thickly faceted eyes. Prothorax compactly covered with common punctures. Coarsely punctured elytra.

17. **Scientific name**: ***Xystrocera globosa* Olivier, 1795**

    Common Name: Monkey pod/samanea round-headed borer
    Description: The average length is 15–32 mm. Body is reddish brown in color and elongated. Very long antennae nearly as long as the body. Elytra are yellow in color with a green or metallic blue longitudinal band.

Family: Cetoniidae

18. **Scientific name**: ***Chiloloba acuta* Wiedmann, 1823**

    Common Name: Chafer beetle
    Description: Shiny metallic green in color with short hairy surfaces on both sides. Commonly seen on grasses. The elytra are raised keen on an edge along the border where they assemble toward the back ending of the carcass.

Family: Chrysomelidae

19. **Scientific name**: ***Spilopyra sumptuosa* Baly, 1860**

    Common Name: Chrysomelid beetle
    Description: Metallic purple color with the reflection of purple green color. 8.5–11.5 mm in length. Undersurface of the body reflects metallic green color with prominent gold color bands bordered with bright golden red color.

20. **Scientific name**: ***Zygogramma bicolorata* Pallister, 1953**

    Common Name: Mexican beetle
    Description: Elongated and oblong body. Strongly convex dorsal surface. Head is black in color. Underneath and pronotum is creamy yellow in color and has hat-shaped black markings. Two elongated spots arranged longitudinally on elytra. Males are smaller than females.

21. **Scientific name**: ***Zygogramma tortuosa* Rogers, 1856**

    Common Name: Parthenium beetle
    Description: Body is elongated and oblong and is creamy yellow in color with brown spots. Head is brown in color. Males are smaller than females.

Family: Cicindelidae

22. **Scientific name**: ***Cicindela sexpunctata***

    Common Name: Tiger beetle
    Description: ½ in. in length. Long antennae and legs. Grayish brown to black in color with white spots. Hairs are present on the face.

23. **Scientific name**: ***Cicindella* sp.**

    Common Name: Tiger beetle
    Description: ½ in. in length. Metallic or iridescent blue, green, and bronze in color. Hairs are present on the face. Antennae are long.

Family: Coccinellidae

24. **Scientific name**: ***Epilachna vigintioctopunctata* Fabricius, 1775**

    Common Name: 28-spotted ladybird beetle
    Description: These species are larger than the other ladybirds. Orange in color. Thirteen black spots found on each wing and two black spots on thorax. Short hairs on the body. 6.7–10.5 mm in length.

25. **Scientific name**: ***Harmonia octomaculata* Fabricius, 1781**

    Common Name: Eight-spotted ladybird
    Description: Elongated and oval in shape, moderately convex. Underneath orange yellow or red in color with black markings on elytra and pronotum. External narrow margins on elytra thickened and rounded. 4.6–7.5 mm in length.

Family: Curculionidae

26. **Scientific name**: ***Alcidodes collaris* Pascoe**

    Common Name: Stem weevils
    Description: Long snout and geniculate antennae with small clubs. 1–40 mm in length. Brown in color.

27. **Scientific name**: ***Alcidodes* sp.**

    Common Name: True weevils
    Description: Adult grows up to 1.4 cm in length. Along the elytra, conspicuous white stripes appear longitudinally. Brownish black in color. Hard and well-sclerotized body. Sometimes the body is covered with scales or hairs.

28. **Scientific name**: ***Hyopolixus truncatulus***

    Common Name: Amaranthus stem weevil
    Description: Shiny bluish black in color. Male is smaller than female. The average length of the female is 14 mm and that of the male is 8 mm. Head is black in color and somewhat rectangular in shape. Face and clypeus are grayish in color. Dark brown eyes. Antennae are shorter than the body. Thorax is black in color with pale markings on the neck. Central dark mark found on the mesonotum. White band over the basal area of the metathorax. Hyaline and large wings. The legs are reddish in color. Abdomen is black in color.

Family: Elateridae

29. **Scientific name**: ***Cardiophorus quadrimaculatus***

    Common Name: Click beetles
    Description: Elongated and parallel-sided body somewhat flattened. Brown to black in color. ¼ to 1½ in. in length. These beetles when to be found on their backs, they normally "click" snapping their prothorax and mesothorax to spin the body in the air to exact themselves.

Family: Hydrophilidae

30. **Scientific name**: ***Hydrophilus triangularis***

    Common Name: Water scavengers
    Description: 38–40 mm in length and 15–16 mm in width. Large in size. Aquatic habitat. Thorax thickness is of 8–9 mm. Elytra are large and glossy black color with a greenish accent. Short and clubbed antennae. Strongly compressed and fringed tarsi of the middle and hind legs with hairs.

Family: Meloidae

31. **Scientific name**: ***Epicauta* sp.**

    Common Name: Blister beetles or meloid beetles
    Description: 3/8 to 1 in. in length. The shape of the beetle is long, cylindrical, and narrow. Head is wider than the pronotum. Solid gray in color with paler wing margin.

Family: Scarabaeidae

32. **Scientific name**: ***Adoretus* Sp.**

    Common Name:
    Description: 10.5–12 mm in length and brownish in color with cream-colored scales. Oblong and oval in shape. Females are larger than males. Posterior is always round in females and quadrate in males. Males have more acute protibial teeth than females.

33. **Scientific name**: ***Agestrata* Sp.**

    Common Name: Cetoniid beetles
    Description: It is largest and most brilliantly metallic in color. Head is distinctive, elongated with prominent eyes and flattened clypeus.

34. **Scientific name**: ***Catharsius molossus*** **Linnaeus, 1758**

    Common Name:
    Description: 15–45 mm in length. Black and opaque in color. Underside reddish hair present. Broad head. Lightly striated elytra.

35. **Scientific name**: ***Cetonia aurata*** **Linnaeus, 1758**

    Common Name: Rose chafer
    Description: 14–22 mm in length. Metallic green in color. Scutellum is in V-shape. Underneath coppery in color. The small triangular region between the wing cases has several other unequal small white lines and marks.

36. **Scientific name**: ***Copris*** **Sp.**

    Common Name:
    Description: These are vigorous beetles. Dull black in color. Middle and behind tibia is slender. No horns on the head and pronotum. Conspicuous striate on the elytra. The length is less than 18 mm.

37. **Scientific name**: ***Gymnopleurus*** **Sp.**

    Common Name:
    Description: Broad and depressed body with slender legs. Moderately convex pronotum. 9 jointed antennae, 3rd joint as long as the 4th and 5th together, and 6th extremely short. Long and narrow mandible.

38. **Scientific name**: ***Holotrichia serrata*** **Hope, 1837**

    Common Name: Root grub
    Description: Adults are dark reddish brown in color. Robust and oval in shape. These species emerge after the first summer showers.

39. **Scientific name**: ***Liatongus radamistus***

    Common Name:
    Description: Bright orange yellow in color. Head is almost semicircular in shape. Smooth metasternum. Flattened rhomboidal area between the eyes. The abdomen, sterna, tibia, and tarsi are the same dark in color.

40. **Scientific name**: ***Odontria xanthosticta*** **White, 1846**

    Common Name: Chafer beetle
    Description: Not descried—specimen lost.

41. **Scientific name**: ***Onitis philemon*** **Fabricius, 1801**

    Common Name:
    Description: Coppery or bronzy black or green in color. Moderately, it is shiny. Oval in shape. Rugulose head with smooth ocular lobes. Unevenly punctured and strong pronotum. The lateral margin is straight in front. The elytra are strongly striate, and the intervals of the elytra are very finely and sparsely punctured.

42. **Scientific name**: ***Onthophagus gazella*** **Fabricius**

    Common Name: Dung beetle
    Description: Yellow in color. Large oval spot upon the lower surface of the four posterior femora. The forehead is finely and closely punctured. The head is semicircular with the front margin.

43. **Scientific name**: ***Oryctes rhinoceros*** **Linnaeus, 1758**

    Common Name: Rhinoceros beetle
    Description: Rounded and convex dorsal surface. In males, horns presented on the head or pronotum. In females, horns are absent. Dark brown in color. Adult body is covered with exoskeleton. A pair of thick wings is present.

44. **Scientific name**: ***Oxycetonia versicolor*** **Fabricius, 1775**

    Common Name: Citrus flower beetle
    Description: Oblong body with slender legs. Slightly hairy on legs. Black in color with dark orange marks. Head is oval in shape. Two black spots on the head.

45. **Scientific name**: ***Scarabaeus*** **sp.**

    Common Name:
    Description: Body flat with slender legs. Tarsi of the middle and hind legs are slender. Long front femur. Strongly oblique middle coxae. Long and narrow femur. Slender tibia. The femora and tibia are long and slender. Slightly convex pronotum. Flat elytra. Very short mesosternum. Moderately long metasternum. Antenna is 9-jointed.

46. **Scientific name**: ***Xylotrupes ulysses*** **Guérin-Méneville, 1830**

    Common Name: Rhinoceros beetle
    Description: Large and stout. Black in color with an average length of 5.5 cm. Male has two horns, one on the head and the other on the thorax. Females lack horns.

Family: Tenebrionidae

47. **Scientific name**: ***Tribolium castaneum*** **Herbst, 1797**

    Common Name: Rust red flour beetle
    Description: Approximately 3–4 mm in length. Reddish brown flattish curved-sided body. Minute punctures are presented on the head and upper part of the thorax. The antennae tip is enlarged. Eyes are reddish black in color.

Order: Dictyoptera
Family: Blattellidae

48. **Scientific name**: ***Blattella germanica*** **Linnaeus, 1767**

    Common Name: German cockroach
    Description: 1½ in. in length. Deep brown in color. Dark brown streaks on the prothorax.

49. **Scientific name**: ***Periplaneta americana*** **Linnaeus, 1758**

    Common Name: American cockroach
    Description: 27–35 mm in length. Reddish brown in color. Wings are well developed.

50. **Scientific name**: ***Prosoplecta*** **sp.**

    Common Name: Cockroach
    Description: Reddish brown in color. Epicranium and above antennal sockets slightly paler. Pronotum and tegmina are reddish brown in color. Abdomen and legs with brownish yellow appendages.

Family: Cimicidae

51. **Scientific name**: ***Cimex lectularius*** **Linnaeus, 1802**

    Common Name: Bedbug
    Description: Small oval flattened reddish brown in color. Wingless. Measuring about ¼ in. in length. Ectoparasites of birds and mammals.

Family: Reduviidae

52. **Scientific name**: ***Acanthaspis*** **sp.**

    Common Name: Assassin bug
    Description: Three segmented. Raptorial forelegs. Head is more or less cylindrical. Brown in color. Elongated head. Short and three-segmented beak. Abdomen is widened in the middle. 25–34 mm in length.

Order: Diptera
Family: Asilidae

53. **Scientific name**: ***Promachus rufipes*** **Fabricius, 1775**

    Common Name: Robber fly
    Description: Head is hollowed out between the eyes. Stout thorax with long strong legs with tapering abdomen. Hairy abdomen.

Family: Calliphoridae

54. **Scientific name**: ***Cochliomyia hominivorax*** **Coqurel, 1858**

    Common Name: Screwworm fly
    Description: 8–10 mm in length. Metallic green in color. Three black longitudinal stripes on the thorax. Filiform-type palps. Anterior spiracles are pale white in color.

Family: Culicidae

55. **Scientific name**: ***Culex*** **sp.**

    Common Name: Mosquito
    Description: Tip of the abdomen is blunt, with the cerci retracted. Color of the thorax is usually dull. The scales will be present along with the wing veins.

Family: Hippoboscidae

56. **Scientific name**: ***Stomoxys calcitrans*** **Linnaeus, 1758**

    Common Name: Stable fly
    Description: Dark brownish in color. Size of the fly is smaller than that of houseflies. Flat in shape and leathery in appearance. Toothed tarsal claws. Widely separated coxae.

Family: Muscidae

57. **Scientific name**: ***Musca domestica*** **Linnaeus, 1758**

    Common Name: Housefly
    Description: ¼ in. in length. Dirty gray in color. Small hairs presented all over the body. Some dark stripes on the thorax and short anal vein.

Family: Sarcophagidae

58. **Scientific name**: ***Sarcophaga*** **sp.**

    Common Name: Flesh fly
    Description: 10–13 mm in length. Grayish in color. Hairy abdomen. Red compound eyes. Three black and gray stripes on the thorax.

59. **Scientific name**: ***Sarcophaga*** **sp.**

    Common Name: Sarcophagid fly
    Description: 8–14 mm in length. Body is black in color with grayish tint. At the end of the abdomen, red-colored genitalia are visible. Long antennae. The fully developed species head has a row of anterior hair close to the inflamed center vertical area.

Family: Simuliidae

60. **Scientific name**: ***Simulium* sp.**

    Common Name: Neirida fly
    Description: Dark in color. Legs are short, and wings are broad.

Family: Syrphidae

61. **Scientific name**: ***Eristalis* sp.**

    Common Name: Syrphid fly
    Description: 7–17 mm in length. Dark in color and less hairy. Head is broader than the thorax. Yellow and black markings on the abdomen.

62. **Scientific name**: ***Ischiodon scutellaris* Fabricius**

    Common Name: Common hoverfly
    Description: 10–11 mm in length. Yellow–black pattern on the abdomen. Elongated basoflagellomere. Bare post-pronotum. Face is partly yellow in color. Parallel-sided abdomen. Enlarged genitalia in males. Yellow to orange color legs.

Family: Tabanidae

63. **Scientific name**: ***Tabanus striatus* Fabricius, 1787**

    Common Name: Horsefly
    Description: Stout body. The eyes are bright in color. The hind tibia lacks apical spurs. Somewhat hemispherical head. Third antennal segment has a toothlike process near the base.

Order: Hemiptera
Family: Belostomatidae

64. **Scientific name**: ***Lethocerus americanus* Leidy, 1847**

    Common Name: Giant water bug
    Description: 45–65 mm in length. Basal segment of beak is short. Short terminal filaments can found on the abdomen.

Family: Cimicidae

65. **Scientific name**: ***Cimex lectularius* Linnaeus**

    Common Name: Bedbug
    Description: Small oval flattened in shape and reddish brown in color. ¼ in. in length. Wingless. Mouthparts consist of a sharp probasas adapted for piercing and sucking the blood.

Family: Coreidae

66. **Scientific name**: ***Anoplocnemis phasiana*** **Fabricius**,

    Common Name: Squash bug
    Description: ½ to 2/3 in. in length. Dark brown in color. Flat wings. Wings held over the back side of the body.

67. **Scientific name**: ***Euthochtha galeator*** **Fabricius, 1767**

    Common Name: Coreid bug
    Description: 13–17 mm in length. Dull brown in color. In males, hind femora are swollen, and underneath, spines will be found. Pronotum in the midst of front part of lateral borders finely and irregularly toothed.

68. **Scientific name**: ***Leptoglossus phyllopus*** **Linnaeus, 1767**

    Common Name: Leaf-footed bug
    Description: ¾ in. in length. Dark brown in color with yellow whitish stripes with central part black in color. Flattened hind legs. On the tibia the hind legs have leaf-like expansions and look flattened.

Family: Eurybrachidae

69. **Scientific name**: ***Eurybrachis tomentosus*** **Blanchard, 1840**

    Common Name: Eurybrachid bug
    Description: Female is greenish in color and is larger than male. Male is with brown-colored tegmina. Body is in green color with white spots.

Family: Gerridae

70. **Scientific name**: ***Gerris*** **sp. (cf. lacustris)**

    Common Name: Pond skaters
    Description: 4–6 mm in length. Dark in color. Long and narrowly shaped body. Live on the surface of the water. Two-segmented tarsi. Ocelli are small in size.

Family: Lygaeidae

71. **Scientific name**: ***Lygaeus equestris*** **Linnaeus, 1758**

    Common Name: Black and red bug
    Description: Abdominal spiracles were found on the dorsal side of the body. Red and black in color. 7–9 mm in length. Moderately swollen front femora.

Family: Notonectidae

72. **Scientific name**: ***Notonecta glauca*** **Linnaeus, 1758**

Common Name: Back swimmer
Description: 8–17 mm in length and little stouter. Four-segmented antennae. Claval commissure is shorter than the hemelytra. Hind legs are oarlike and long. Hind tarsi are without claws.

Family: Pentatomidae

73. **Scientific name**: ***Halyomorpha halys*** **Stal, 1855**

Common Name: Stinkbug
Description: 12–17 mm in length. Ovoid in shape and five-segmented antennae. Dark and light bands take place on the last two segments of antennae. Coppery or bluish metallic-colored patches cover the pronotum. And these pronotum margins are smooth.

74. **Scientific name**: ***Nezara viridula*** **Linnaeus, 1758**

Common Name: Green stinkbug
Description: Bright in color and clearly marked. Produce disagreeable odor. Five-segmented antennae.

Family: Pyrrhocoridae

75. **Scientific name**: ***Dysdercus cingulatus*** **Fabricius, 1775**

Common Name: Red cotton bug
Description: 11–17 mm in length. Elongated and oval in shape. Brightly marked with red and black. Hemelytra have many veins with branches and cells.

Family: Reduviidae

76. **Scientific name**: ***Zelus renardii*** **Kolenati, 1856**

Common Name: Assassin bugs
Description: Elongated head. Constricted and necklike area behind the eye. Three-segmented and short beak was found. Abdomen is widened in the middle. Raptorial forelegs.

Family: Scutelleridae

77. **Scientific name**: ***Chrysocoris stolli*** **Wolf**

    Common Name: Green jewel bug
    Description: Very large scutellum and it extends to the apex of the abdomen. Wings visible only at the edge of the scutellum. 8–10 mm in length. Brown in color.

Order: Hymenoptera
Family: Formicidae

78. **Scientific name**: ***Camponotus compressus*** **Fabricius, 1787**

    Common Name: Black common ant
    Description: Tibiae of legs prismatic and abdomen covered with sparse erect hairs. Petiole with single node. 12-segmented antennae without club and faraway from the clypeus. Legs are black in color, and antennae are reddish brown in color.

79. **Scientific name**: ***Monomorium indicum*** **Forel, 1902**

    Common Name: Small black ant
    Family: Formicidae
    Description: Small in size. Black in color. Antennal segments are unique. Eggs are white in color.

80. **Scientific name**: ***Solenopsis geminata*** **Fabricius, 1804**

    Common Name: Small red ant
    Family: Formicidae
    Description: 10–11-segmented antennae. Hairs are found sparsely on the body. The number of antennal segments is unique in this group, 10 with 2-segmented club.

Family: Apidae

81. **Scientific name**: ***Apis cerana*** **Fabricius, 1793**

    Common Name: Indian bee
    Description: About 10 mm in length. Smaller in size and less hairy. Well-defined and even have black and yellow stripes on the abdomen. It is bigger than the common honeybee.

82. **Scientific name**: ***Apis dorsata* Fabricius, 1793**

    Common Name: Rock bee
    Family: Apidae
    Description: 17–20 mm in length. Golden brown in color. Abdomen black in color. Transparent wings with golden brown color. Hairs found on the head.

83. **Scientific name**: ***Apis florea* Fabricius, 1787**

    Common Name: Little bee or red dwarf honeybee
    Family: Apidae
    Description: Dwarf bee. Small in size. White hairs on the head. Wings are transparent. Black stripes on the body with white hairs.

84. **Scientific name**: ***Xylocopa* sp**.

    Common Name: Carpenter bee
    Family: Apidae
    Description: 25 mm in length. Dorsum of their metasoma is largely bare, and the second submarginal cell is triangular in shape.

Family: Evaniidae

85. **Scientific name**: ***Evania appendigaster* Linnaeus**

    Common Name: Ensign wasp
    Description: 10–25 mm in length. Metasoma is short and oval in shape and it is attached by the base of the hind coxae.

Family: Megachilidae

86. **Scientific name**: ***Megachile* sp**.

    Common Name: Leaf-cutter bee
    Description: Moderate in size and stout-bodied. Forewing has two submarginal cells of about equal length. Longer labrum is present.

Family: Sphecidae

87. **Scientific name**: ***Podium luctuosum* Smith, 1856**

    Common Name: Sphecid wasp
    Description: Posterior margin of the pronotum is always straight in dorsal view. Unbranched body hairs. Relatively bare body. 1 in. in length.

Family: Vespidae

88. **Scientific name**: ***Polistes dominulus*** **Christ, 1791**

    Common Name: Paper wasp
    Description: Elongate and slender. Metasoma is spindle in shape. Brown marked with yellow in color, 19–26 mm in length.

89. **Scientific name**: ***Vespa cincta*** **Degeer, 1773**

    Common Name: Yellow band wasp
    Description: Black and yellow in color. Antenna is clavate. Posterior margin of the pronotum is in U-shape.

Order: Isoptera
Family: Termitidae

90. **Scientific name**: ***Microtermes*** sp.

    Common Name: Termites
    Description: Chewing-type mouthparts. Front and hind wings almost similar in size. Saddle-shaped pronotum.

91. **Scientific name**: ***Odontotermes obesus*** **Rambur, 1842**

    Common Name: Termites
    Description: Bristles are clearly seen on the thoracic tergites.

Order: Lepidoptera
Family: **Ammatidae**

92. **Scientific name**: ***Amata*** **sp**.

    Common Name: Amatid moth
    Description: Black wings with white spots. Brick red in color with black lines on the abdomen. Black hairs found on the marginal side of the thorax. Rounded abdomen.

Family: Arctiidae

93. **Scientific name**: ***Creatonotos gangis*** **Linnaeus, 1763**

    Common Name: Tiger moth
    Description: Wingspread about 40 mm. Brownish in color. Brightly spotted. Nocturnal. Labial palps are short. 2 rounded basal abdominal segments.

Family: Cossidae

94. **Scientific name**: ***Endoxyla leucomochla* Turner, 1915**

    Common Name: Witjuti grub
    Description: Heavy-bodied with medium size. Much elongated front wing. Thinly scaled wings.

Family: **Eupractidae**

95. **Scientific name**: ***Nisaga simplex* Walker, 1855**

    Common Name:
    Description: Wingspan is 4 cm in length. Light brownish in color. Brownish color hairs on the thorax. Antenna is of Callosamia type.

Family: Hesperiidae

96. **Scientific name**: ***Caprona ransonnetti* Felder, 1868**

    Common Name: Golden angle
    Description: This butterfly is ochreous brown in color. In male, before the apex three small semitransparent white spots can be found. Gray brown outer discal area. Cilia are white in color. In female, forewing has marginal lunules and spots. Underside of forewing is paler brown with light back spots and outer black borders. Hind wing is more prominent with the macular band.

Family: Lycaenidae

97. **Scientific name**: ***Jamides celeno* Cramer, 1775**

    Common Name: Common cerulean
    Description: 27–40 mm in length. Underside grayish white or brown in color. Upper side of the male is pale blue in color with a thin black border. Female is duller than male with broad black borders.

98. **Scientific name**: ***Freyeria trochilus* Freyer, 1845**

    Common Name: Grass jewel
    Description: 15–22 mm in length. Underside grayish brown in color with 2 black spots along costa. Black spots along margin crowned with green and orange. Upper side dark brown with 3–4 orange crowned spots along the termen.

Family: Noctuidae

99. **Scientific name**: ***Sesamia inferens*** **Walker, 1556**

Common Name:
Description: Light in color and antennae are drasterian type. Wingspread about 20 mm in length. Forewing is dark brown in color. Hairs on thorax.

Family: Nymphalidae

100. **Scientific name**: ***Acraea terpsicore*** **Linnaeus, 1758**

Common Name: Tawny coster
Description: Upper side red in color. Wings bordered with black. Upper forewing has thin dark border. Hind wing is broad with white spots. Both wings are spotted with black.

101. **Scientific name**: ***Ariadne ariadne*** **Linnaeus, 1763**

Common Name: Angled castor
Description: Wingspan is 45–60 mm in length. Rounded wing edges. Orange brown in color. Cilia are white in color and are alternated by brown. Underside much brown in color. It much resembles the common castor. Ochraceous rufous-colored abdomen, thorax, head, and antennae.

102. **Scientific name**: ***Byblia ilithyia*** **Drury, 1773**

Common Name: Joker butterfly
Description: Narrow black markings in male and dark brown markings in female. Underside, it has white basal and central band. White spots in discal and marginal band.

103. **Scientific name**: ***Charaxes solon*** **Fabricius, 1793**

Common Name: Black rajah
Description: 70–80 mm in length. Upper side black, with yellow spots in discal area on both wings that are broken into separate spots at the apex of the forewing. Underneath, silvery gray with black wavy lines and submarginal and marginal yellowish green spots.

104. **Scientific name**: ***Danaus genutia*** **Cramer, 1779**

Common Name: Striped tiger
Description: 70–100 mm in length. Prominent blackened veins and with white apical spots on the black apex of the forewing.

105. **Scientific name**: ***Danaus chrysippus* Linnaeus, 1758**

Common Name: Plain tiger
Description: 70–80 mm in length. Chestnut orange with black borders. Apex black with white subapical white band of linked spots. Four black discal spots on the underside of the hind wing.

106. **Scientific name**: ***Ergolis merione* Cramer**

Common Name: Castor butterfly
Description: Wingspan is 45–60 mm in length. Rusty brown in color. Curly black lines can be found on the wings. Front margin of the forewings has white dot. Much dark in color underside of the wings.

107. **Scientific name**: ***Euploea core* Cramer, 1780**

Common Name: Common Indian crow
Description: 85–95 mm in length. Dark velvety butterfly with two rows of marginal cream-colored spots, with the inner row of spots being larger than the outer.

108. **Scientific name**: ***Hypolimnas misippus* Linnaeus, 1764**

Common Name: Danaid egg fly
Description: Wingspan is 70–85 mm in length. Upper wing of the male is purple-black in color with a white patch. Above of the female upper wing is orange in color. Near the apex, smaller spot is present. Larger white spot on the hind wing. Thorax, antennae, and head are dark brown in color. Lower side of the thorax and palpi is white in color. Chestnut-colored abdomen.

109. **Scientific name**: ***Junonia almana* Linnaeus, 1758**

Common Name: Peacock pansy
Description: 60–65 mm in length. Orange in color with prominent peacock eyespots. Two smaller eyespots on the upper forewing. The largest eyespot is on hind wing. Margins of the wings on the upper side bordered by 3 parallel black lines.

110. **Scientific name**: ***Junonia lemonias* Linnaeus, 1758**

Common Name: Lemon pansy
Description: 40–60 mm in length. Brown in color. With prominent peacock eyespot on each wing. Pale yellow spotting on the upper forewing. Margins of the wings on the upper side bordered by 3 parallel black lines.

111. **Scientific name**: ***Junonia orithya* Linnaeus, 1758**

Common Name: Blue pansy
Description: 45–60 mm in length. Upper hind wing brilliant blue in color and upper forewing velvet black, while apex pale with white bands. A pair of blue-centered red eyespots on all wings.

112. **Scientific name**: ***Tirumala limniace* Caramer, 1775**

Common Name: Blue tiger
Description: 90–100 mm in length black in color with pale blue markings. The cell-spot of the hind wing is paler.

Family: Papilionidae

113. **Scientific name**: ***Graphium Agamemnon* Linnaeus, 1758**

Common Name: Tailed jay
Description: 85–100 mm in length. Blackish in color with bright green spots and markings. Tail is short and stout.

114. **Scientific name**: ***Graphium doson* C.R. Felder, 1864**

Common Name: Common jay
Description: The body, thorax, and head are black in color. Bottom of the body white or yellowish in color. Both forewings and hind wings are black in color. Large greenish spots found in the middle of the forewing. Curved-edged hind wing.

115. **Scientific name**: ***Pachliopta aristolochiae* Fabricius, 1775**

Common Name: Common rose
Description: 80–110 mm in length. Red-bodied swallowtail with elongated forewings with pale gray stripes in the distal part of the cell and between veins. Five elongated spots were present on the hind wing. In that 5 spots, 4 are white, while the lower one that is near the body is pink. Seven red crescents present on the margin series.

116. **Scientific name**: ***Pachliopta hector* Linnaeus, 1758**

Common Name: Crimson rose
Description: 90–110 mm in length. Black swallowtail with bright crimson spots on the hind wings. Prominent white discal and apical bands on the forewing.

117. **Scientific name**: ***Papilio demoleus* Linnaeus, 1758**

Common Name: Lime or citrus butterfly
Description: 80–100 mm in length. Yellow butterfly with black spots. Upper forewing has red tornal spot and apical black and blue spots.

118. **Scientific name**: ***Papilio polytes* Linnaeus, 1758**

Common Name: Common Mormon
Description: 85–100 mm in length. Brownish black in color. Forewings are dark brown in color. The head, abdomen, thorax, and antennae are black in color. Female is larger than male.

Family: Pieridae

119. **Scientific name**: ***Catopsilia pomona* Fabricius, 1775**

 Common Name: Common emigrant
 Description: 55–80 mm in length. Yellow to translucent greenish white in color. Antennae are black or red in color. Underside of the wings may be unmarked or with red-ringed silver spots in the center.

120. **Scientific name**: ***Catopsilia pyranthe* Latrelle, 1758**

 Common Name: Mottled emigrant
 Description: 50–70 mm in length. White butterfly with black apical and marginal border on the upper forewing. Underside mottled with brown lines and red-ringed silver spots in the center of the wings.

121. **Scientific name**: ***Cepora nerissa* Fabricius, 1775**

 Common Name: Common gull
 Description: 40–65 mm in length. Upper wings white with blackened veins. Upper forewing has black terminal border with white spots. Underside of the hind wing pale or dark yellow, with veins heavily outlined in dark green.

122. **Scientific name**: ***Eurema hecabe* Linnaeus, 1758**

 Common Name: Common grass yellow
 Description: 35–50 mm in length. Male is yellow in color, with apex and termen on upper–forewing and terminal border of upper hind wing being broadly black. In female, black borders are wider. Underside of the forewing has two black spots in cell.

123. **Scientific name**: ***Pieris rapae* Linnaeus, 1768**

 Common Name: Cabbage butterfly
 Description: Body is black in color with white wings. Tip of the upper wing has black band and black spot in the center. Females have two spots, and males have one spot. Wingspan is 30–50 mm. Yellowish green underneath the wings.

Family: Satyridae

124. **Scientific name**: ***Melanitis leda* Linnaeus**

 Common Name: Common evening brown
 Description: 60–80 mm in length. Two prominent black eyespots surrounded by orange patches at the apex of the forewing, each with large eye pupil. Underside gray, with prominent dark striation and a series of white-pupilled eyespots, large, and small.

Family: Sesiidae

125. **Scientific name**: ***Melittia cucurbitae*** **Harris, 1828**

    Common Name: Pumpkin vine borer
    Description: Hind legs are heavily clubbed. With a long border of orange scales. Long and narrow forewings and hind wings are broad. Bright in color.

Order: Mantophasmatidae
Family: Mantidae

126. **Scientific name**: ***Mantis*** **sp.**

    Common Name: Praying mantid
    Description: 2–3 cm in length. Wingless. Chewing-type mouthparts. Antenna is long and filiform. Tarsi are five-segmented.

Order: Neuroptera
Family: Chrysopidae

127. **Scientific name**: ***Chrysoperla*** **sp.**

    Common Name: Green lacewing
    Description: Greenish in color. Eyes are copper in color. Wings with many veins. Prothorax is normal in size. Antenna is not clubbed.

Order: Odonata
Family: Aeshnidae

128. **Scientific name**: ***Anaciaeschna jaspidea*** **Burmeister, 1839**

    Common Name: Rusty darner
    Description: Above bluish gray and below pale yellow eyes. Thorax is reddish brown in color, and on each side, greenish yellow stripes can be observed. Transparent wings. Abdomen is reddish brown in color and 43–48 mm in length. Female is smaller than male and is also similar to male.

129. **Scientific name**: ***Anax guttatus*** **Burmeister, 1839**

    Common Name: Blue-tailed green darner
    Description: Eyes are blue with yellow and black behind. The first and second segment of abdomen is pale green, second segment is turquoise blue, the third segment green with a pair of dorsal triangular turquoise blue spots. 4–7th segment has 3 pairs of bright orange spots. In 8 and 9th segments, two pairs of orange spots, and 10th segment is entirely yellow. Size of the abdomen is 56–62 mm in length. Wings are transparent and 50–58 mm in length with long, narrow, and reddish wing spot.

Family: Gomphidae

130. **Scientific name**: ***Ictinogomphus rapax*** **Rambur, 1842**

Common Name: Common clubtail
Description: Bluish gray eyes. Abdomen is 52 mm in length with black color and yellow marks. Black legs. Transparent wings with black wing spot and 40–44 mm in length. Thorax is black-marked with yellow.

131. **Scientific name**: ***Paragomphus lineatus*** **Selys, 1850**

Common Name: Lined hooktail
Description: This dragonfly is yellow in color either black and brown markings on the body. Medium in size. Male has a peculiar clawlike hook. Eyes are blue gray in color. Three parallel brown stripes were found on the sides of the thorax.

Family: Libellulidae

132. **Scientific name**: ***Brachythemis contaminata*** **Fabricius, 1793**

Common Name: Ditch jewel
Description: Eyes are olivaceous brown above bluish gray below in color. Thorax is in olivaceous brown to reddish brown in color with two reddish brown lateral stripes. Male face is olivaceous in color, and female face is yellowish white in color. Male wings are transparent with reddish venation with rusty wing spot. Female wings are transparent with bright orange wing patches with rusty wing spot. Male abdomen is bright red in color and 18–21 mm in length, and female abdomen is olivaceous brown with a black middorsal stripe and 18–20 mm in length.

133. **Scientific name**: ***Crocothemis servilia*** **Drury, 1773**

Common Name: Ruddy marsh skimmer
Description: Red or reddish yellow colored dragonfly. At the wing base amber colored patch can be observed. Male face is red in color, with above side of eyes are red and sides are purple in color. Legs are reddish in color. Abdomen is 24–25 mm in length with red color. Female face is pale yellow in color. Eyes are brown in color. Thorax is dark brown in color. Transparent wings with pale yellow wing spot. Abdomen is 25–32 mm in length with yellowish brown in color.

134. **Scientific name**: ***Diplacodes trivialis*** **Rambur, 1842**

Common Name: Ground skimmer
Description: Light pale green in color. Wings are transparent. Eyes are light blue in color. Black wing spot. Black line on the abdomen. Habitat is wet rice fields and shallow lakes.

135. **Scientific name**: ***Neurothemis tullia* Drury, 1773**

Common Name: Pied paddy skimmer
Description: Male face is black in color with blackish brown eyes. Legs are black in color. Abdomen is 16–20 mm in length with black color and creamy white stripe on the upper side of the abdomen. Wing tips are transparent. Dull brown wing spot. Female face is olivaceous yellow in color. Eyes are pale brown in color. The outer surface of the legs is yellow in color and the inner surface black. Abdomen is 16–19 mm in length with bright yellow color with broad black band above the abdomen.

136. **Scientific name**: ***Orthetrum sabina* Drury, 1770**

Common Name: Green marsh hawk
Description: Male face is yellowish green in color. Abdomen is green in color and 30–36 mm in length with broad black rings on 1–3 segments. Transparent wings. It has black with reddish brown wing spot. Female is almost alike male. Abdomen is 32–35 mm in length.

137. **Scientific name**: ***Pantala flavescens* Fabricius, 1798**

Common Name: Wandering glider
Description: Abdomen is yellow in color and 29–35 mm in length. Thorax is rusty in color. In males, eyes are reddish brown above side and below the sides are bluish gray. In females, eyes are olivaceous in color. Transparent wings. The base of the hind wing is amber yellow in color. Bright reddish brown wing spot.

138. **Scientific name**: ***Rhyothemis variegata* Linnaeus, 1763**

Common Name: Common picture wing
Description: Medium in size. Yellow and brown wings. Abdomen is black in color with a length of 23–25 mm in males and bluish black abdomen with a length of 20–22 mm in females. Forewing is transparent and golden yellow in color with black wing spot in males. Tips of the forewing are transparent. A dark brown unclear region extends to the center of the forewing with black wing spot in females.

Family: Macromiidae

139. **Scientific name**: ***Epophthalmia vittata* Burmeister, 1839**

Common Name: Blue-eyed cruiser
Description: Black in color. Thoracic part with yellow bands. Head is blue in color. Six light reddish yellow spots on the abdomen. Wings are transparent. Both wings are similar in size with black wing spot.

140. **Scientific name**: ***Macromia cingulata* Rambur, 1842**

Common Name: Macromia cingulata
Description: It is large in size. Yellow markings are absent behind the eyes. Wings are transparent. Black wing spot on the edge of the wing.

Order: Orthoptera
Family: Acrididae

141. **Scientific name**: ***Acrida exaltata* Walker, 1859**

Common Name: Short-horned grasshoppers
Description: Body greenish in color. Lateral carinae with black line. Tegmina are exclusive of distinctly colored margin. Hind legs long and slender.

142. **Scientific name**: ***Acrida*** sp.

Common Name: Short-horned grasshoppers
Description: Large in size. It resembles grass. Head is conical-shaped with narrow and long hind femora. Elongated body.

143. **Scientific name**: ***Anacridium flavescens* Fabricius, 1793**

Common Name: Short-horned grasshoppers
Description: Head is yellowish in color with two blackish bands slanting backward from between the eyes. Front side, it is in black and depressed with a red mark in the middle. Frontal ridge is black in color. Head and pronotum are shorter than the antennae. Antennae are black in color. Strongly carinated black pronotum. Body is greenish brown in color. Long tegmina with yellowish gray in color. At the base of the wings pale greenish yellow in color. Hind femora are whitish in color. Legs are olive green in color.

144. **Scientific name**: ***Atractomorpha crenulata* Fabricius, 1793**

Common Name: Short-horned grasshoppers
Description: Body is long, compressed, and slender. Conical-shaped head. Fastigium almost as stretched as the eye. In front, it is very slanted. Frontal ridge compressed between the antennae. Triquetral and short antennae. Subfiliform antennae. In females, antennae are slightly depressed and widened at the base. Oblong eyes. Wings almost as long as the tegmina.

145. **Scientific name**: ***Brachycrotaphus longiceps***

Common Name: Short-horned grasshoppers
Description: More brown on the back side of the species with a pale band running after the eyes as far as the periphery of the outer lobe of the pronotum. Body is pale in color. Antennae rather longer than the head and pronotum together. Eye is shorter than the fastigium of the vertex.

146. **Scientific name**: ***Chrotogonus oxypterus* Blanchard, 1836**

Common Name: Short-horned grasshoppers
Description: Short, rugose, sub-apterous, and yellowish brown in color. Head is red in color. Antennae are yellow in color with small black slanting stripes. Large and prominent eyes. These eyes cover the sides of the head. Pronotum is short and broad humped, small black dots appear front side of it. Sternum is black with spots. Tegmina are brownish yellow in color with dark spots. Upper side of the abdomen is yellow in color with small and rather unclear slanting black lines.

147. **Scientific name**: ***Dioabolocatantops pinguis* Stal, 1860**

Common Name: Short-horned grasshoppers
Description: Reddish brown in color and somewhat stout. Filiform-type antennae. Frontal ridges lightly punctured. Pronotum is closely punctured. Behind, it is obtusely angulated. Tegmina extending beyond the abdomen, slightly narrowed at the tip, which is rounded, darker toward the base, and sub-hyaline toward the tip. Wings dull hyaline, or slightly greenish toward the base and slightly clouded toward the tip. Undersurface of the body and legs pale in color. Hind femora stout, with two transverse black spots above hind tibia and tarsi red in color.

148. **Scientific name**: ***Epistaurus sinetyi* Bolivar, 1902**

Common Name: Short-horned grasshoppers
Description: Pale reddish in color. Median carina of the vertex very unclear. Hind femora are brown in color, and lower outer area is brown in color. Tibia is clothed with long gray hairs, brown at the base and pale ringed. Abdomen is red in color with a row of spots on the back interrupted by the median line. Short and slightly compressed cerci. It is incurved toward the tip.

149. **Scientific name**: ***Euprepocnemis alacris* Serville, 1838**

Common Name: Short-horned grasshoppers
Description: Yellowish in color. A black streak below each eye. A broad velvetty black sub-parallel-sided stripe runs over the vertex and pronotum. The median carina narrowly, and the lateral borders rather broadly yellow. Deflexed lobes yellow, with an oblique blackish bar, and suffused pale markings above and below it. Wings hyaline, slightly greenish toward the base. Abdomen is shining greenish brown in color. Femora yellowish, with a longitudinal black streak on the outer surface. Tibia and tarsi are purplish brown in color.

150. **Scientific name**: ***Gastrimargus africanus* Saussure, 1888**

Common Name: Short-horned grasshoppers
Description: Green in color with red antennae. Behind each eye, a pale spot runs to the back of the head, bordered above and below with black color, which is generally continuous on the front of the pronotum. Abdomen is greenish in color. Long and narrow tegmina with green inner margin and rest of the tegmina brown in color. Tegmina are longer than the wings and wings beneath are bright sulpher yellow.

151. **Scientific name**: ***Oedaleus abruptus* Thunberg, 1815**

Common Name: Short-horned grasshoppers
Description: Green or testaceous, with brown and white markings. Scutellum of vertex is long, narrow, sub-triangular, and pointed. Upper side of the pronotum brown with white cross marks. On the sides of the head and pronotum, it has white and brown oblique stripes. Tegmina are brown in color with three pale bands. Base of the wings pale greenish yellow in color. Reddish tibia.

152. **Scientific name**: ***Oxya fuscovittata* Marschall, 1836**

Common Name: Short-horned grasshoppers
Description: Large head, which is broader than the pronotum. Obtuse, transverse fastigium. It is not contracted above the eyes. Slightly slanting face which is almost vertical. Flattened and smooth pronotum. Narrow tegmina and expanded toward the base of the costa as long as the abdomen. Broad wings rounded at the tips. Long and slender legs.

153. **Scientific name**: ***Phlaeoba panteli* Bolivar, 1902**

Common Name: Short-horned grasshoppers
Description: Ferruginous brown in color with spotted black dots. Narrowly ensiform antennae. It is shorter than the head and pronotum and brown in color at the tips. Broad fastigium which is rounded at the front. The median carina continued on the head and pronotum. Moderately broad and narrowed tegmina, which are longer than the abdomen. Narrow wings.

154. **Scientific name**: ***Trilophidia annulata* Thunberg, 1815**

Common Name: Short-horned grasshoppers
Description: Brown or gray in color with black margins. Slightly thickened antennae which are pale at the base. Rugose pronotum with a high median carina. Tegmina are gray in color. Marked tibia and femora. Wing is yellow at the base, and beyond, it is black in color.

155. **Scientific name**: ***Tristria pulvinata* Uvarov, 1921**

Common Name: Short-horned grasshoppers
Description: Lower surface is broad and bowl-shaped, with fully developed wings. Elytra short and it not reach to the abdomen. Strong prosternal tubercle.

156. **Scientific name**: ***Truxalis indica* Bolivar, 1902**

Common Name: Short-horned grasshoppers
Description: Hind wings are so dark brown in color. This species is easily identified by pronotum sellate.

157. **Scientific name**: ***Tylotropidius varicornis* Walker, 1870**

Common Name: Short-horned grasshoppers
Description: Pronotum is brown in color. Lateral carina is pale in color. Compressed, truncated prosternal tubercle. Castaneous tegmina, with row of triangular white spots on the pale longitudinal stripe in the coastal area. Very slender hind femora which are thickened at the base. Hind tibia is dull blue in color. Tarsi are dull blue in color. Tibia and tarsi very pilose. Straight, rounded, or slightly compressed cerci.

Family: Gryllidae

158. **Scientific name**: ***Gryllus domesticus* Linnaeus**

Common Name: House crickets
Description: 16–21 mm in length. Brown in color. In females, ovipositor is present and is absent in males.

159. **Scientific name**: ***Gryllus* sp.**

Common Name: Field cricket
Description: 15–26 mm in length. Dark brown to black in color. Ovipositor is brown in color. Males and females show difference in coloration, and females are light in color with reddish legs. Male species play song.

Family: Pyrgomorphidae

160. **Scientific name**: ***Chrotogonus trachypterus* Blanchard, 1836**

Common Name: Short-horned grasshoppers
Description: Brown, rugose, and tuberculate. Short and broad head. Fulvous antennae, ringed with black. Prominent, brown, and shiny eyes. Eyes occupy three-quarters of the length of the head. Short and broad pronotum. Sternum is yellowish in color with black spots. Brown tegmina are shorter than the abdomen. Hyaline wings, which are nearly as long as the tegmina. Upper side of the abdomen is brown in color, and beneath, it is

pale in color. Four forelegs are slender and short and yellow with brown spots. Short hind legs. Femora as long as abdomen. Tibia with short pale spines.

Family: Tettigoniidae

161. **Scientific name**: ***Conocephalus fasciatus* De Geer, 1773**

    Common Name: Slender meadow katydid
    Description: 17–18 mm in length. Abdomen is shorter than forewings. Hind wings go beyond forewings by 2–3 mm. Cerci are green in color.

162. **Scientific name**: ***Conocephalus indicus* Redtenbacher**

    Common Name: Meadow katydid
    Description: 17–18 mm in length. Greenish with slender body. Dorsal surface of the front tibia does not have spines.

163. **Scientific name**: ***Mecopoda elongata* Linnaeus, 1758**

    Common Name: Long-horned grasshoppers
    Description: Some species are brown in color and some green or pale green in color. It produces high sound.

164. **Scientific name**: ***Neoconocephalus ensiger* Harris, 1841**

    Common Name: Sword bearer
    Description: 45–64 mm long and conical-shaped long head. Ovipositor is sword-shaped.

165. **Scientific name**: ***Scudderia furcata* Brunner, 1878**

    Common Name: Fork-tailed bush katydid
    Description: Parallel-sided front wings. Green in color. Supra-anal plate is large and inflated.

b. **Spiders**

Order: Araneae
Family: Araneidae

1. **Genus *Araneus***: Opisthosoma and carapace are covered with hairs. Longitudinal fovea in the male and transverse fovea in the female. Unequal-sized median eyes. These spiders are orb web builders.

   Species recorded from the study region: (1) *Araneus* sp.1

2. **Genus *Argiope***: Female is larger than male. Posterior lateral eyes are larger than posterior median eye. Flat carapace with short white small and weak chelicerae.

Species recorded from the study region: (1) *Argiope anasuja*, (2) *Argiope minute, and* (3) *Argiope* sp.1

3. **Genus *Cyclosa***: Drably colored body. Carapace is pointed in obverse and opisthosoma stuffed extending posteriorly outside the level of the spinnerets in a pointed tubercle. Posterior median eyes are smaller than anterior median eyes and are close to each other. Small epigynum.

   Species recorded from the study region: (1) *Cyclosa hexatuberculata and* (2) *Cyclosa mulmeinensis*

4. **Genus *Cyrtophora***: Orb web builders. Dorsally flat carapace with long cephalic region. Sub-equal lateral eyes separated from each other. Moderately long and stout legs.

   Species recorded from the study region: (1) *Cyrtophora cicatrosa*

5. **Genus *Gibbaranea***: Females are larger than males. It is brownish in color with whitish shade. Quite broad prosoma with hair. Female has spherical and large abdomen. Middle eyes are large. Short and powerful chelicerae. Moderately long eggs.

   Species recorded from the study region: (1) *Gibbaranea bituberculata*

6. **Genus *Neoscona***: Epigynal scape is broad and spoon-shaped. Carapace is by longitudinal thoracic groove. Smallest posterior lateral eye. Largest anterior median eye. Usually vertical web.

   Species recorded from the study region: (1) *Neoscona excelsus*, (2) *Neoscona mukerjei*, (3) *Neoscona theisi, and* (4) *Neoscona* sp.1

Family: Corinnidae

7. **Genus *Castianeria***: Heavily sclerotized and shiny opisthosoma and the carapace. White scalelike setae with transverse bands are present on the dorsal side of the opisthosoma. Oval-shaped carapace. Short and wide labium. Thin and long legs.

   Species recorded from the study region: (1) *Castianeria* sp.1

Family: Eresidae

8. **Genus *Stegodyphus***: White setae are presented on carapace. Sub-triangular-shaped cephalic region. It is longer than the width. Leg 1 is thick and elongated. Many white setae presented on legs. Several white setae present on the abdomen.

   Species recorded from the study region: (1) *Stegodyphus sarasinorum*

Family: Filistatidae

9. **Genus *Pritha***: Elongated prosoma and opisthosoma. Bunch of hairs are present on the prosoma. Absence of thoracic groove. A tarsal claw does not have spines. Clumped eyes. Divided and triangular cribellum.

   Species recorded from the study region: (1) *Pritha* sp.1

Family: Hersiliidae

10. **Genus *Hersilia***: Tangentially angulated and flat carapace and head. Very high and prominent clypeus. Long posterior spinnerets.

    Species recorded from the study region: (1) *Hersilia savignyi*

Family: Lycosidae

11. **Genus *Arctosa***: Long carapace. Light wide longitudinal median stripe with dark brown color. Four large posterior eyes are arranged in quadrangle shape, facial area is verticle and sides of the face are slant. The labium is always longer than width, and the basal excavation is prominent, usually one-third or more of the length of the labium. Clypeus is not vertical. Abdomen overall grayish with a dark median strip, often breaking into chevrons, bars, or dots. Members of this genus spin no web. The larger forms live in a silk-lined burrow and under stones, with the entrance covered by a thin sheet of silk. The female carries their eggs in egg sac. Egg sac attached to her spinnerets.

    Species recorded from the study region: (1) *Arctosa* sp.1

12. **Genus *Hippasa***: Wide carapace. Long thin legs. These legs are covered with spines and hairs. Front side of the cephalic region is pronouncedly narrowed. Strong chelicera is strong. Slender pedipalp in male.

    Species recorded from the study region: (1) *Hippasa pisaurina*

13. **Genus *Lycosa***: Long carapace. Dark brown in color with light wide longitudinal median stripe. Vertical facial area and sides are slant with four posterior eyes. Longer labium than the width. Clypeus is not vertical. Opisthosoma is overall grayish in color with dark median stripe.

    Species recorded from the study region: (1) *Lycosa bistriata*, (2) *Lycosa madani*, (3) *Lycosa pictula*, (4) *Lycosa prolific, and* (5) *Lycosa* sp.

14. **Genus *Pardosa***: Medium and small in size. Vertical clypeus. Elevated cephalic region. Wider labium than the length. Labium is with basal articular notches. Pale or dark legs and moderately long and slender. Scopulae are

sparse. Opisthosoma is ovoid in shape with dark to pale in color. Spiders of these genera are found near water bodies.

Species recorded from the study region: (1) *Pardosa birmanica*, (2) *Pardosa altitudes*, (3) *Pardosa mysorensis*, (4) *Pardosa oakleti*, (5) *Pardosa pseudoannulata*, (6) *Pardosa pusiola*, (7) *Pardosa sangosa*, (8) *Pardosa shyamae*, (9) *Pardosa sumatrana, and* (10) *Pardosa* sp.1

Family: Miturgidae

15. **Genus *Cheiracanthium***: Medium-to-large-sized and slender spider. First pair of legs are long and thinner legs. Robust body. Front end of the abdomen has clump of hairs. Pale in color. 5–10 mm in length. Abdomen is yellow in color.

    Species recorded from the study region: (1) *Cheiracanthium saraswati*, (2) *Cheiracanthium danieli, and* (3) *Cheiracanthium* sp.

Family: Oxyopidae

16. **Genus *Oxyopes***: Body enclosed with short scalelike recumbent shining setae. Anteriorly carapace is vertical, and margins were found laterally. Cephalic area is slightly prominent. Long legs with spinous and longitudinal gray bands present on femora. Elongated opisthosoma.

    Species recorded from the study region: (1) *Oxyopes ratnae and* (2) *Oxyopes* sp.1

17. **Genus *Peucetia***: These spiders are called as green lynx spiders. The carapace of cephalic region is narrow. Bright green in color. Procurved posterior eye row. Vertical face. Elongated opisthosoma. **Species recorded from the study region**: (1) *Peucetia viridana and* (2) *Peucetia* sp.1

Family: Philodromidae

18. **Genus *Philodromus***: These species are extremely alert. Flattened carapace. Smoothly curved on the lateral side. Longer than the width. Small eyes are present. All eyes are uniform in size. Legs are long and slender. Opisthosoma is oval in shape, and it is laterally angulated. Flat but dorsally it has heart-shaped markings.

    Species recorded from the study region: (1) *Philodromus shillongensis and* (2) *Philodromus* sp.1

19. **Genus *Tibellus***: Plump-bodied in shape. Legs are short to medium in length. 4–7 mm in length. Eight dark eyes are present in two horizontal rows. Metatarsus IV of females does not have calamistrum. Two tarsal claws. The abdomen noticeably attractive dorsally. Female abdomen does not have cribellum.

    Species recorded from the study region: (1) *Tibellus* sp.1

Family: Pholcidae

20. **Genus *Crossoprise***: Circular carapace with slightly raised and forwardly projected eye region. Short opisthosoma is ovate in shape with dark patches and yellow spots. Striped legs with spots. Two pairs of distinctive apophysis in male chelicerae.

    Species recorded from the study region: (1) *Crossopriza lyoni*

Family: Pisauridae

21. **Genus *Tinus***: Carapace is longer than width. Relatively long and prograde legs. Opisthosoma is oval-to-elongate-shaped with longitudinal bands or spots. Six spinnerets.

    Species recorded from the study region: (1) *Tinus* sp.1

Family: Salticidae

22. **Genus *Marpissa***: Gray to brown in color. Flat in shape with furry appearance. 6–11 mm in length (males 6–8 mm and females 8–11 mm). They have an elongated and gray-colored abdomen. Abdomen has two complete longitudinal black stripes. Carapace is oval in shape. Elongate oval-shaped abdomen and pointed at the back. Carapace is yellowish brown in color. Pale yellowish-colored legs. Forelegs are darker.

    Species recorded from the study region: (1) *Marpissa dhakuriensis and* (2) *Marpissa* sp.1

23. **Genus *Phidippus***: Eight eyes. Anterior median eyes are large. 6–15 mm in length. Hairy. Abdomen and cephalothorax are black in color with white hairs. Distinctly marked abdomen. Triangular white spot in the middle of the abdomen. Males are smaller than females.

    Species recorded from the study region: (1) *Phidippus* sp.1

24. **Genus *Plexippus***: Robust spiders. The average length of this species is 10 mm. Carapace covered with short grayish hairs. Female is slightly larger than male. Male has black abdomen and carapace. Center of the abdomen

has broad central stripe and another broad white stripe on each side of the abdomens. A pair of white spots near the posterior end of the abdomen. Female is brownish gray in color.

Species recorded from the study region: (1) *Plexyppus payakullii and* (2) *Plexyppus* sp.1

25. **Genus *Rhene***: Pale yellowish color abdomen. Cephalothorax is metallic brown in color and it is wider than the length. Dark brown-colored legs. Brown eyes, with anterior row of eyes recurved. Chelicerae are dark brown in color. Oval-shaped sternum. Posterior eyes located far from the anterior group of eyes.

    Species recorded from the study region: (1) *Rhene* sp.1

Family: Scytodidae

26. **Genus *Scytodes***: Carapace is black in color. The second leg is longer than the fourth. The average body length is 8 mm. **Species recorded from the study region**: (1) *Scytodes* sp.1

Family: Selenopidae

27. **Genus *Selenops***: It is called as wall crab spider. Gray in color. Female is larger than male.

    Species recorded from the study region: (1) *Selenops* sp.

Family: Sparassidae

28. **Genus *Olios***: Flattened body. Gray in color. All legs bend forward. Forelegs are much longer than the hind. It is called as huntsman spider.
    Family: Tetragnathidae

    Species recorded from the study region: (1) *Olios* sp.

29. **Genus *Larinia***: Legs are completely bright in color. Opisthosoma has bright longitudinal bands. Dorsally opisthosoma consists of 3 pairs of dark dots. Male pedipalp bent toward cymbium. Epigyne is short and it is not reaching away from side parts posterior.

    Species recorded from the study region: *Larinia* sp.1

30. **Genus *Leucauge***: Broadly elliptical silvery opisthosoma and the transverse depression anterior to the dorsal groove. Opisthosoma is twice longer than the width and blunt at the both ends with bands. Soft epigastric plate.

    Species recorded from the study region: *Leucauge celebesiana*

31. **Genus *Tetragnatha***: Prominently long and narrow body. Longer than the width. Carapace is oval in shape and flattened above with a conspicuous thoracic groove. Longer prosoma than the width. Long chelicerae in the males. Long and narrow opisthosoma and spinnerets near the end. Each eye is surrounded by a black ring. Well-developed chelicerae.

    Species recorded from the study region: (1) *Tetragnatha fletcheri*, (2) *Tetragnatha javana*, (3) *Tetragnatha mandibulata*, (4) *Tetragnatha maxillosa*, (5) *Tetragnatha moulmeinensis*, (6) *Tetragnatha okumae*, (7) *Tetragnatha sutherlendi, and* (8) *Tetragnatha* sp.1

32. **Genus *Zygiella*** : Opisthosoma is elliptical in shape with flattened paracymbium. Thoracic region is lower than the carapace with very few hairs. Large anterior median eyes. Opisthosoma with paired dark patches without humps.

    Species recorded from the study region: (1) *Zygiella indica*, (2) *Zygiella malanocrania*, and (3) *Zygiella* sp.1

Family: Theridiidae

33. **Genus *Argyrodes***: Flat carapace. Chelicera bears one or two retromarginal teeth and two or three promargin teeth. Small colulus. Opisthosoma is subtriangular to very long. In both sexes, anterior border with stridulating edges. Pedipalp in male with poorly sclerotized median apophysis. Female epigynum covered with resinous material.

    Species recorded from the study region: (1) *Argyrodes cyrtophore*, (2) *Argyrodes* sp.1, and (3) *Argyrodes* sp.2

34. **Genus *Theridula***: Colulus large to small or reduced to pair of setae only. Carapace without transverse groove in the thoracic area. Opisthosoma with tubercle on each side.

    Species recorded from the study region: (1) *Theridula* sp.1 and (2) *Theridula* sp.

35. **Genus *Theridon***: Longer carapace than the width. Indistinct fovea. Eyes are sub-equal in size. Enlarged chelicerae in males. In female, chelicerae with one or two teeth. Long legs. Carapace is shorter than patella-I and tibia-I. Leg IV is shorter than leg II in males. Opisthosoma is spherical in shape.

    Species recorded from the study region: (1) *Theridion manjithar*, (2) *Theridon* sp.1, and (3) *Theridion* sp.2

Family: Thomisidae

36. **Genus *Mesumenoides***: Small-to-medium-sized smooth carapace. Eyes are present in two transverse recurved rows. Legs are long and powerful. Legs III and IV are shorter and more robust than legs I and II. Several

prolateral spines and ventral spines are present, and dorsal spines are absent. Broad opisthosoma which is moderately flat with yellowish white longitudinal bands. Embolus is short and twisted in males.

Species recorded from the study region: (1) *Mesumenoides* sp.1

37. **Genus *Thanatus***: Dark brown abdomen. It is longer than the width. Yellowish brown color legs and cephalothorax. The average length of this species is 8 mm. Cephalothorax is longer than the width and broad in middle with covered hairs. A little deep groove is there in the center of cephalothorax. Eight black eyes are present in two rows. Yellowish brown chelicerae with dark brown fangs. Heart-shaped sternum which is covered with hairs. Long and stout legs. Legs with spines and brownish dots. Two pale brown dots are present at the posterior end of the abdomen.

    Species recorded from the study region: (1) *Thanatus stripatus*

38. **Genus *Thomisus***: Front side of the carapace is truncated, and upper corners are strongly diverged. Without setae. Eyes are poorly developed and small and sub-equal in size. Chelicerae are without teeth. Opisthosoma is wider than the length. Female has simple and less developed epigynum without hood. Male is darker than the female.

    Species recorded from the study region: (1) *Thomisus andamanensis and* (2) *Thomisus pugilis*

39. **Genus *Xysticus***: The average length of this species is 3–100 mm. Carapace is gray with brown in color with black stripes. Abdomen is brown in color with various matching markings. Males are smaller than females and dark brown in color. Males have different color patterns. Sometimes prosoma is smaller than the opisthosoma.

    Species recorded from the study region: (1) *Xysticus* sp.1

Family: Uloboridae

40. **Genus *Uloborus***: Broader carapace in dorsal view. Slightly longer than the width. Sides are moderately curved. Eight small eyes are present. Longer spear-shaped opisthosoma than the width. Opisthosoma is rounded at front and pointed at the rear.

    Species recorded from the study region: (1) *Uloborus krishnae and* (2) *Uloborus danolius*

c. **Zooplankton**

Family: Asplanchnidae

1. **Scientific name**: ***Asplanchna priodonta*** **(Gosse, 1850)**

   Description: 250–1500 micrometers in length. Cylindrical shaped. Jaw is hard, curved and sharply pointed. Pedal glands are absent. Four flame cells on protonephridium.

Family: Brachionidae

2. **Scientific name**: ***Brachionus bidentata*** **(Anderson, 1889)**

   Description: Well-developed anterointermediate spines. 170–579 μm in length. 115–265 μm in width. Anterolateral dorsal spine (16–97 μm) is smaller than the posterolateral spine (10–34 μm).

3. **Scientific name**: ***Brachionus calyciflorus*** **(Pallas, 1766)**

   Description: Anteromedian spines have extensive bottom, and posterior spines are frequently present. Transparent and smooth lorica. 180–570 μm in length. 124–300 μm in width.

4. **Scientific name**: ***Brachionus falcatus*** **(Zacharias, 1898)**

   Description: Occipital border with six spines. Ventrally curved and longer intermediates. Medians and laterals are short and equal in length. Long posterior spines. Six spines on the anterior dorsal margin. 185–405 μm in length.

5. **Scientific name**: ***Brachionus urceolaris*** **(Muller, 1773)**

   Description: Gently curved juncture between anteromedian and anterointermediate spines. Posterior spines are absent.

6. **Scientific name**: ***Keratella tropica*** **(Apstein, 1907)**

   Description: Dorsoventrally compressed lorica with a length of 150–240 μm. Lateral and generally slightly longer than intermediate. Lorica with two unequal posterior spines. Right spine is longer than the left.

7. **Scientific name**: ***Platyias quadricornis*** **(Ehrenberg, 1832)**

   Description: Moderately rigid and segmented foot. Two anterior and posterior median spines.

Family: Chydoridae

8. **Scientific name**: ***Chydorus sphaericus*** **(O.F. Muller, 1785)**

   Description: Compound eye. Body is spherical in shape. Prominent preanal angle.

Family: Cyclopidae

9. **Scientific name**: ***Mesocyclops leuckarti (*****Claus, 1857)**

   Description: Not descried—specimen lost.

10. **Scientific name**: ***Tropocyclops prasinus*** **(Fischer, 1860)**

    Description: Small and 12-segmented antennae. Elongated and narrow urpsome. The average length of this species is 0.5–0.9 mm.

Family: Cyprididae

11. **Scientific name**: ***Hemicypris fossiculata***

Family: Daphniidae

Description: Belong to ostracod, which is one of the important groups of zooplankton and known as seed shrimp and occupies the important position in the food web.

12. **Scientific name**: ***Ceriodaphnia cornuta*** **(Sars, 1885)**

Family: Diaptomidae

Description: Depressed and small head. Cervical sinus is present. Post-abdominal scrape with mid pectin. Rostrum is absent.

13. **Scientific name**: ***Heliodiaptomus viduus*** **(Gurney, 1916)**

    Description: Common Indian diaptomid genera and typically found in standing water bodies seasonal and permanent in nature.

Family: Filinidae

14. **Scientific name**: ***Filinia longiseta***

    Description: 315–515 μm in length. Spinelcts are found on the bristles in the summer. Habitat of this species is lake.

Family: Habrotrochidae

15. **Scientific name**: ***Habrotrocha bidens* (Gosse, 1851)**

    Description: Spindle-shaped body. Mouth furnished with two different teeth. It is two-teethed.

Family: Lecanidae

16. **Scientific name**: ***Lecane luna* (Muller, 1776)**

    Description: A narrower lorica and straight anterior–dorsal margin. The average length of this species is less than 150 μm.

Family: Lecanidae

17. **Scientific name**: ***Monostyla bulla* (Gosse, 1851)**

    Description: A narrower lorica and straight anterior–dorsal margin. The average length of this species is less than 150 μm.

Family: Moinidae

18. **Scientific name**: ***Moina brachiata* (Jurine, 1820)**

    Description: Large head and approximately one-fourth of the body. Large eye situated in front of the long antenna.

Family: Mytilinidae

19. **Scientific name**: ***Mytilina ventralis* (Gosses, 1886)**

    Description: Body is compressed laterally with dorsal sulcus. The front margins of the side lorica plates are curved. Small thorns like structures presented on the borders of each frontal lorica.

20. **Scientific name**: ***Mytilina acanthophora* (Hauer, 1938)**

    Description: This species is compressed laterally and dorsally arched. Long toes. Anterior ventral curves with triangular cusps. Granulated lorica.

Family: Notommatidae

21. **Scientific name**: ***Cephalodella gibba* (Ehrenberg, 1830)**

    Description: Short and stout body. Laterally compressed abdomen. Short foot. Short, thin, and straight toes. 80–160 μm in length.

Family: Philodinidae

22. **Scientific name**: ***Rotifer tardus* (Ehrenberg, 1830)**

    Description: Spurs are thick. Longer toes. The last joint of the foot comes down to the toes and it acts as a sucker. During early stage it has only two teeth each side and in the latter stage the three teeth on one side are found, if not on both.

Vertebrates

a. **Fishes**

Family: Channidae

1. ***Channa striata* Bloch, 1793**

   Common Name: Snakehead murrel
   Local Name: Haal, Shawl, Shol (Assam); Shol (West Bengal); Morrul, Morl, Soura (Bihar); Sowl, Dhoali, Carrodh (Punjab); Dolla (Jammu); Sola (Orissa); Korramennu, Korra-matta (Andhra Pradesh); Sowrah, Veralu, Kaunan (Kerala); Pooli-kuchi, Koochinamarl (Karnataka); Sohr, Dekhu (Maharashtra).
   Family: Channidae
   Habitat: Inhabits in ponds, streams, and rivers. With stagnant and muddy water.
   Characters: Adults grow up to 75 cm; the average length of this fish is 41 cm. Coloration be likely greenish with spots and stripes, mainly on the fins. Elongated head, white belly, and dark color.
   Geographic range: Found commonly in Pakistan, India, Nepal, Bangladesh, and Myanmar till China and Sri Lanka.

Family: Cichlidae

2. ***Tilapia mossambica* Peters, 1852**

   Common Name: Tilapia
   Local Name: Jilabi meen (South India), Chilabi, Tilapi (Maharashtra), Tilapia.
   Habitat: Inhabits in ponds, streams, and rivers. With stagnant and muddy water.
   Food: Small fish, frogs, insects, earthworms, tadpoles.
   Description: Blackish blue in color males, generally grayish in color with large black blotches along the sides. Fins are bordered with red. It has prolonged and pointed dorsal and anal fins. The tail is round in shape. Fish grows up to a maximum length of 25 cm.
   Geographic range: Native to East Africa. Widely introduced in India, Pakistan, and Sri Lanka.

Family: Cyprinidae

3. ***Puntius denisonii* FDay, 1865**

   Common Name: Denison barb
   Habitat: Inhabits in ponds, streams, and rivers. With stagnant and muddy water.
   Food: Small fish, frogs, insects, earthworms, tadpoles.
   Description: Torpedo-shaped body with silver scales, through the eye a red line runs from the snout. A black line runs below the red line till the tail. As they mature, in matured fish, distinctive green/blue marking develops on the top of the head. This fish grows up to a length of 15 cm.
   Geographic range: India (Karnataka and Kerala).

4. ***Catla catla* F. Hamilton, 1822**

   Common Name: Catla
   Local Name: Thoppa meen, Japan kendai, Koora kendai (Tamil Nadu), Karakatla (Kerala), Katala (Maharashtra), Thambra (Gujarat), Catla (West Bengal and Bihar), Bacha, Botchee (Andhra Pradesh).
   Habitat: Inhabits in ponds, streams, and rivers. With stagnant and muddy water.
   Food: Small fish, frogs, insects, earthworms, tadpoles.
   Description: This is a large silvery fish with large head. Lips are thick and scales are large. Dark gray in color in the upper side of the body and turning silver in color along the sides. Pectoral fins are long and lengthen up to the base of the pelvic fins. The maximum length is 150–200 cm and weight 70 kg.
   Geographic range: India, Pakistan, Bangladesh and Nepal and Myanmar. It has introduced recently in Sri Lanka and China.

5. ***Labeo rohita* F. Hamilton, 1822**

   Common Name: Rohu
   Local Name: Bocha-gandumeenu (Andhra Pradesh), Kennadi kendai (Tamil Nadu), Tambada-massa (Maharashtra), Rohitham (Kerala), Rahu (West Bengal), Rohu (Bihar), Rohi (Orissa).
   Habitat: Inhabits in ponds, streams, and rivers with low elevation (<600 m ASL). With stagnant and muddy water.
   Food: Small fish, frogs, insects, earthworms, tadpoles.
   Description: It is a large carp/barb, with bulging belly with bluish/greenish dorsal coloration. Its sides and ventral surface are pale/silver, while the fins are brown/red. Thick and fringed lips. Back is arched and eyes larger and reddish in color. It grows up to a maximum length of 100 cm and weight more than 15 kg.
   Geographic range: India, Pakistan, Bangladesh, Nepal, and Myanmar. Introduced in Andaman Islands and Sri Lanka.

6. ***Cirrhinus cirrhosus* Bloch, 1795**

Common Name: Mrigal
Local Name: Aruzu (Andhra Pradesh), Vencandi (Tamil Nadu), Mrigal (Maharashtra), Venkata (Kerala).
Habitat: Inhabits in ponds, streams, and rivers. With stagnant and muddy water.
Food: Small fish, frogs, insects, earthworms, tadpoles.
Description: Barbels are well developed. In the dorsal side, this fish shows dark brown color, and on the lower surface, it shows traces of yellow. Fins are darker than the body. Elongated body. The maximum length of this fish is 60 cm. The average length is around 26 cm.
Geographic range: India. Endemic to Peninsular India.

7. ***Cyprinus carpio* Linnaeus, 1758**

Common Name: Common carp
Local Name: Siprinus
Habitat: Inhabits in ponds, streams, and rivers. With stagnant and muddy water.
Food: Small fish, frogs, insects, earthworms, tadpoles.
Description: Stout-bodied carp. It is olivaceous with yellowish fins, and during the breeding season, it turns into bright red. Head is moderate in size. Triangular and rounded at the snout. Pectoral fins are large and rounded.
Geographic range: Original natural distribution is narrow belt in Europe, along the Danube river. Introduced in Sri Lanka and India.

8. ***Puntius sarana* F. Hamilton, 1822**

Common Name: Peninsular olive barb
Local Name: Gidpakke (Karnataka), Kannaku, Goodha paraga (Andhra Pradesh), Pungella (Tamil Nadu), Kuruka, Pullan, Mundutte, Kuruva (Kerala).
Habitat: Rivers and backwaters.
Food: Small fish, frogs, insects, earthworms, tadpoles.
Description: Medium-sized silvery fish. Body is deep, and fins are orange in color. The lower half of the body is to a large extent pale in color. Bases of the scales are black. The tail fin is black-tipped. The maximum length of this fish is 25 cm. Barbels are two pairs, in that one pair is long and noticeable.
Geographic range: Occurs only in Peninsular India. Mainly found in Krishna and Cauvery river systems of Karnataka, Andhra Pradesh, and Tamil Nadu. And common in backwaters of Kerala.

9. ***Rasbora daniconius*** **F. Hamilton, 1822**

Common Name: Rasbora
Local Name: Needean-jabbu, Kolkane, Pattai kunju, Ogari, Vaina paruva, Kolainjan kande (Karnataka), Narangi (Andhra Pradesh), Patta kanju (Tamil Nadu), Darikhana (Assam), Kokanutchee (Kerala), Dandwan (Maharashtra) Jilo (Orissa).
Habitat: Inhabits in ponds, streams, and rivers. With stagnant and muddy water.
Food: Small fish, frogs, insects, earthworms, tadpoles.
Description: This is small and very active fish. Slender and elongated body. Body is in golden color, and a lateral line is gold-bordered black in color. Fins are short and reddish orange in color. It grows up to a length of 10 cm.
Geographic range: India, Pakistan, Nepal, Bangladesh, Sri Lanka, Thailand, and Myanmar.

Family: **Siluridae**

10. ***Wallago attu*** **Bloch and Schneider, 1801**

Common Name: Wallago/boal
Local Name: Bahle, Chate, Godlay (Karnataka), Walagh, Vazhai (Tamil Nadu), Attu vaalay (Kerala), Balu shivada (Maharashtra), Valuga (Andhra Pradesh), Bayari, Paran (Bihar, Madhya Pradesh), Ballai, Boabe (Orissa).
Habitat: Inhabits in ponds, streams, reservoirs, tanks, channels, and rivers. With stagnant and muddy water.
Food: Small fish, frogs, insects, earthworms, tadpoles.
Description: Large in size. Elongated and laterally flattened fish. Small eyes and wide mouth and deeply forked tail. Gold and silvery white in color, with a yellow lateral band. Fins are yellow in color. It grows to a maximum length of 200 cm and weight 45–55 kg. It has two pairs of barbels.
Geographic range: India.

11. ***Ompok bimaculatus*** **Bloch, 1794**

Common Name: Indian butter catfish
Local Name: Godla, Kembari (Karnataka), Silaivalai, Chaithavelai (Tamil Nadu), Dukaduma, Theenuva (Andhra Pradesh), Goongware, Moone (Maharashtra), Pabho (Assam), Puffta (West Bengal).
Habitat: Inhabits in ponds, streams, and rivers. With stagnant and muddy water.
Food: Small fish, frogs, insects, earthworms, tadpoles.

Description: Medium-sized, laterally flattened with a wide mouth, and tail forked. Dorsal fin is a small knifelike structure, with an anal fin extending over more than three-fourths of the length of the body. Overall coloration is silver brown, back being greenish in color. Black spot is present on the shoulder.
Geographic range: India, Pakistan, Afghanistan, Nepal, Sri Lanka, Bangladesh, and Myanmar.

b. **Amphibians**

Family: Bufonidae

12. ***Duttaphrynus melanostictus*** **Schneider, 1799**

Order: Anura
Common Name: Common Indian toad
Local Name: Kappe.
Habitat: Human-dominated agriculture ecosystems and urban areas.
Food: They have been noted to feed on a wide range of invertebrates including scorpions.
Description: The top of the head has several bony ridges. The snout is short and blunt. The space between the eyes is broader than the width of the upper eyelid. The tympanum is very distinct. The dorsal side is yellowish in color and covered with spiny warts. The spines and ridges are black. The first finger is longer than the second, and toes are webbed.

Family: Ranidae

13. ***Hylarana erythraea*** **Schlegel, 1837**

Order: Anura
Common Name: Green paddy frog
Local Name: Kappe
Habitat: Agriculture fields, ditches along the road, irrigation ponds.
Description: Dorsal side light to dark in color and ventral side whitish in color. It has cream-colored dorsolateral folds bordered with black. Yellow limbs with irregular spotting. Smooth skin and long free fingers that expand into tiny disks with grooves. Hind limbs are long. Males are smaller than females.
Distribution: This species is widespread in Asia from Myanmar, east through mainland Southeast Asia, and then south to Peninsular Southeast Asia including Singapore and Indonesia and Borneo. It occurs at elevations up to 1200 m asl.

14. ***Euphlyctis cyanophlyctis* Schneider, 1799**

Order: Anura
Common Name: Skittering frog
Local Name: Kappe
Habitat: Found in marshes, pools, and various other wetlands within a variety of habitat types.
Description: Narrow interorbital space. Distinct tympanum, about two-third the size of the eye. Slender finger which is pointed or slightly swollen at the tips. Completely webbed toes. Long inner metatarsal tubercle. Male has vocal slits under the lower jaw. Smooth ventram. Light gray, olive green, or light brown dorsum. Sometimes dorsum is black in color with irregular black spots. Posterior dark thighs with one or two yellow or white irregular longitudinal stripes. Light brown vocal sacs. Frequently seen at the edges of water and eyes put above the water. And hardly ever seen outside of the water.
Distribution: This species ranges throughout much of South Asia including southern Afghanistan and Sri Lanka. It is present from the sea level to at least 2,500 m asl. This species is native to Afghanistan, Bangladesh, India, Iran, Nepal, Pakistan, and Sri Lanka.

15. ***Hoplobatrachus tigerinus* Daudin, 1803**

Common Name: Indian bull frog
Local Name:
Family: Ranidae
Habitat: The frog does not stay in water for a long time; it spends most of its time hiding and feeding in the surrounding vegetation.
Description: Slightly longer head. Pointed snout. Fingers obtusely pointed. Sides of the throat vocal sacs are present which are blue in color. Body is dorsum olive green in color or gray with dark blotches. Vertebral streak is light yellow in color. Dark bar limbs. White ventram.
Distribution: Native to India, Afghanistan, Bangladesh, Myanmar, Nepal, Pakistan, and Sri Lanka. Introduced to Madagascar and Maldives.

Family: Rhacophoridae

16. ***Polypedates maculatus* J.E. Gray, 1834**

Order: Anura
Common Name: Common Indian tree frog
Local Name: Kappe
Habitat: A typical frog of moist deciduous forest. This frog has become semiurban, especially in the cities with extensive gardens.
Description: Snout–vent length of mature males is about 34–57 mm and females 44–89 mm. Dorsum smooth, chin and chest smoothly granular,

venter and underside of thigh granular. Males with single internal vocal sac and nuptial pad at the base of the first finger. Dorsum olivaceous to chestnut, sometimes brownish yellow or gray with scattered dark spots; loreal and temporal regions dark brown or black; limbs with dark crossbars. Hinder side of thighs with round yellow spots which are usually separated by a dark brown network.
Distribution: India, Nepal, Bangladesh, Sri Lanka

c. **Reptiles**

Family: **Agamidae**

1. ***Sitana ponticeriana*** **(Cuvier, 1829)**

   Common Name: Fan throat lizard
   Local Name: Tonda (Telugu)
   Habitat: Found in open ground patches in thin forests.
   Description: Small upper head scales. Ventral scales are smaller than the dorsal scales, with sharp keels which form instantly longitudinal lines. Foot is longer than the lower thigh. Brown color, dark spots along the middle of the back. Dark spot is present on the neck. Gular appendages are blue, black, and red in color. Round tail. The maximum length is 8 in. In that, tail takes 5 in.
   Distribution: India and in some parts of Pakistan and Sri Lanka.

2. ***Calotes versicolor*** **(Daudin, 1812)**

   Common Name: Indian garden lizard
   Local Name: Tonda (Telugu).
   Habitat: Found in parks and gardens, lower shrubs, and bushes where it feeds on insects and other prey.
   Description: Brownish buff to grayish color. The throat of the male becomes red and black in color during breeding season. Short crest above the neck. Small spines are present above the tympanum. Swollen cheeks are present in male. Total size of this species is 37 cm (snout to vent: 10 cm).
   Distribution: Iran to India, and from southern China to Thailand.

3. ***Psammophilus dorsalis*** **(Griffith and Pidgeon, 1837)**

   Common Name: Peninsular rock agama
   Local Name:
   Habitat: These lizards lie on the bare land.
   Description: Elongated and depressed long head (with swollen cheeks in male). Eye cavity is longer than the snout. Slightly flattened body. Adult has small, uniform, smooth, dorsal scales. Dorsal scales are smaller than ventral scales. Ventral (underside) scales are slightly bigger than the gular (under chin). Strong legs which are uniformly covered by keeled scales. The tail is

to some extent compressed and enclosed with keeled scales, which are larger below than above. Young and females are olive brown, spotted, speckled, or marbled with dark brown with elongated white spots along each side of the back. In male, upper surface of the head and body has pale brownish color. Yellowish brown lips, yellow strip extends beyond the ear. Behind the eye, a dark brown or black lateral stripe begins and expands to cover the lower sides. The lower side is yellow, and throat mottled with gray. From snout to vent, its length is 135 mm; the tail is 200 mm. Males are bigger than females.
Distribution in world and India: Peninsular India.

Family: Colubridae

4. ***Ptyas mucosus* (Linnaeus, 1758)**

Common Name: Rat snake
Local Name: Dhaman (Marathi, Hindi, Gujarati, Oriya), Dhanaraas (Bengali), Sara Pamboo (Tamil), Kere haavu (Kannada), Jerri pothu (Telugu).
Habitat: Generally, it will be found nearby rat burrows. All types of habitats such as urban, rural, forest, arid, coastal, and marshy.
Food: Rats and also feeds on toads, frogs, lizards, geckos, small birds, other snakes, squirrels, and bats.
Description: The average length is 200 cm. The maximum length will be around 345–355 cm. The body color can be yellowish brown, moss green, creamy, or black. Tail is very long and tapering. Underside is yellowish. Vertical black lines are present on the lip scales. Big round eyes with yellow borders around the pupils. Mating season is March to May.
Distribution: All over India except Kashmir.

5. ***Ahaetulla nasuta* (Lacepede, 1789)**

Common Name: Green snake
Local Name: Hasiru Haavu (Kannada), Lata saanp (Hindi), Harantol (Marathi), Malan leelavan (Gujarati), Lawdoga (Bengali), Pachay pamboo (Tamil), Pasara paamu (Telugu).
Habitat: Bushes and trees in forests, coastal areas, paddy fields.
Food: Small birds, eggs, hatchlings. It also eats lizards and frogs.
Description: Green, long thin body resembling a vine. Unique elongated, pointed head. Capable of binocular vision. Large golden yellow eyes with a horizontal pupil. Body color varies from bright green to dark green with a two white lines.
Distribution: All over India except in the northwest and much of the Gangetic basin.

Family: Elapidae

6. ***Naja naja* (Linnaeus, 1758)**

Common Name: Spectacled cobra/Indian cobra
Local Name: Nag (Hindi, Marathi, Gujarati), Gokho (Bengali), Gokhura (Oriya), Naagar havo (Kannada), Naagu paamu (Telugu).
Habitat: Forests, grasslands, paddy fields, urban areas, villages. It also resides in anthills and old houses.
Food: Toads, frogs, mice, small birds, and occasionally small snakes.
Description: It is found in shades of brown, yellow, black, and gray. The head is distinct with large nostrils and black eyes having round pupils. The spectacle mark is visible when it spreads the hood.
Distribution: All over India.

7. ***Bungarus caeruleus* (Schneider, 1801)**

Common Name: Krait
Local Name: Manyaar (Marathi), Manir, Karait (Hindi), Kalotaro (Gujarathi), Kalaj (Bengali), Chitti (Oriya), Kahuvirat (Tamil).
Habitat: Spotted in forests, near human habitations, paddy fields, marshy as well as arid and rocky areas.
Food: Other snakes as well as other kraits. Occasionally, it has been known to eat mice, frogs, and even lizards.
Description: Slender bluish black to dark brown. Paired white bands across the entire length. Body has smooth glossy scales of which the ones on the uppermost row are distinctly enlarged and hexagonal. Head black, while the lower lip and underside are white.
Distribution: All over India.

Family: Gekkonidae

8. ***Hemidactylus brookii* (Gray, 1845)**

Common Name: Brook's gecko
Local Name: Balli (Telugu), Chipakli (Hindi), Alli (Kannada).
Habitat: Open forested areas and human habitations.
Description: Above it is yellowish brown with irregular dark spots. On the side of the head, it has one or two dark lines which are passing through the eye. Lips with dark bars. White lower parts. Dark brown scales with light dots. Length of head and body 58 mm and tail 60 mm. Head is oval in shape with small scales. Flattened body with granular scales.
Distribution in world and India: Africa, Asia (India, Bhutan, Thailand, Maldives, Peninsular Malaysia, Hong Kong, Philippine Islands, Sri Lanka, Myanmar, Pakistan, Bangladesh, Nepal, Indonesia, Borneo), and America.

9. ***Hemidactylus frenatus* (Dumeril & Bibron, 1836)**

Common Name: Common house gecko
Local Name: Balli (Telugu), Chipakli (Hindi), Alli (Kannada).
Habitat: Human habitation, in both rural and urban areas. Terrestrial habitat.
Description: It is in mottled dark brown to pale brown in color. Tail possesses a row of tiny spikes on each side and oval in cross section. The total length is 13.5 cm (snout to vent: 6.5 cm). Large head with smooth dorsal scales. Fingers not have webbing. Mainly nocturnal.
Distribution in world and India: Throughout mainland Asia and Southeast Asia, South and East Africa, and across the Pacific Ocean to Mexico.

Family: Scincidae

10. ***Eutropis carinata* (Schineider, 1801)**

Common Name: Common skink
Local Name: Haavu raani
Habitat: Forest biome, scrubland (terrestrial habitat).
Description: Moderate thick snout. Scaly lower eyelid. Above it is brown or olive brown in color with small black spots or longitudinal lines. Darker sides with or without lighter spots. Yellowish lower surface. Males have a scarlet band from the shoulder to the thigh in the breeding season.
Distribution in world and India: Throughout India, Nepal, Bangladesh, Maldives, and Sri Lanka.

11. ***Eutropis macularia* (Blyth, 1853)**

Common Name: Bronze grass skink
Local Name: Haavu raani
Habitat: Deciduous forests, evergreen forests, and plantations.
Description: Slender body with 5–9 keels. Bronzy brown dorsum. Short snout. Scaly lower eyelid. No postnasal is absent. Tail is larger than the head and body. Above it is olive brown in color. Darker sides with white–black-edged spots. Back is uniform or black-spotted or two black longitudinal scales. Two light well-marked lateral lines only on the neck. Lower surfaces yellowish. From snout to vent, the length is 2.5 in. and tail 3.6 in.
Distribution in world and India: Asia (India, Pakistan, Nepal, Bhutan, Sri Lanka, Bangladesh, Myanmar, Thailand, Laos, Cambodia, Vietnam, NW Malaysia.

d. **Birds**

Family: Accipitridae

1. ***Accipiter badius*(Gmelin, 1788)**

   Common Name: Shikra
   Habitat: Light deciduous forest and village groves.
   Geographic range: Indian subcontinent.
   Description: The shikra (*Accipiter badius*) is a small BOP (bird of prey). This is resident bird. Adult male pale blue gray above; white, closely barred with rusty brown below; indistinct mesial stripe. Female larger, dark smoky brown washed with gray above; Juvenile scaly, dark brown above; heavily streaked below.

2. ***Aquila nipalensis* (Hodgson, 1833)**

   Local Name: Steeped eagle
   Habitat: Open plains, often near water.
   Description: Deep blackish brown to pale almost buffish brown; yellow gape extends till the rear of the eye.
   Geographic range: It is reported to be in Pakistan, Nepal, and central India. And also it is reported to be from the hills of Assam, Manipur, and Meghalaya. Rare in south peninsula till Karnataka.

3. ***Elanus axillaris* (Latham, 1802)**

   Common Name: Black-shouldered kite
   Habitat: This is resident species. Deciduous biotype: thin savannah forest, grassland, cultivation; plains and hills up to 1300 m.
   Description: A dainty ashy gray and black raptor, the black shoulder patch conspicuous at rest, and black primaries showing in overhead flight.
   Geographic range: Subcontinent; vagrant (an occasional stray) to Sri Lanka and Lakshadweep.

4. ***Haliastur indus* (Boddaert, 1783)**

   Common Name: Brahminy kite
   Habitat: Seacoast and inland waters.
   Description: This is resident bird. It is chestnut kite with white head, neck, and breast. Outer primaries black tipped. Tail is round in structure. Immature is overall brown, showing dark-tipped creamish primaries in overhead flight.
   Geographic range: Subcontinent (except Balochistan); Andaman and Sri Lanka.

Family: Aegithinidae

5. ***Aegithina tiphia* (Linnaeus, 1758)**

Common Name: Common iora
Habitat: Open forest, cultivation, and scrub jungle.
Description: This is resident species. Breeding male is in black, green, and yellow with two white wing bars; tail is entirely black above. Female is yellowish green in color with greenish brown wings; two white wing bars. Non-breeding male has black tail like female.
Geographic range: Himalayan foothills, from Himachal Pradesh to Arunachal Pradesh, and most of the subcontinent from the east of Gujarat to the southern tip; Sri Lanka.

Family: Alaudidae

6. ***Eremopterix griseus* (Scopoli, 1786)**

Common Name: Ashy-crowned sparrow lark
Habitat: Sparsely scrubbed wasteland, stubble and plowed fields.
Description: This is resident bird species. A small, squat, thick-billed crestless sparrow-like lark. Crown is in sandy brown with ashy color, with whitish cheeks and a black streak from chin to eye. Below brownish black. It has an undulating song flight. Female overall sandy brown.
Geographic range: Endemic to sub-Himalayan subcontinent; Sri Lanka.

7. ***Mirafra erythroptera*(Blyth, 1845)**

Common Name: Indian bush lark
Habitat: Sparse scrub jungle and fallow land.
Description: The chestnut in wings more intense and noticeable than the other bush larks. Brown cheeks bordered by white; distinct white supercilium; breast spotted; whitish underparts.
Geographic range: Endemic bird. Plains of Pakistan, northwest India, and most of the subcontinent.

Family: Anatidae

8. ***Anas poecilorhyncha* (Forster, 1781)**

Common Name: Spot-billed duck
Habitat: Jheels and shallow reservoirs.
Description: Size of this bird is more or less than 61 cm. Scaly patterned gray and dark brown. White border to metallic green speculum. Legs in coral red color. Two orange red spots at the base of yellow-tipped bill.
Geographic range: Subcontinent up to 1800 m. Andaman and north Sri Lanka (occasional).

Family: Ardeidae

9. ***Ardea cinerea* (Linnaeus, 1758)**

Common Name: Gray heron
Local Name:
Family: Ardeidae
Habitat: Swamps, estuaries, and rocky offshore islets, and also large inland waters.
Description: Size of this bird is bigger than 98 cm. Ashy gray with white crown and neck and long back occipital crest. A black dotted line running down the middle of fore neck till breast.
Geographic range: Subcontinent; Andaman and Nicobar, Sri Lanka, Lakshadweep, and Maldives.

10. ***Ardea purpurea* (Linnaeus, 1766)**

Common Name: Purple heron
Habitat: Jheels, reedy lakes, and rivers.
Description: Size bigger than 98 cm. Purplish blue or slaty above. Crown and crest slaty black, rest of head and neck ferruginous; black stripe on the fore neck and sides of neck. Chin and throat white. Plumes on upper breast buffy white with chestnut streaks. Rest of the underparts black and rich chestnut.
Geographic range: Indian subcontinent, Andaman and Nicobar, Sri Lanka, recorded from Lakshadweep and Maldives.

11. ***Ardeola grayii* (Sykes, 1832)**

Common Name: Indian pond heron
Habitat: Shallow marshes, streams, paddy fields, and ponds.
Description: Resident to India, size more or less than 46 cm. Snow-white in color, which is prominent during flight but in resting time it is camouflaged with earth by light brown color. It looks very handsome in breeding plumage with head and neck yellowish-brown with long recumbent white or buff occipital crest.
Geographic range: Indian subcontinent, Andaman and Nicobar, Lakshadweep, Maldives, Sri Lanka.

12. ***Bubulcus ibis* (Linnaeus, 1758)**

Common Name: Cattle egret
Habitat: Associated with grazing cattle.
Description: Size more or less than 51 cm. Heavy yellow bill and dark legs. During breeding season, golden buff on head, neck, and back.
Geographic range: Indian subcontinent, Sri Lanka, Andaman and Nicobar, Lakshadweep, and Maldives.

13. ***Casmerodius albus* (Linnaeus, 1758)**

Common Name: Great egret
Habitat: Marshes, rivers.
Description: Large, lanky, snow-white egret with black legs. During breeding season, legs are tinged reddish in color. Yellow bill, while breeding black in color.
Geographic range: Throughout Indian subcontinent; Sri Lanka, Maldives, and south Andamans.

14. **Egretta garzetta(Linnaeus, 1766)**

Common Name: Little egret
Habitat: Inhabits in inland wetlands, marshes, and estuaries.
Description: This is small and snow-white egret. Bill is black in color. The yellow feet contrasting with black legs. Bunch of ornamental plumes over the breast, nape, and back.
Geographic range: Throughout the subcontinent, Sri Lanka, Maldives, Lakshadweep, and Andaman and Nicobar.

Family: Caprimulgidae

15. ***Caprimulgus indicus* (Latham, 1790)**

Common Name: Indian jungle nightjar.
Habitat: Dry and moist deciduous jungle.
Description: Broad, black streaks on head and scapulars diagnostic. Throat band broken. Two small white spots were present on the wings in males. In the case of females, buff is present on wings. Forehead has broader black streaks, and tail bands are broader. Tail bands are broader.
Geographic range: Subcontinent, Himalayas, northeast hill states, Assam, Bangladesh, Sri Lanka, and Andamans.

Family: **Cerylidae**

16. ***Ceryle rudis* (Linnaeus, 1758)**

Common Name: Lesser pied kingfisher
Habitat: Jheels and slow-flowing streams.
Description: This is black-billed kingfisher. Bird is black and white in color. Eye band is black in color. Gorget is single, and in the case of females, it is broken in the middle. Size of this bird is above 31 cm.
Geographic range: India and Sri Lanka.

Family: Charadriidae

17. ***Vanellus malabaricus* (Boddaert, 1783)**

Common Name: Yellow-wattled lapwing
Habitat: Wastelands, stubbles, and fallow fields in dry regions.
Description: Size is less than 27 cm. Sandy brown with white belly, black cap, and yellow lappets of skin above and in front of the eyes.
Geographic range: In India, throughout the Peninsular India, Gujarat north to Himachal Pradesh, east to Bihar, Sri Lanka, and Sind in Pakistan.

18. ***Vanellus indicus* (Boddaert, 1783)**

Common Name: Red-wattled lapwing
Habitat: Near water in open and cultivation lands.
Description: Size is more than 33 cm. Above it is bronze brown in color, and below it is white with black breast, head, and fore neck. Bill red and tipped black.
Geographic range: Sub-Himalayan subcontinent and Sri Lanka.

Family: Ciconiidae

19. ***Mycteria leucocephala* (Pennatit, 1769)**

Common Name: Painted stork
Habitat: Large marshes.
Description: Largely white. Upper wing is coverts black with white barring. Tail and wing quills are black in color. Bill is yellow, long, and slightly decurved at the tip. Legs are orange red in color.
Geographic range: Subcontinent and Sri Lanka.

20. ***Ciconia episcopus* (Boddaert, 1783)**

Common Name: White-necked stork
Habitat: Flooded grasslands, irrigated fields, marshes, and dry grasslands.
Description: Size is above 107 cm. A red-legged black stork with noticeable white neck and undertail coverts. Black crown.
Geographic range: Subcontinent and Sri Lanka.

Family: Cisticolidae

21. ***Prinia socialis* (Sykes, 1832)**

Common Name: Ashy prinia
Habitat: Grass reeds, scrub, and grass around the cultivation lands and gardens.

Description: Red eyes, upperparts and crown is dark ashy gray in color. Whitish tips and black subterminal spots are present on the tail. Whitish buff throat. This bird is smaller than 13 cm in size.
Geographic range: India, Pakistan, Bangladesh, and Sri Lanka.

Family: Columbidae

22. ***Columba livia* (Gmelin, 1789)**

Common Name: Blue rock pigeon
Habitat: Around the human habitation and cultivation land.
Description: Bluish gray with shining. In breast and around the neck, it has metallic green–purple–magenta shine. Tail band is black terminal. Two dark bars were found on the wings. It is 33 cm in size.
Geographic range: Subcontinent, Sri Lanka, Maldives, and Lakshadweep. Introduced in Nicobars.

23. ***Streptopelia senegalensis* (Linnaeus, 1766)**

Common Name: Laughing dove
Habitat: Open semiarid biotype.
Description: Colored with pinkish brown and gray. Head and neck are lilac pink in color and black and rufous on the sides of the fore neck. More or less than 27 cm.
Geographic range: Peninsular plains, subcontinent east to Nepal, Bangladesh, and Tripura.

24. ***Streptopelia chinensis* (Scopoli, 1786)**

Common Name: Spotted dove
Habitat: Dry and moist deciduous biotype.
Description: This dove is pinkish brown in color. Black-and-white chess-board-like dots at the base of the hind neck. Less than 30 cm in length.
Geographic range: Subcontinent except the arid northwest regions and Sri Lanka.

Family: Corvidae

25. ***Corvus splendens* (Vieillot, 1817)**

Common Name: House crow
Habitat: Human habitations.
Description: Above 43 cm. Black in color. This is a well-known bird in all the villages and cities.
Geographic range: Subcontinent, south of Himalayan foothills, and Sri Lanka.

Family: Cuculidae

26. ***Centropus sinensis* (Stephens, 1815)**

Common Name: Greater coucal
Habitat: Deciduous scrub jungle and gardens.
Description: Crow like glossy, black bird with chestnut wings and long broad black tail. Size of this bird is above 48 cm.
Geographic range: Subcontinent, south of the Himalayas, and east of the Indus valley except Thar Desert and Sri Lanka.

27. ***Eudynamys scolopaceus* (Linnaeus, 1758)**

Common Name: Asian koel
Habitat: Lightly wooded country, towns, and country.
Description: Male is shiny metallic black in color. Bill is in yellowish green with crimson eyes. Female: Above it is dark brown in color with white spots. Below it is white, barred with blackish brown. Size is more or less than 43 cm.
Geographic range: Subcontinent, Lakshadweep, Maldives, Andaman and Nicobars, and Sri Lanka.

28. ***Phaenicophaeus viridirostris* (Pennant, 1769)**

Common Name: Blue-faced malkoha
Habitat: Deciduous and semievergreen scrub and secondary jungle.
Description: Bill is bright green in color, eye patch is blue in color, and tail is long and broad, with noticeable white tips.
Geographic range: Peninsular India, from Gujarat to east Orissa and to the south till Kanyakumari.

Family: Dicruridae

29. ***Dicrurus macrocercus* (Vierllot, 1817)**

Common Name: Black drongo
Habitat: Open deciduous forest and cultivation lands.
Description: Size is above 31 cm. Glossy, jet black in color. Long, deeply branched tail with slim body.
Geographic range: Subcontinent and Sri Lanka.

Family: Estrildidae

30. ***Lonchura malabarica* (Linnaeus, 1766) or *Euodice malabarica***

Common Name: Indian silverbill
Habitat: Grassland, cultivation lands, and sparsely scrubbed and secondary jungle.

Description: Pale bluish bill. Underparts and rump are white in color. Long and pointed blackish tail.
Geographic range: Peninsular and continental India except northeast hill states and planes of Pakistan. Bangladesh and Sri Lanka.

31. ***Lonchura Malacca* (Linnaeus, 1766)**

Common Name: Black-headed munia or tricolored munia
Habitat: Cultivation fields, tallgrass, and swampy areas.
Description: The size of the bird is below 10 cm. Bill is bluish gray in color. Hood is black in color. Center of belly, vent, coverts under tail and thighs are black. Males and females show similar appearance.
Geographic range: India and Bangladesh.

32. ***Lonchura punctulata* (Linnaeus, 1758)**

Common Name: Spotted munia
Habitat: Open woodland, grassland, gardens, and cultivation lands.
Description: Head, throat, upper breast, and upperparts of the bird are chestnut brown in color. Lower breast, flanks, and rump are speckled black and white in color. White belly. Size of this bird is less than 10 cm.
Geographic range: Subcontinent, Sri Lanka, and narrow foothill zone of Pakistan.

Family: Halcyonidae

33. ***Halcyon smyrnensis* (Linnaeus, 1758)**

Common Name: White-breasted king fisher
Habitat: Water bodies, cultivation lands, and forest edges.
Description: Head, neck, and lower underparts are chocolate brown in color. Remaining upperparts are in turquoise blue color. Chin, throat, and center of the breast are white in color. Bill is bright coral red in color. Size is above 28 cm.
Geographic range: Most of the subcontinent, Andaman and Nicobar, Lakshadweep, and Sri Lanka.

Family: Hirundinidae

34. ***Hirundo daurica* (Laxmann, 1769)**

Common Name: Red-rumped swallow
Habitat: Open, cultivated countryside land.
Description: It is fulvous white in color. Below of the bird finely streaked with dark brown in color. Sides of the neck, hind neck, and rump are rufous orange in color. Size of this bird is more or less than 20 cm.
Geographic range: Subcontinent, Andamans, Car Nicobar, Lakshadweep, and Sri Lanka.

35. ***Hirundo smithii* (Leach, 1818)**

Common Name: Wire-tailed swallow
Habitat: Open countryside, cultivation land, and near water.
Description: Above side of this bird glossy steel blue with a bright chestnut cap. Underparts of this bird are pure shiny white in color. Two long tail streamers. Size of this bird is more or less than 14 cm.
Geographic range: Subcontinent and Sri Lanka.

36. ***Hirundo concolor* (Sykes, 1832)**

Common Name: Dusky crag martin
Habitat: Hilly or mountainous country with rock face, gorges and caves, and stone buildings (old forts).
Description: This bird is a small-sized bird in the swallow family. Size of this bird is less than 13 cm in length. Body and wings are broad. Short square tail with small white patches near the tips. Upperparts are sooty brown, and underparts are slightly paler.
Geographic range: Indian subcontinent, southwestern China, and the northern parts of Thailand, Vietnam, and Laos. Sri Lanka and Borneo.

Family: Megalaimidae

37. ***Megalaima haemacephala* (Statius muller, 1776)**

Common Name: Coppersmith barbet
Habitat: Dry and moist deciduous biotype.
Description: Yellow on the throat and a patch around the eye. Forehead and breast in crimson color. Underparts are yellowish in color with green streaked. Green on above side of the bird. Size is above 17 cm.
Geographic range: Most of the subcontinent and Sri Lanka.

Family: Meropidae

38. ***Merops leschenaulti* (Vieillot, 1817)**

Common Name: Chestnut-headed bee-eater
Habitat: Mixed moist deciduous forest near the streams.
Description: Bright cinnamon–chestnut-colored crown, hind neck, and mantle. Yellow throat bordered by a rufous and black gorget. Above it is grass green in color, and below it is pale. Bluish rump and central tail features. Size of this bird is more or less than 21 cm.
Geographic range: India (Himalayan foothills, Uttarakhand to northeast hill states, Peninsular India, and Eastern and Western Ghats), Bangladesh, Andaman and Nicobar, and Sri Lanka.

39. ***Merops orientalis* (Linnaeus, 1766)**

Common Name: Small bee-eater
Habitat: Open countryside and cultivation land.
Description: Overall body is grass green in color. Pin tail. Throat is verditer blue bordered below by a black gorget. Size of this bird is above 21 cm.
Geographic range: Sub-Himalayan subcontinent and Sri Lanka.

Family: Motacillidae

40. ***Anthus campestris* (Linnaeus 1758)**

Common Name: Tawny pipit
Habitat: Sparsely scrubbed semidesert, fallow land, pastures, and plowed fields.
Description: Above lightly marked and pale brown. Outer rectrices white in color. Below it has whitish buff and without stripes on breast. Size of this bird is more or less than 15 cm.
Geographic range: Subcontinent and Sri Lanka.

41. ***Anthus rufulus* (Vieillot, 1818)**

Common Name: Paddy field pipit
Habitat: Grasslands and marshy ground. Dark brown tail with white outer rectrices. Breast is brown in color. Size is more or less than 15 cm.
Geographic range: Subcontinent and Sri Lanka.

Family: Muscicapidae

42. ***Copsychus saularis* (Linnaeus, 1758)**

Common Name: Oriental magpie robin
Habitat: Dry and moist deciduous forest, near human habitations and secondary jungles.
Description: Head, breast, and upperparts are black. Large white wing patch. Below white from the breast. Size is more or less than 20 cm.
Geographic range: Subcontinent, south of the Himalayas and east of Indus valley and Kachchh, Andamans, and Sri Lanka.

43. ***Saxicoloides fulicatus* (Linnaeus, 1766)**

Common Name: Indian robin
Habitat: Arid and stony areas, semidesert with scattered bushes, cultivation fields, and around the habitations.

Description: Male: Shining black with white wing patch. Female: Above brown. Lacks white wing patch. Below gray brown with chestnut vent. Size is more or less than 16 cm.
Geographic range: Subcontinent, from Indus valley east to West Bengal and Bangladesh, and south to Sri Lanka.

44. ***Saxicola caprata* (Linnaeus, 1766)**

Common Name: Pied bushchat
Habitat: Cultivated fields and sparsely scrubbed hill sides.
Description: Male: Pied, black feathers fringed with brown. Female: Above brown in color. Rump, vent, upper and under tail coverts buff are brownish-yellow in color.
Geographic range: Most of the subcontinent, south of the Himalayas, and Sri Lanka.

Family: Nectariniidae

45. ***Nectarinia asiatica* (Latham, 1790)**

Common Name: Purple sunbird
Habitat: Light deciduous forest and gardens.
Description: Pale olive brown. Ventrally yellow in color, with a broad blue-black band down middle of throat and breast.
Geographic range: Subcontinent, except high Himalayas.

46. ***Cinnyris asiaticus* (Latham, 1790)**

Common Name: Purple sunbird
Habitat: Light deciduous forest and gardens.
Description: Above olive brown and uniform yellow below. Size is less than 10 cm.
Geographic range: Subcontinent, except high Himalayas.

47. ***Leptocoma zeylonica* (Linnaeus, 1766)**

Common Name: Purple-rumped sunbird
Habitat: Scrubs, light secondary jungle, dry cultivated land near by countryside, and gardens.
Description: *Male*: Crown is metallic green in color. Rump and throat metallic purple color. A green shoulder patch. *Female*: Brown above, rufous on wings. Grayish white throat and flanks. Breast and center of belly are lemon yellow color.
Geographic range: Entire Peninsular India. Bangladesh and Sri Lanka.

Family: Oriolidae

48. ***Oriolus kundoo* (Sykes, 1832)**

Common Name: Indian golden oriole
Habitat: Semievergreen and deciduous biotype.
Description: Male: Golden yellow with black color in wings and tail. Female: Greenish yellow and duller. Underparts of this bird are paler and streaked with dull brown. Size of this bird is above 25 cm.
Geographic range: Subcontinent, except northwest arid regions, northeast hill states, Bangladesh, and Sri Lanka.

49. ***Oriolus oriolus* (Linnaeus, 1758)**

Common Name: Eurasian golden oriole
Habitat: Semievergreen and deciduous biotype.
Description: Underparts of this bird are paler and streaked with dull brown. The black eye stripe does not extend behind the eye. Bill is shorter. Size of this bird is above 25 cm.
Geographic range: Subcontinent, except northwest arid regions, northeast hill states, Pakistan, Bangladesh, and Sri Lanka.

Family: Passeridae

50. ***Passer domesticus* (Linnaeus, 1758)**

Common Name: House sparrow
Habitat: Human habitations, cultivated lands, and secondary jungles.
Description: *Male*: Gray crown. Black borders of eyes and throat. Rufous wing with a broad, white wing bar. Gray brown rump. Dark brown tail. *Female*: Above the sparrow contains grayish brown. Supercilium is pale in color.
Geographic range: Subcontinent, Sri Lanka, introduced to Maldives and Andaman and Nicobar.

Family: Phalacrocoracidae

51. ***Phalacrocorax fuscicollis* (Stephens, 1826)**

Common Name: Indian cormorant
Habitat: Jheels, reservoirs, rivers, and estuaries.
Description: Above shining bronze black and below jet black with white tufts of feathers behind eye-coverts. In breeding season, eyes are in emerald green. Size of this bird is above 80 cm.
Geographic range: Subcontinent (except Himalayas) and Sri Lanka.

Family: Phasianidae

52. ***Francolinus pondicerianus* (Gmelin, 1789)**

Common Name: Gray francolin
Habitat: Thorny scrub, light dry jungle, and cultivation fields.
Description: Overall gray. Upperparts are rusty. Yellowish rufous throat patch enclosed in a black prominent necklace. Size of this bird is around 33 cm.
Geographic range: India from base of the Himalayas, east to West Bengal and south to entire peninsula. Northwest Sri Lanka and south Pakistan.

53. ***Pavo cristatus* (Linnaeus, 1758)**

Common Name: Indian peafowl
Family: Phasianidae
Habitat: Moist and dry deciduous forest, cultivation and village boundaries.
Description: *Male*: Crest is fan-shaped. Breast and neck are blue in color. Black-centered coppery disks or eyespots. Sweeping metallic, bronze green train present. Size of the male is 110 cm without train (2–2.5 m in full plumage). *Female*: Smaller than male, crested, and train absent. Upperparts are brown. Abdomen buffy white. Female size is 86 cm.
Geographic range: Subcontinent and Sri Lanka. Introduced in Sindh and Andaman.

Family: Picidae

54. ***Dinopium benghalense* (Linnaeus, 1758)**

Common Name: Lesser golden-backed woodpecker
Habitat: Dry and moist deciduous biotype.
Description: Rump black. Throat and shoulders are spotted. In males, crown and crest are crimson in color. A female has crimson only on crest. Size of this bird is above 29 cm.
Geographic range: Entire subcontinent and Sri Lanka.

Family: Ploceidae

55. ***Ploceus philippinus* (Linnaeus, 1766)**

Common Name: Baya weaver
Habitat: Open cultivation land, grassland, and secondary scrub
Description: *Male*: Breeding male has both the crown and the breast or only the yellow crown. Blackish face. Upperparts are dark brown and broadly streaked. Either blackish or creamy buff on throat. *Female*: Mantle, crown and wings yellowish buff with dark brown streaks. Face and supercilium have yellowish buff. White throat. Upper belly and breast yellowish buff with

brown streaks on flanks. Lower belly cream buff. Size of this bird is more or less than 15 cm.
Geographic range: Subcontinent (except desert region of northwest) and Sri Lanka.

Family: Psittaculidae

56. ***Psittacula krameri* (Scopoli, 1769)**

Common Name: Rose-ringed parakeet
Habitat: Moist and dry deciduous biotype, thin jungle, orchards, cultivation, and human environments.
Description: Grass green parakeet. Lower mandible is black. The male has a rose pink and black collar. Size of this bird is above 42 cm.
Geographic range: Subcontinent and Sri Lanka. Introduced to Lakshadweep, Maldives, and Andamans.

Family: Pycnonotidae

57. ***Pycnonotus cafer* (Linnaeus, 1766)**

Common Name: Red-vented bulbul
Habitat: Cultivation and scrublands.
Description: Black head and throat. Upperparts and breast have scaly markings. White rump and white-tipped blackish tail. Size is less than 20 cm.
Geographic range: Subcontinent and Sri Lanka.

Family: Rallidae

58. ***Amaurornis phoenicurus* (Pennant, 1769)**

Common Name: White-breasted water hen
Habitat: Marshes, ponds, and rice fields.
Description: Above it has dark slate gray. White facial mask and underparts. Vent and under tail coverts are reddish brown in color. Size of this bird is more or less than 32 cm.
Geographic range: Subcontinent, Andaman and Nicobar, Lakshadweep, Maldives, and Sri Lanka.

59. ***Fulica atra* (Linnaeus, 1758)**

Common Name: Common coot
Habitat: Lakes, large jheels, and reservoirs.

Description: Slate black color. Dumpy and tail-less. The stout is pointed; bill is white in color and horny shield covering on forehead. Size is more or less than 32 cm.
Geographic range: Subcontinent, Andamans, Maldives, and Sri Lanka.

60. ***Porphyrio porphyrio* (Linnaeus, 1758)**

Common Name: Purple moorhen
Habitat: Marshy reed beds and flooded rice fields.
Description: Long red legs and enormous toes. Large purplish blue. The huge red bill with a frontal shield. White patch under the short tail. Size of this bird is more or less than 43 cm.
Geographic range: Subcontinent and Sri Lanka.

Family: Recurvirostridae

61. ***Himantopus himantopus* (Linnaeus, 1758)**

Common Name: Black-winged stilt
Habitat: Lagoons, jheels, and reservoirs.
Description: A delicate, lanky, pied wader. Bill is thin, black, and long. Legs are long, thin and reddish. Size of this bird is less than 25 cm.
Geographic range: Subcontinent, Sri Lanka, Lakshadweep, and Maldives.

Family: Sternidae

62. ***Sterna aurantia* (Gray, JE, 1831)**

Common Name: River tern
Habitat: Reservoirs and large rivers.
Description: Deeply forked tail, yellow orange bill, and red legs.
Geographic range: Subcontinent.

Family: Sturnidae

63. ***Acridotheres tristis* (Linnaeus, 1766)**

Common Name: Common myna
Habitat: Neighborhood of homesteads, cities, and villages.
Description: Glossy black head. Yellow legs. Dark color body. A large white wing patch seen in flight. Size of this bird is 23 cm.
Geographic range: Subcontinent, Sri Lanka, Andaman and Nicobar, Lakshadweep, and Maldives.

Family: Sturnidae

64. ***Sturnia malabarica*** **(Gmelin, 1789)**

    Common Name: Chestnut-tailed starling
    Habitat: Secondary jungle.
    Description: Silvery gray and rufous in color. Whitish eye. Silvery gray hood. Mantle and wings are in gray rufous color. Ferruginous large tail. Size of this bird is less than 21 cm.
    Geographic range: Continental and Peninsular India.

65. ***Sturnus pagodarum*** **(Gmelin, 1789)**

    Common Name: Brahminy starling
    Habitat: Open deciduous forest and scrub jungle, near cultivation lands and habitations.
    Description: Crested starling. Upperparts are gray brown in color and below orange buff. Forehead, crown, nape, and long pointed crest glossy black. Brown tail with white edges. Size of this bird is less than 22 cm.
    Geographic range: India (except northeast), north Sri Lanka.

66. ***Sturnus roseus*** **(Linnaeus, 1758)**

    Common Name: Rosy starling
    Habitat: Open cultivation fields and semidesert areas.
    Description: Rose pink in color. Shiny black hood, crest, wings, thighs, vent, and tail. Bill and legs are reddish in color. Size of this bird is more or less than 23 cm.
    Geographic range: Subcontinent, Sri Lanka, and Andamans.

Family: Threskiornithidae

67. ***Plegadis falcinellus*** **(Linnaeus, 1766)**

    Common Name: Glossy ibis
    Habitat: River banks and marshes.
    Description: A small, blakish ibis. Head is feathered and slender with down-curved bill. Breeding birds are rich dark. Nonbreeding birds have white-streaked, brown head and neck. Size of this bird is more or less than 52 cm.
    Geographic range: India (south to Kerala and Tamil Nadu), Pakistan, Bangladesh, and Sri Lanka.

68. ***Pseudibis papillosa*** **(Terminck, 1824)**

    Common Name: Black ibis
    Habitat: Jheel margins, stubble fields, river banks, and grasslands.
    Description: Glossy black in color. Bill is slender and curly. White patch near the shoulder. Brick red legs. Size of this bird is more or less than 68 cm.
    Geographic range: Southeast Sind, eastward to Bangladesh, and south especially along the eastern peninsula.

Family: Timaliidae

69. ***Turdoides striata* (Dumont, 1823)**

Common Name: Jungle babbler
Habitat: Deciduous forest, cultivation fields and scrub jungle and bamboo plantation area.
Description: Earth brown color. Creamy white eyes. Bill and legs are yellow in color. Streaked with pale brown breast. Gray throat. Creamy buff belly. Size is more than 25 cm.
Geographic range: Subcontinent.

70. ***Turdoides affinis* (Jerdon, 1845)**

Common Name: White-headed babbler
Habitat: Secondary jungles, around cultivation fields and villages.
Description: Above dark brown, fringed with gray giving a scaly appearance. Pale cream crown. Grayish brown ear coverts. Below buff. Throat and breast have mottled. Size of this bird is more than 23 cm.
Geographic range: Southern Peninsular India and Sri Lanka.

Family: Upupidae

71. ***Upupa epops* (Linnaeus, 1758)**

Common Name: Common hoopoe
Habitat: Light forest and near villages.
Description: Fawn-colored bird with black-and-white markings on wings. Black-and-white-tipped crest. Long and gently curved beak. Size of this bird is more than 31 cm.
Geographic range: Subcontinent, Andamans, Sri Lanka, Maldives, and Lakshadweep.

e. **Mammals**

Family: Canidae

1. ***Vulpes bengalensis* (Sharo, 1800)**

Common Name: The Indian fox
Local Name: Sanna nari (Kannada), Konka nakka or Gunta nakka (Telugu), Lomri (Hindi).
Habitat: Common in wasteland and scrublands. Most of the species found in cultivated lands and borders of irrigation channels.

Distribution: The whole of India from the foothills of the Himalayas to Cape Comorin.
Food: Small mammals, reptiles, and insects, rather than birds.
Description: Common fox of the Indian plains, a pretty, slender-limbed animal, smaller and slimmer in build than the red fox; distinctive in the black tip to its tail. The backs of its ears are generally of the same tone as the head and nape, never jet black and strongly contrasted as in the red fox.

Family: Cercopithecidae

2. ***Macaca radiata* (E. Geoffroy, 1812)**

Common Name: Bonnet macaque
Local Name: Bandar (Hindi), Makad, Lal manga (Marathi), Vella manthi (Malayalam), Korungoo (Tamil).
Habitat: It is the common species in village and jungle, both in the foothills and in the plains.
Distribution: The Indian peninsula as far north as Bombay on the west and the Godavari river on the east. Bonnet macaques from Travancore are considered a distinct race.
Food: Eat anything eatable, fruits, berries, leaves, and shoots, and also insects, grubs, and spiders.
Description: A medium-sized long-tailed macaque. A bonnet of long dark hairs radiates in all directions from a whorl on its crown. The bonnet does not quite cover the forehead, where the hairs are short and neatly parted in the center. The coat of bonnet macaque is variable both among individuals and with the season. In the cold weather, it is usually lustrous olive brown: the underparts whitish. With the onset of the hot weather, the coat loses its luster, turns harsh and scraggy, and fades to a buffy gray.

Family: Cervidae

3. ***Axis axis* (Erxleben, 1777)**

Common Name: Chital
Local Name: Nil, Nilgai, Rojh, Roz, Rojra;
Habitat: Nilgai avoid dense forests. Their usual haunts are hills sparsely dotted with trees, or level or undulating plains covered with grass and patches of scrub. They freely enter cultivation lands.
Distribution: The nilgai is found only in the Indian peninsula from the base of the Himalayas to Mysore. It does not occur in eastern Bengal, or Assam, or on the Malabar Coast.

Food: They both graze and browse, feeding on the leaves and the fruit of the ber (*Ziziphus*) and other trees.
Description: A great ungainly animal somewhat horselike build, with high withers and low rump. The adult bull has a coarse iron gray coat, a white ring below each fetlock, and two white spots on each cheek. His lips, chin, the inside of his ears, and the undersurface of his tail are white.

Family: Herpestidae

4. ***Herpestes edwardsii* (E. Geoffroy saint, Hilaire, 1818)**

Common Name: Indian gray mongoose.
Local Name: Mangus, Newal, Newara (Hindi), Mungoos (Marathi), Nurulia (Gujarathi), Keeree (Tamil, Malayalam and Kannada), Yentawa mangisa (Telugu).
Habitat: This is not a creature of forest, but of open lands, of scrub jungle, and of cultivation.
Distribution: The whole of India from the Himalayan foothills to Cape Comorin, extending westward to Persia and Mesopotamia and southward to Ceylon.
Food: They prey on the rats and mice, snakes, lizards and frogs, insects, scorpions, and centipedes, and in fact on anything that comes on their way.
Description: A tawny yellowish gray mongoose with no stripe on the side of its neck. The alternate light and dark rings on its hairs give its coat a grizzled look. The tail which is as long as its body is tipped with white or yellowish red, never black. With a shorter, finely speckled winter coat.

Family: Leporidae

5. ***Lepus nigricollis* (F. Cuvier, 1823)**

Common Name: The Indian hare
Local Name: Khargosh (Hindi), Sasa (Marathi), Mola (Kannada), Musal (Tamil), Choura pilli (Telugu).
Habitat: Large tracts of bush and jungle alternating with cultivated plains afford them ideal conditions. Many hares live in the neighborhood of villages and cultivation.
Distribution: This black-napped hare ranges from southern India to the Godavari on the east and in the west as far north as Khandesh and Berar and the adjoining districts of Madhya Pradesh.
Food: Grass.
Description: This is distinctive in having a dark brown or black patch on the back of its neck from the ears to the shoulder and upper surface of tail black.

Family: Muridae

6. ***Mus booduga* (Gray, 1837)**

   Common Name: Little Indian field mouse
   Local Name: Cuha (Hindi), Ili (Kannada), Tella Yeluka (Telugu).
   Family: Muridae
   Habitat: Found in compounds and gardens and may even venture into a house.
   Distribution: Throughout the India
   Food: Cereals and it will cause damage to the crops also.
   Description: It is about 2–3 in. 5–8 cm in body length with a tail slightly over 2 in. The dorsal fur varies from pale sandy in the desert and thorn forest zone to brown or dark grayish brown in moist country.

7. ***Mus musculus* (Linnaeus, 1758)**

   Common Name: Common house mouse
   Local Name: Cuha (Hindi), Ili (Kannada), Yeluka (Telugu).
   Habitat: It lives chiefly in houses but sometimes in gardens and fields near villages and towns.
   Distribution: It is found over the greater part of the Indian peninsula from the Himalayas to Cape Comorin.
   Food: It is omnivorous and feeds on any food that it can get.
   Description: It measures about 2–3 in. in body length and has a tail about equally long. The general color varies from dark to light brown, paler below.

Family: **Pteropodidae**

8. ***Pteropus giganteus* (Eschscholtz, 1831)**

   Common Name: Indian flying fox
   Local Name: Gadal, Badur (Hindi). Watwagul (Marathi), Sikat Yella (Kannada), Sikurayi (Telugu), Vowval (Tamil and Malayalam).
   Habitat: The largest of Indian bats, usually seen flying with slow wing beats at dusk.
   Distribution: India, Ceylon, Burma. Rare in west Rajasthan, Kachchh, and Sind. Does not occur as a resident species in the higher hills.
   Food: Feeds only on the juice of fruits, extracted by chewing dry the fleshy pulp, which is spat out.
   Description: Head and body about 9 in. (229 mm); wingspread about 4 feet (122 cm); weight 568–625 gm. The large size of this bat makes it unmistakable. Head reddish brown with a darker, sometimes blackish, snout. Hind neck and shoulders pale brownish yellow to straw; behind shoulders dark brown or black. Ventrally yellowish brown. Chin, neck, vent, and flanks darker. Wings black.

Family: Sciuridae

9. ***Funambulus palmarum* (Linnaeus, 1766)**

Common Name: Three-striped palm squirrel
Local Name: Gilheri (Hindi), Beral (Bengali), Khadi khar (Marathi), anna pilli (Tamil), Gili (Kannada), Udutha (Telugu).
Habitat: Inhabits in the Indian peninsula from the base of the Himalayas southward. This species is predominating in the south and in the moist parts of western and eastern India.
Distribution: Inhabit the Indian peninsula from the base of the Himalayas southwards. This species is predominating in the south, and in the moister parts of western and eastern India.
Food: Food is the usual food of this species is Fruits, Nuts, Young shoots, Buds, and Bark. When the silk cotton trees bloom, these squirrels drink the nectar, thus helping them in their pollination.
Description: Head and body 5 in. (13 cm) to 6 in. (15 cm); tail, slightly longer. Three dorsal stripes. Light or dark coat. It has a particularly shrill birdlike call which it repeats again and again.

Family: Suidae

10. ***Sus scrofa* (Linnaeus, 1758)**

Common Name: Wild boar
Local Name: Suar, Barba, Bad janwar, bura janwar (Hindi), Ran dukkar (Marathi), Punni (Tamil), Kad hundi (Kannada), Kartu punni (Malayalam), Adavi pandi (Telugu).
Habitat: Grass or scanty bush jungle, sometimes in forest; commonly found high crops after the rains.
Distribution: Widely distributed. It ranges over nearly the whole of India, Burma, and Tenasserim, Siam, and part of the Peninsular Malaysia. It is very common in Ceylon.
Food: Omnivorous animal. Living on crops, roots, tubers, insects, snakes, offal, and carrion. Feed in the early morning and late in the evening.
Description: A well-grown male stands 36 in. (90 cm) high at the shoulder, and its weight may well exceed 500 lb (230 kg). Lower tusks 12 in. (32.1 cm.) on the outside curve.

Photographs of some of the faunal species are given in photoplates (Photoplates 4.1, 4.2, 4.3, 4.4, 4.5, 4.6, 4.7 and 4.8)

Common rose
Pachliopta aristolochiae

Crimson rose
Pachliopta hector

Common Cerulean
Jamides celeno

Danaid Egg fly (Male)
Hypolimnas misippus

Common grass yellow
Eurema hecabe

Blue tiger
Tirumala limniace

Blue Pansy
Junonia orithya

Peacock pansy
Junonia almana

Plain Tiger
Danaus chrysippus

Photoplate 4.1 Butterflies

Common Picture Wing
Rhyothemis variegata

Pied paddy skimmer (male)
Neurothemis tullia

Ruddy Marsh Skimmer (Female)
Crocothemis servilia

Ruddy Marsh Skimmer (male)
Crocothemis servilia

Ditch Jewel
Brachythemis contaminata

Blue Tailed Green Darner
Anax guttatus

Common Club tail
Ictinogomphus rapax

Photoplate 4.2 Dragonflies

Certacanthacris tatarica

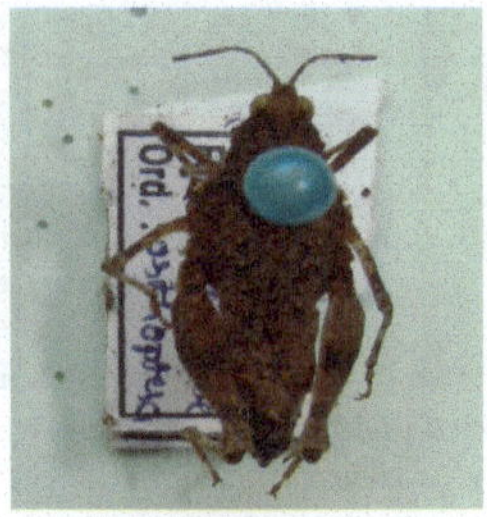

Chrotogonus oxypterus

Chrotogonus trachypterus

Anacridium flavescence

Eyprepocnemis alacris

Tylotropidius varicornis

Atractomorpha crenulata

Oedaleus abruptus

Gastriamrgus africanus

Oxya fuscovittata

Dioabolocatantops pinguis

Tristria pulvinata

Photoplate 4.3 Grasshoppers

Snakehead murrel
Channa striata

Tilapia
Tilapia mossambica

Denison barb
Puntius denisonii

Catla
Catla catla

Rohu
Labeo rohita

Mrigal
Cirrhinus cirrhosus

Common carp
Cyprinus carpio

Peninsular olive barb
Puntius sarana

Rasbora
Rasbora daniconius

Wallago/Boal
Wallago attu

Indian Butter catfish
Ompok bimaculatus

Photoplate 4.4 Fishes

Peninsular rock agama
Psammophilus dorsalis

Indian cobra
Naja naja

Green snake
Ahaetulla nasuta

Indian garden lizard
Calotes versicolor

Fan-throated lizard
Sitana ponticeriana

Rat Snake
Ptyas mucosus

Common Krait
Bungarus caeruleus

Brook's gecko
Hemidactylus brookii

Common house gecko
Hemidactylus frenatus

Common skink
Eutropis carinata

Bronze grass skink
Eutropis macularia

Green Paddy Frog
Hylarana erythraea

Skittering frog
Euphlyctis cyanophlyctis

Indian Bull frog
Hoplobatrachus tigerinus

Common Indian Toad
Duttaphrynus melanostictus

Common Indian tree frog
Polypedates maculatus

Photoplate 4.5 Herpatofauna

Black drango
Dicrurus macrocercus

Yellow wattled lapwing
Vanellus malabaricus

Red wattled lapwing
Vanellus indicus

House sparrow
Passer domesticus

Red vented bulbul
Pycnonotus cafer

Small bee eater
Merops orientalis

Spot-billed duck
Anas poecilorhyncha

Grey heron
Ardea cinerea

Purple heron
Ardea purpurea

Indian pond heron
Ardeola grayii

Cattle egret
Bubulcus ibis

Great egret
Casmerodius albus

Photoplate 4.6 Birds

Little egret
Egretta garzetta

Baya weaver
Ploceus philippinus

Rose ringed parakeet
Psittacula krameri

Indian peafowl
Pavo cristatus

White breasted kingfisher
Halcyon smyrnensis

Goldenbacked woodpecker
Dinopium benghalense

Common hoopoe
Upupa epops

Spotted munia
Lonchura punctulata

Indian silver bill
Lonchura malabarica

Tricolored munia
Lonchura malacca

Red rumped swallow
Hirundo daurica

Greater coucal
Centropus sinensis

Photoplate 4.6 (continued)

Blue faced malkoha
Phaenicophaeus viridirostris

Asian koel (Male)
Eudynamys scolopaceus

Asian koel (Female)
Eudynamys scolopaceus

House crow
Corvus splendens

Spotted dove
Streptopelia chinensis

Laughing dove
Streptopelia senegalensis

Blue rock pigeon
Columba livia

White necked strock
Ciconia episcopus

Painted stork
Mycteria leucocephala

Ashy prinia
Prinia socialis

Indian jungle nightjar
Caprimulgus indicus

Lesser pied kingfisher
Ceryle rudis

Photoplate 4.6 (continued)

Shikra
Accipiter badius

Steeped eagle
Aquila nipalensis

Black-shouldered kite
Elanus axillaris

Brahminy kite
Haliastur indus

Common iora
Aegithina tiphia

Ashy-crowned sparrow lark
Eremopterix griseus

White headed babbler
Turdoides affinis

Jungle babbler
Turdoides striata

Indian bush lark
Mirafra erythroptera

Wire tailed swallow
Hirundo smithii

Grey francolin
Francolinus pondicerianus

Common myna
Acridotheres tristis

Photoplate 4.6 (continued)

Rosy starling
Sturnus roseus

Chestnut tailed starling
Sturnia malabarica

Brahminy starling
Sturnus pagodarum

Glossy ibis
Plegadis falcinellus

Black ibis
Pseudibis papillosa

Indian cormorant
Phalacrocorax fuscicollis

Indian robin
Saxicoloides fulicatus

White-breasted water hen
Amaurornis phoenicurus

Common coot
Fulica atra

Purple moorhen
Porphyrio porphyrio

Black winged stilt
Himantopus himantopus

River tern
Sterna aurantia

Photoplate 4.6 (continued)

Dusky crag martin
Hirundo concolor

Coppersmith barbet
Megalaima haemacephala

Chestnut-headed bee eater
Merops orientalis

Tawny pipit
Anthus campestris

Paddy field pipit
Anthus rufulus

Oriential magpie robin
Copsychus saularis

Pied bush chat
Saxicola caprata

Purple sunbird
Nectarinia asiatica

Eurasian golden oriole
Oriolus oriolus

Indian golden oriole
Oriolus kundoo

Photoplate 4.6 (continued)

Black naped hare
Lepus nigricollis

Wild Boar
Sus scrofa

Indian Flying fox
Pteropus giganteus

Bonnet Macaque
Macaca radiata

Three striped palm squirrel
Funambulus palmarum

The Indian fox
Vulpes bengalensis

Indian gray mongoose
Herpestes edwardsii

Common House mouse
Mus musculus

Indian Field mouse
Mus booduga

Photoplate 4.7 Mammals

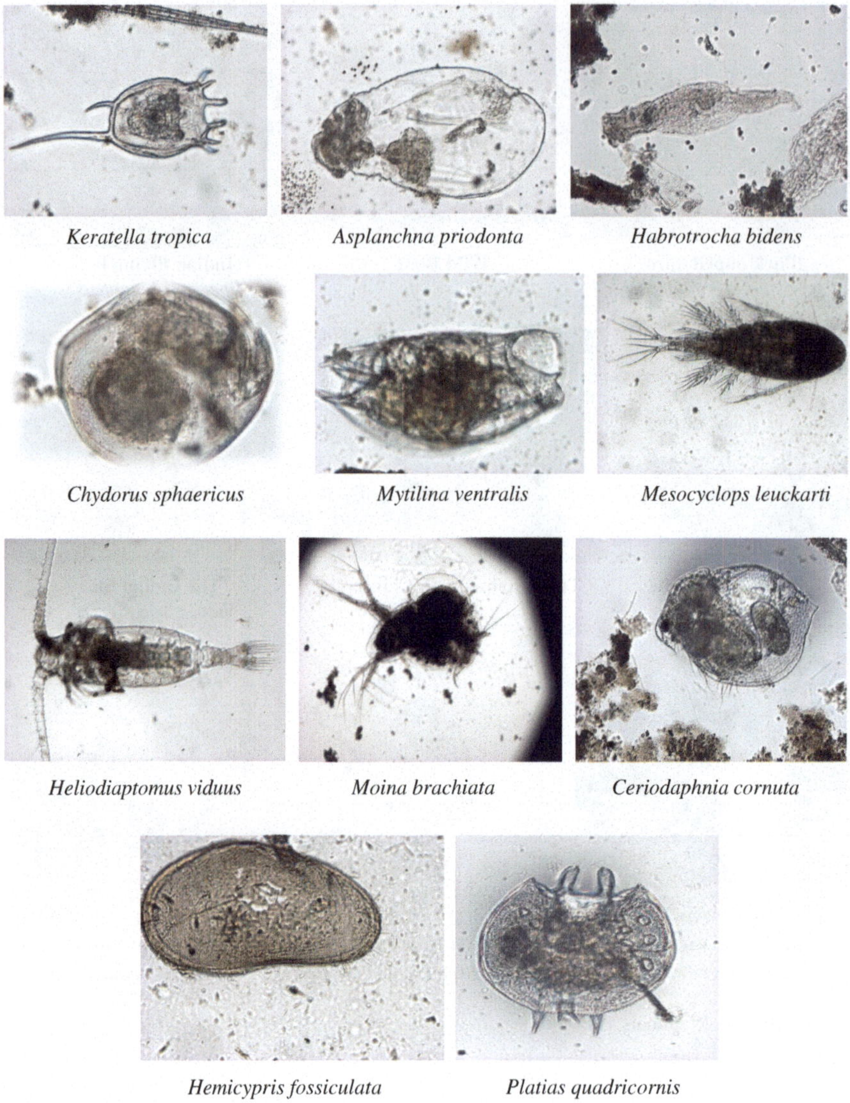

Photoplate 4.8 Zooplankton

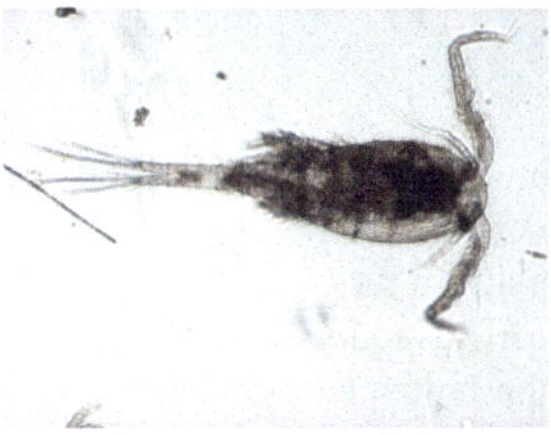

Tropocyclops Prasinus

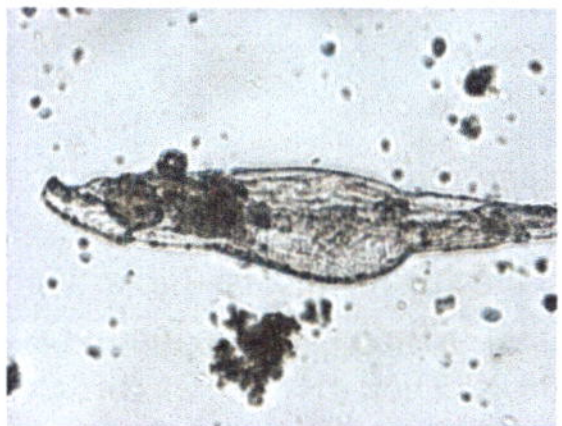

Rotifer tardus

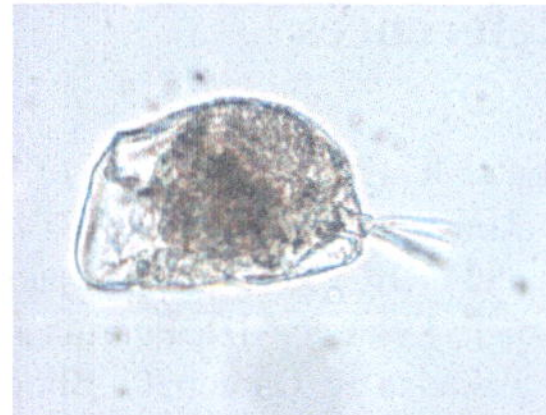

Mytilina acanthophora

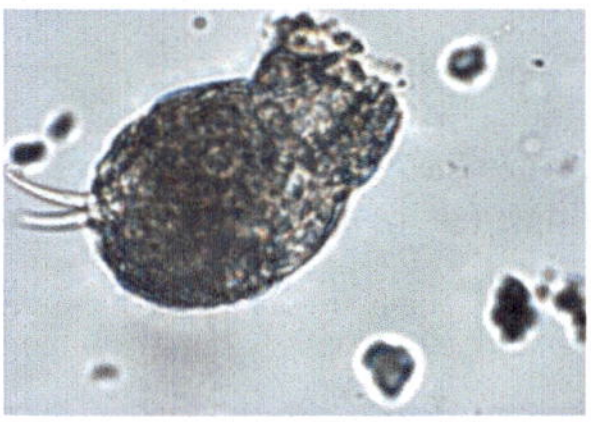

Cephalodella gibba

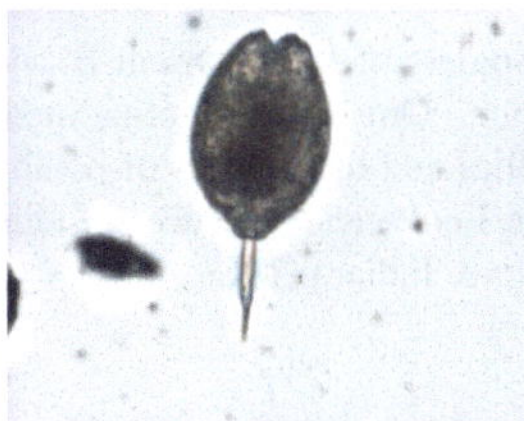

Monostyla bulla

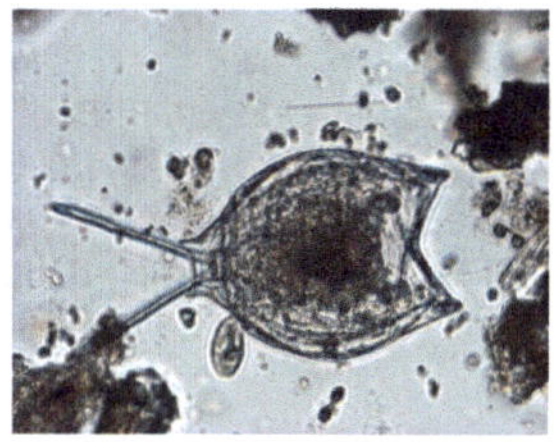

Lecane luna

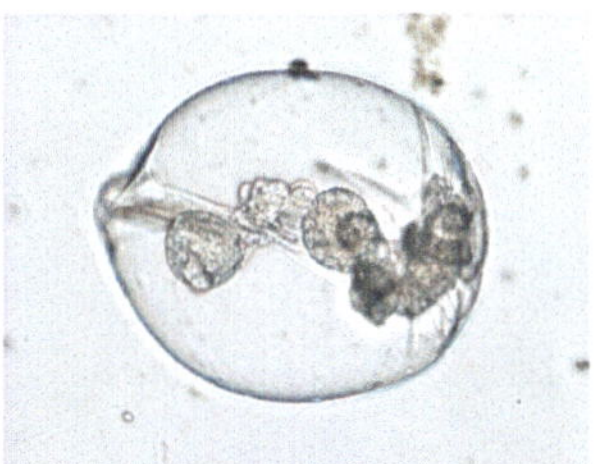

Horlla brehmi

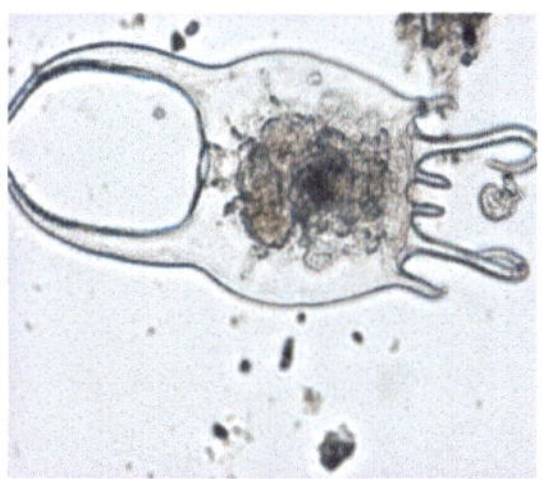

Brachionus falcatus

Paracy final

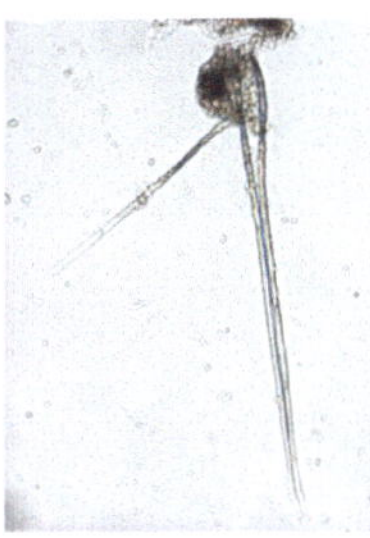

Fillinia longicepa

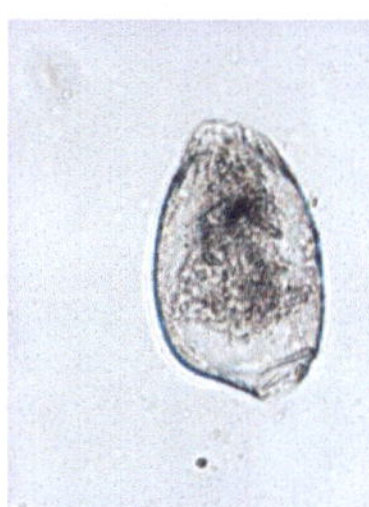

Brachionusurceolaris

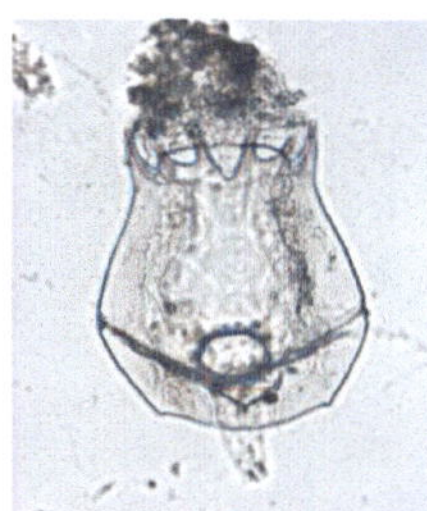

Brachionus bidentata

Photoplate 4.8 (continued)

References

Daniel JC (2002) The book of Indian reptiles and amphibians. Bombay Natural History, Oxford University Press

Daniels RJR (2002) Fresh water fishes of peninsular India. Universities Press

Gay T, Kehimkar ID, Punrtha J (2009) Butterflies of India. Oxford University Press

Manakadan R, Daniel JC, Bhopale N, Dick JH (2012) Birds of the Indian subcontinent: a field guide. Bombay Natural History, Oxford University Press

Prater SH (2005) The book of Indian animals. Bombay Natural History Society, Oxford University Press

Sebastian PA, Peter KV (2008) Spiders of India. Orient Blackswan publication

Singh AP (2011) Butterflies of India. Om Books International, India

Subramanian KA (2005) Dragonflies and damselflies of peninsular India-A field guide e-book of Project Lifescape. Centre for Ecological Sciences, Indian Institue of Science and Indian Academy of Sciences, Bangalore, India, p 118

Encyclopedia of Life: http://eol.org/

Chapter 5
Status of Biodiversity as per IUCN, IWPA, and CITES in Studied Semiarid Region, Karnataka, India

5.1 Introduction

The variety and variability among living organisms from all sources, terrestrial, marine, and other aquatic ecosystems and the ecological complexes of which they are a part; is diversity within species, between species and of ecosystems (Convention on Biological Diversity). In addition, the biodiversity in simplest way can be defined as "sum total species richness of a given area".

In recent times, our earth is facing an umpteen number of environmental problems. The major problem is the loss of biodiversity, which is a global issue in this era of scientific and technological world. The main reasons for this problem are extreme changes in the environment and habitat loss due to population explosion and unsustainable developmental practices. These are all anthropogenic-driven processes, where the species are not getting liberty of time and space for their survival; therefore, it leads to biodiversity loss. Each and every species plays an important role in the maintenance of equilibrium in the ecosystem. Biodiversity loss leads to the disruption in food chains and food webs of the ecosystem. Between the 1600s and 1950s, the rate of extinction of species was one species in a ten-year period and the current situation indicates that one species every year is vanishing from the planet earth (Kumar and Asija 2000). The main reasons for declining biodiversity is the human activities such as shrinking of habitat due to the development of agriculture, development of cities, industries, unabated pollution, and climate change. There are four main threats for biodiversity loss, i.e., (1) habitat loss, (2) over-exploitation, (3) pollution, and (4) pesticides (Khan and Shishodia 1998).

5.2 Biodiversity of Study Region

Among flora, 73 trees, 59 shrubs, 134 herbs, 45 climbers, 4 pteridophytes, 1 bryophyte, and 1 gymnosperm species were recorded from the study area.

S. Nautiyal et al., *Biodiversity of Semiarid Landscape*,
Environmental Science, DOI 10.1007/978-3-319-15464-0_5

In the faunal studies from invertebrates, 164 species of insects (13 families), 82 spiders (19 families), and 17 mollusca (9 families) are documented. From vertebrates, 11 species of fishes, 5 amphibians, 12 reptiles, 71 aves, and 11 mammals are documented.

5.3 Major Threats to Biodiversity

Following are the major threats to biodiversity of the studied semiarid region:

1. Spraying of pesticides to crops is the main threat for Insects and Spiders
2. Land-use change
3. Hunting for food
4. Eroding traditional knowledge related to use and conservation of biodiversity
5. Road accidents
6. Developmental activities
7. Unplanned urbanization
8. Habitat destruction and deforestation
9. Climate change

Wildlife conservation is the method of conservation and protection of threatened plants and animals and their habitats. The endangered species in India were identified by various national and international organizations such as International Union for Conservation of Nature (IUCN), The World Wide Fund for Nature (WWF), The Convention on International Trade in Endangered Species of Wild Fauna and Flora (CITES), and The Indian Wildlife Protection Act (IWPA).

5.4 International Union for Conservation of Nature (IUCN)

IUCN founded in 1948 is the world's first global environmental organization (https://www.iucn.org). The IUCN Red List of threatened species founded in 1963 is the world's most comprehensive inventory of the global conservation status of biological species.The IUCN provides the threatened species taxonomic, conservation status and global distribution, availability of information on flora and fauna that were evaluated using the IUCN Red List categories and criteria. The main objective of the IUCN is to highlight those species of plants and animals that are facing high risk of global extinction, i.e., critically endangered, endangered, and vulnerable. The IUCN provides the data on extinct species of plants and animals and in the data deficient taxa where there is not sufficient information, which can be evaluated. IUCN also evaluates the least concern species of flora and fauna for the sake of transparency, with all least concern species are now included in IUCN Red List. Because of this, the IUCN not only focuses on threatened species, but also

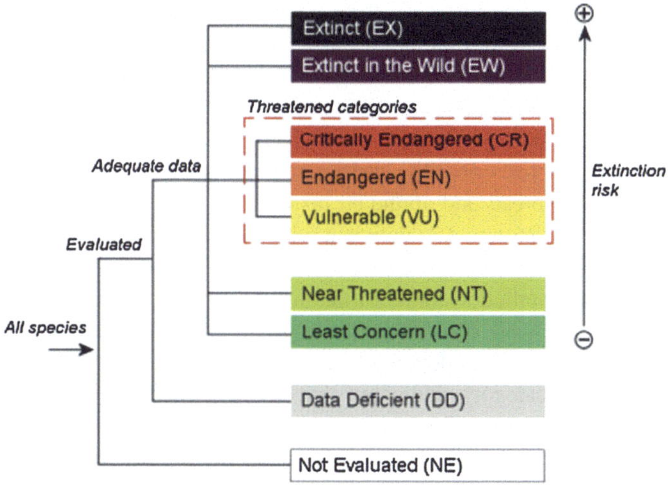

Fig. 5.1 IUCN categories. *Source* http://www.iucnredlist.org/about

considers the status of all the species of plants and animals, which includes the amphibians, birds, mammals, fresh water crabs, corals for increasing the number of taxonomic groups (http://www.iucnredlist.org/about).

There are nine clearly defined categories into which every taxon in the world can be classified (Fig. 5.1). Extinct and extinct in the wild are relatively self-explanatory. The category near threatened is applied to taxa that do not qualify as threatened now, but may be close to qualifying as threatened. The category least concern is applied to taxa that do not qualify (and are not close to qualifying) as threatened or near threatened. The category data deficient highlights taxa for which sufficient information is lacking to make a sound status assessment. The category not evaluated applies to taxa that have not yet been evaluated against the Red List Criteria (IUCN Standards and Petitions Subcommittee 2010).

The status of the species under IUCN categories based on the evaluation is given Fig. 5.1. IUCN assigns the species status according to their geographical distribution, availability, and number of individuals and availability. And accordingly categorized into categories, viz., critically endangered (CR); endangered (EN); vulnerable (VU); near threatened (NT); least concern (LC); data deficient (DD); not evaluated (NE).

The IUCN categories:

- Critically endangered: A taxon is critically endangered when it is facing an extremely high risk of extinction in the wild in the immediate future.
- Endangered: A taxon is endangered when it is not critically endangered but is facing a very high risk of extinction in the wild in the near future.
- Vulnerable: A taxon is vulnerable when it is not critically endangered or endangered but is facing a high risk of extinction in the wild in the medium-term future.

- Near Threatened: A taxon is near threatened when it has been evaluated against the criteria but does not qualify for critically endangered, endangered, or vulnerable now, but is close to qualifying for or is likely to qualify for a threatened category in the near future.
- Least concern: A taxon is least concern when evaluated against the criteria and does not qualify for critically endangered, endangered, vulnerable, or near threatened. Widespread and abundant taxa are included in this category.
- Data deficient: A taxon is data deficient when there is inadequate information to make a direct or indirect assessment of its risk of extinction based on its distribution and/or population status. A taxon in this category may be well studied, and its biology is well known, but appropriate data on abundance and/or distribution are lacking. Data deficient is therefore not a category of threat. Listing of taxa in this category indicates that more information is required and acknowledges the possibility that future research will show that threatened classification is appropriate. It is important to make positive use of whatever data are available. In many cases, great care should be exercised in choosing between data deficient and a threatened status.
- Not evaluated: A taxon is not evaluated when it is not yet been assessed against the criteria (IUCN).

5.4.1 Conservation Status of Plant Species in Study Region (as per IUCN categories)

Out of the total species, three species from the study area are vulnerable and 17 species are least concern as per the IUCN category. According to IUCN status, the three vulnerable species are *Acacia ferruginea, Chloroxylon swietenia,* and *Santalum album.* The species *Acacia auriculiformis, Erythrina indica, Couroupita guianensis, Punica granatum, Euphorbia tirucalli, Alternanthera sessilis, Bacopa monnieri, Commelina benghalensis, Commelina diffusa, Hydrilla verticillata, Lemna gibba, Phyla nodiflora, Saccharum spontaneum, Sphaeranthus amaranthoides, Typha angustifolia. Marsilea quadrifolia,* and *Thuja occidentalis* belonging to the pteridophytes and gymnosperms, respectively, are recorded least concern species as per IUCN categories from the study area. In the total recorded species from the study region, 159 species are data deficient and 114 species are not evaluated. None of the species from study area is found under the categories of CITES and IWPA.

5.4.2 Conservation Status of Animal Species in the Study Region (as per IUCN categories)

Among 164 species of insects, 117 species have been evaluated for conservation status at global scale under IUCN. Of the 117 species, 13 species are listed under the category of least concern and 104 under the category of data deficient of IUCN. And 47 species are not evaluated for their conservation status. Eighty-two (82) species of spiders have been documented from the study area, of which 23 species are not evaluated for the conservation status and remaining 59 species are listed under data-deficient category of IUCN. Seventeen (17) species of mollusks were recorded from the study region, of which 11 are not evaluated and six are listed under least concern category of IUCN.

A total of 11 species of fishes have been found in the study region. One species is not evaluated and remaining 10 species are evaluated and categorized under IUCN specifications. Two species are in data deficient category, four are least concern, two are near threatened, one species is vulnerable, and one is endangered (Box 1).

Box 1: Fish species under IUCN categories

Puntius denisonii (Denison Barb)—Endangered
Cyprinus carpio (Common carp)—Vulnerable
Ompok bimaculatus (Indian Butter-catfish)—Near Threatened
Wallago attu (Wallago)—Near Threatened

Among the 5 species of reported amphibians, 3 species are in the category of not evaluated, and remaining 2 species are evaluated and found to be categorized under least concern category of IUCN.

Among the 13 species of reptiles, two are not evaluated, 10 are evaluated, five are categorized under data deficient, and five are as least concern. The main threat for reptiles in the area is road accidents due to running vehicles on the road. However, accidental killing of snakes by the people is also reported in the area.

Among 71 species of birds, two species are not evaluated for IUCN and remaining 69 species are evaluated and considered under various categories of IUCN. Five species were lacking the adequate data put under data deficient category, 62 species are listed under least concern, and two species are reported to be near threatened (Box 2).

Box 2: Bird species under IUCN categories

Mycteria leucocephala (Painted stork)—Near Threatened
Sterna aurantia (River tern)—Near Threatened

A total of 10 mammalian species were found in the study region. According to IUCN, all species are evaluated from study area of which eight species are categorized under least concern, 1 species categorized under near threatened, and 1 species listed as endangered (Box 3). The main threat for common rabbit and wild boar is hunting. The habitat destruction was reported to the threat for Giant Golden-crowned flying fox.

Box 3: Mammalian species under IUCN categories

Oryctolagus cuniculus (Common Rabbit)—Near Threatened.
Acerodon jubatus (Giant Golden-Crowned flying fox)—Endangered.

5.5 Indian Wildlife Protection Act (IWPA)

India has a strong legal mechanism and policy framework to regulate wildlife trade in the market by framing a special piece of legislation for the protection and conservation of wildlife in the year 1972. As framed, Wildlife Protection Act 1972 prohibits the trade of 1800 wild species of animals and plants to meet its objectives (**WWF**). The main objective of this act is providing protection to the wild animals and plants by prohibiting hunting, poaching, picking, and uprooting specified wild species and this Act empowers central government to declare some areas as wildlife sanctuaries, national parks, and zoos (http://www.haryanaforest.gov.in and IWPA 1972).

- **Prohibition of hunting**: In this Act, hunting and smuggling of wild animals is strictly prohibited. Hunting is the major threat for loosing wildlife animals.
  Prohibition of picking, uprooting specified plants: This Act of 1972 lays emphasis on specified forests where people are not allowed to pick, uproot, damage, and collect any specified plant from the wild. Nevertheless, with the special permission from the forest department, scientists and researchers can collect the specimens for the purpose of education.
- **Sanctuaries, National Parks, and Closed Areas**: Under this act, any suitable area can be declared as wildlife sanctuary, national parks, and closed area by the central government to develop wildlife. In these places, no persons are allowed without any permission from the forest department.
- **Recognition of Zoos**: Zoos are constituted by the Central Zoo Authority (CZA) and Central government will specify the minimum standards for housing, upkeep, and care of the animals kept in the zoo.
- **Trade or commerce in wild animals and plants**: No person is allowed to sell and offer for sale or export any wild animals and plants to other places except written permission from the forest department. Wild animals and plants are the property of government, and killing, smuggling, and teasing wild animals will be treated as crime.

The Act contains 66 Sections divided into seven chapters and six schedules. Chapter I contains short title and definitions. Chapter II deals with Authorities under the Act. Chapter III deals with the protection of Specified Plants. Chapter IV provides for declaration of sanctuaries, national parks, and closed areas. Chapter IV-A deals with the Central Zoo Authority and recognition of zoos. Chapter V deals with trade or commerce in wild animals, animal articles, and trophies. Chapter V-A deals with prohibition of trade or commerce in trophies, animal articles, etc. Chapter VI relates to prevention and detection of offenses and, finally, Chapter VII contains miscellaneous provisions.

It has six schedules, which give varying degrees of protection. Schedule I and part II of Schedule II provide absolute protection—offenses under these are prescribed the highest penalties. Species listed in Schedule III and Schedule IV is also protected, but the penalties are much lower. Schedule V includes the animals that may be hunted. The plants in Schedule VI are prohibited from cultivation and planting. For hunting, the enforcement authorities have the power to compound offenses under this Schedule.

5.5.1 Fauna Listed Under IWPA from Study Area

Under IWPA, none of the plant species is listed from the study area. However, 65 faunal species were listed under IWPA Schedules. Among the species, one insect species is listed in IWPA. From vertebrates, 1 reptile, 60 bird species, and 3 mammals are listed under IWPA from the study area. In Box 5, species names are given as per various schedules of under IWPA. Spiders, Molluscans, fishes, and amphibians are not listed in the Indian Wildlife Protection Act from the study area. *Sus scrofa* (Wild boar) listed in IWPA Schedule III, however, as per local people this species being hunted by poachers for various reasons (Box 4).

Box 4: IWPA Listed Faunal species from study Area

Hypolimnas misippus Family—Nymphalidae (Insects)—(Schedule-I)
Veranus bengalensis (Indian monitor lizard)—Reptile (Schedule-I)
Streptopelia senegalensis (Laughing dove)—Aves (Schedule-IV)
Streptopelia chinensis (Spotted dove)—Aves (Schedule-IV)
Pavo cristatus (Indian peafowl)—Aves (Schedule-I)
Vulpes bengalensis (Indian fox)—Mammal (Schedule-II)
Macaca radiata (Common monkey)—Mammal (Schedule-II)
Sus scrofa (Wild boar)—Mammal (Schedule-III)
Remaining 57 birds in Schedule V and Schedule IV are given in Appendix.

5.6 The Convention on International Trade in Endangered Species of Wild Fauna and Flora (CITES)

CITES is an international contract among governments of different countries. The aim of the CITES is to internationally trade in specimens without damaging of wild animals, plants, and their survival. Currently there are 176 parties having membership in CITES by having conservation agreements. Since 1976, India also became a member of the CITES. Many countries are facing the problem of illegal trade in wildlife including India. The main reason for this is the lack of enforcement and implementation of the laws, which relate to wildlife protection.

Approximately 29,000 species of plants and 5000 species of animals and wildlife products (e.g., Food products, exotic leather, and medicines) are confined by CITES against over-exploitation through international trade. Every protected species is listed in CITES under one of the lists, called Appendices.

All the plants and animal species listed in three Appendices of CITED are based on their protection given to them (CITES).

- **Appendix-I**: Most endangered animals and plants are listed under Appendix-I. CITES prohibits international trade in these species with some exceptions, e.g., specimens can be used for scientific research. As on December 22, 2011, under Appendix-I 669 faunal species and 305 plant species were listed.
- **Appendix-II**: The species that may become extinct unless trade is controlled is listed under this Appendix and as such 4693 species of fauna and 22,105 species of plants listed from December 22, 2011.
- **Appendix-III**: Species at the demand of a country that previously regulates trade in the species, when the country needs support of other countries to make compulsory the trade convention (e.g., to prevent exploitation of the species). 160 faunal species and 120 plant species are listed under Appendix-III as on December 22, 2011.

5.6.1 Species Listed Under CITES from the Study Area

None of the plants and invertebrates are listed in CITES from the study area. From the vertebrates, three reptiles, three aves, and three mammals are listed in the Appendices I, II, and III to the CITES (Box 5). However, trading of wild animals in any form the region was not seen during the study period.

Box 5: CITES Listed Faunal species from study Area

Ptyas mucosus (Rat snake)—Reptile—(Appendices II)
Naja naja (Indian cobra)—Reptile—(Appendices II)
Varanus bengalensis (Indian monitor Lizard)—Reptile—(Appendices-I)
Accipiter badius (Shikara)—Aves—(Appendices-II)
Aquila nipalensis (Stepped eagle)—Aves (Appendices-II)
Haliastur Indus (Brahminy kite)—Aves (Appendices-II)
Vulpes bengalensis (Indian fox)—Mammal (Appendices-III)
Macaca radiate (Common monkey)—Mammal (Appendices-II)
Herpestes edwardsii (Indian gray mongoose)—Mammal (Appendices-III)

5.7 Conservation Methods

Conservation methods: In the study area, threatened species can be conserved using the following methods.

- In situ methods
- Ex situ methods

In situ conservation is onsite conservation; it is a process of conserving species in their own natural habitats and natural conditions. Ex situ conservation is called offsite conservation. It is the process of conserving threatened species of plants and animals outside its natural habitat. In this method, human involvement will play a more important role in botanical gardens and zoos. Here are the few conservation methods that need to be implemented in the study area:

- Natural regeneration
- Botanical gardens development
- Nursery development
- Forest conservation
- Agro forest conservation
- Threatened species conservation
- Implementation of Protective Laws
- Educating the Public
- Habitat improvement

1. **Natural regeneration**: Generally, in all the plant species natural regeneration occurs in a life cycle and increases their population.
2. **About four species listed under IUCN categories are from** the study area. *S. album* is naturally growing in several places such as forests, wastelands, and nearby villages. This species is also growing in the boundaries of the agricultural fields. This species are growing naturally and abundantly throughout Karnataka and other states of South India. For this species, compared to southern and western parts of the Karnataka, the distribution in this region is very low. Species namely *A. ferruginea* and *C. swieteniaa* are planted by the Forest department in the forestlands and wastelands. Status of these three plant species and their distribution in India and elsewhere are given in Table 5.1.
3. **Botanical garden development**: In the study area, botanical gardens have to be established for conservation and regeneration of plant species. Not many botanical gardens are found to be established in the region except in Gulbarga University, Gulbarga. Therefore, small botanical gardens can be established at premises of the various institutions, for example in the premises of various educational and other institutions (colleges, agriculture and horticulture departments etc.).
4. **Nursery development**: A nursery is a place where plants are propagated and grown to usable size. Nurseries are one of the best examples of ex situ method of conservation. Nurseries need to be established for conserving the

Table 5.1 Status of all plant species listed in IUCN Red Data Book, the current status and distribution

S. no.	Botanical name	Vernacular name	IUCN status	Distribution in the world	Distribution in India
1	*Acacia ferruginea* DC.	Banni gida	Vulnerable	India, Nepal, Sri Lanka	Peninsular India
2	*Chloroxylon swietenia* DC.	Masivala	Vulnerable	Sri Lanka, India, Africa—Nigeria, and Madagascar	Central and southern India
3	*Santalum album* L.	Gandadamara	Vulnerable	India, China, Sri lanka, Indonesia, Malaysia, Philippines, and Australia	Karnataka, Maharashtra, Tamil Nadu, Kerala, Andhra Pradesh, Uttar Pradesh Less than, Madhya Pradesh, Orissa

species. Through nurseries, plants can easily propagate using various methods. Nurseries are helpful to create awareness and produce resource on threatened species, medicinal species, timber species, and other ornamental species. Villagers of the study area need to be trained for this purpose, and they should be motivated to develop nurseries in the villages.

5. **Agroforestry development**: Agroforestry is the system of land use that combines growing and rising of crops and livestock along with multipurpose plant species. The land can be used to raise agricultural crops and trees, and to rear animals. The trees can be planted along the boundaries of the agricultural land for conservation and this will provide production of food, fodder, and fire wood for the people. This kind of land-use development should be taken as priority basis.
6. **Forest conservation**: Forest conservation is one of the primary steps needed to be taken for the conservation of the threatened species and development of the region. This region is an arid zone in southern India where plant growth is limited as well as sparsely distributed. The Forest Department should involve people in various conservation programmes.
7. **The creation of protected areas**: Protected areas are important for conservation of biodiversity. Identification of areas for the establishment of protected areas in the study region has to be given priority.
8. **Awareness among the people**: There is a need to conduct time-to-time awareness programs biodiversity conservation and sustainable socioecological development.
9. **Habitat improvement**: Habitats of wildlife should be improved by constructing water bodies and plantation and nourishment of grasses and trees.

The summary of the animal species as per their status under CITES, IWPA, and IUCN is given in Table 5.2. According to International Union for the Conservation

Table 5.2 Animal species as the status of CITES, IWPA, and IUCN

	No. of species	No. of families	CITES	IWPA	IUCN								
					NE	DD	LC	NT	VU	EN	CR	EW	EX
Insects	164	67	–	1	47	104	13	–	–	–	–	–	–
Spiders	82	19	–	–	22	55	–	–	–	–	–	–	–
Molluscans	17	9	–	–	11	–	6	–	–	–	–	–	–
Fishes	11	4	–	–	1	2	4	2	1	1	–	–	–
Amphibians	5	3	–	–	3	–	2	–	–	–	–	–	–
Reptiles	13	6	3	1	2	5	5	–	–	–	–	–	–
Aves	71	37	3	60	2	5	62	2	–	–	–	–	–
Mammals	10	9	3	3			8	1	–	1	–	–	–

of Nature (IUCN), there are four fishes, two birds, and two mammals listed under endangered, vulnerable, and near threatened categories in the study area.

Among the four fishes, *Puntis denisonii* is the endangered, *Cyprinus carpio* vulnerable, and *Ompok bimaculatus* and *Wallago attu* are near threatened species as categorized by IUCN. The main habitat of these fishes is river, lakes, and tanks.

The major threats confronted in the conservation of these fishes are anthropogenic activities such as resource pollution, habitat destruction, indiscriminate use of pesticides, and over-exploitation as a source of food.

Conservation of these fishes demands an appropriate strategy and approach to deal with. First of all, protection and conservation of their habitat is a priority whereby water bodies such as rivers, lakes, and tanks have to be free from pollution and unabated use of pesticides. There is an immense need to check the over-exploitation of these fishes by humans as a source of food. The guidelines of IUCN should be properly followed to protect these species of fishes at micro- and macro-levels.

Two birds, *M. leucocephala* and *S. aurantia,* are listed as near threatened under IUCN. The main habitats for *M. leucocephala* are river banks, lakes, reservoirs, and paddy fields and for *S. aurantia* are rivers and fresh water lakes. The major threat for these birds is habitat loss, pollution, and disturbance in their habitats. The conservation of these birds is very vital to save them from extinction. The main strategies are protecting their habitat, restricting human interference, controlling pollution, and checking hunting of these birds.

Two mammalian species are categorized under IUCN as near threatened and endangered from our study area. One is *O. cuniculus* whose habitat is in burrows near grasses and cereal fields. *A. jubatus* lives in terrestrial habitat preferring long trees such as fig trees. Habitat destruction, pesticides used in agriculture, and hunting of rabbit for food are the main threats for this mammalian species in our study area.

For the conservation of these fishes, birds, and mammals, the focus should be on the habitat conservation of all the species in danger of becoming extinct according to IUCN. There should be sustainable growth, which will protect the environment and save their habitats, and there is intense need to prevent hunting of rabbits in the study area. Educational and awareness programmes have to be carried on for the conservation of rivers and lakes. Laws and policies are effective in the conservation of biodiversity; thus, there is a need for the authorities to make effective use of them. The distribution of animal species in India and elsewhere from the study region cited under IWPA, CITES, and IUCN is described in detail in Tables 5.3, 5.4, and 5.5, respectively.

Table 5.3 Status of species under IWPA (Indian Wildlife Protection Act) from the study region and their distribution in India and elsewhere

S. no.	Family	Scientific name	Common name	IWPA	Distribution in world	Distribution in India
Insects						
1	Nymphalidae	*Hypolimnas misippus* Linnaeus, 1764	Danaid egg fly	Sch-IV	Found across Africa, Asia, and Australia	It occurs throughout India
Reptiles						
1	Colubridae	*Ahaetulla nasuta* Lacepede, 1789	Green snake	Sch-IV	India, Bangladesh, Sri Lanka, Myanmar, Thailand, Cambodia, Vietnam	It occurs throughout India
2	Colubridae	*Ptyas mucosus* Linnaeus, 1758	Rat snake	Sch-II	The Oriental Rat Snake has an extensive geographical distribution in Asia	It occurs throughout India including the islands of Andaman and Nicobar
3	Elapidae	*Naja naja* Linnaeus, 1758	Indian cobra	Sch-II	The Indian cobra is native to the Indian subcontinent which includes present-day Nepal, Pakistan, India, Bangladesh, and Sri Lanka	It occurs throughout India
4	Elapidae	*Bungarus Caeruleus* Schineider, 1801	Krait	Sch-IV	This species is found in Peninsular India from Sindh (Pakistan), to the West Bengal plains. It occurs throughout South India and Sri Lanka at elevations up to about 1600 m	Found in most of India including the Andaman and Nicobar, up to 1700 m above sea level. They are scarce in Bengal, Assam, and Orissa, where Banded Kraits are found
5	Varanidae	*Varanus bengalensis* Daudin, 1802	Indian monitoring Lizard	Sch-I	Afghanistan; Bangladesh; Cambodia; China; India; Indonesia (Jawa, Sumatera); Iran, Islamic Republic of; Lao People's Democratic Republic; Malaysia; Myanmar; Nepal; Pakistan; Sri Lanka; Thailand; Vietnam	It occurs throughout India

(continued)

Table 5.3 (continued)

S. no.	Family	Scientific name	Common name	IWPA	Distribution in world	Distribution in India
Aves						
1	Accipitridae	*Accipiter badius*	Shikra	Sch-IV	Widely distributed in Asia and Africa	Throughout India
2	Accipitridae	*Aquila nipalensis*	Stepped eagle	Sch-IV	The Steppe Eagle breeds from Romania east through the South Russian and Central Asian steppes to Mongolia. The European and Central Asian birds winter in Africa, and the eastern birds in India. It lays 1–3 eggs in a stick nest in a tree. Throughout its range it favors open dry habitats, such as desert, semidesert, steppes, or savannah	Central Part of India, Karnataka and Andhra Pradesh
3	Accipitridae	*Elanus axillaris*	Black shouldered kite	Sch-IV	Sub-Saharan Africa and tropical Asia, but it has a foothold within Europe in Spain and Portugal	Throughout subcontinent
4	Accipitridae	*Haliastur indus*	Brahminy kite	Sch-IV	This kite is a familiar sight in the skies of Sri Lanka, Nepal, India, Pakistan, Bangladesh, and southeast Asia and as far south as New South Wales, Australia, through which region it is widespread and resident	Throughout India
5	Aegithinidae	*Aegithina tiphia*	Common iora	Sch-IV	India to Southwest China, Southeast Asia	Himalayan foothills, from Himachal to Arunachal, and most of subcontinent from east of Gujarat to the southern tip
6	Alaudidae	*Eremopterix griseus*	Ashy-crowned sparrow lark	Sch-IV	India, Pakistan, Bangladesh, Sri Lanka	Endemic to sub-Himalayan subcontinent
7	Alaudidae	*Mirafra erythroptera*	Indian Bush lark	Sch-IV	India, Pakistan, Bangladesh, Sri Lanka	North west India and most of the subcontinent
8	Anatidae	*Anas poecilorhyncha*	Indian spot bill duck	Sch-IV	Bangladesh; Cambodia; China; Hong Kong; India; Lao People's Democratic	Throughout subcontinent

(continued)

Table 5.3 (continued)

S. no.	Family	Scientific name	Common name	IWPA	Distribution in world	Distribution in India
					Republic; Myanmar; Nepal; Pakistan; Sri Lanka; Thailand; Vietnam	
9	Ardeidae	*Ardea cinerea*	Grey heron	Sch-IV	Throughout Asia, Europe	Throughout India
10	Ardeidae	*Ardea purpurea*	Purple heron	Sch-IV	Africa, central and southern Europe, and southern and eastern Asia	Indian subcontinent
11	Ardeidae	*Ardeola grayii*	Indian pond heron	Sch-IV	Southern Iran and east to India, Burma, Bangladesh, and Sri Lanka	Throughout Indian subcontinent
12	Ardeidae	*Bubulcus ibis*	Cattle egret	Sch-IV	Nearctic, Palearctic, Oriental, Ethiopian, Neotropical, Australian, Oceanic Islands	Throughout Indian subcontinent
13	Ardeidae	*Casmerodius albus*	Great egret	Sch-IV	Widely distributed egret. Distributed across most of the tropical and warmer temperate regions of the world, in southern Europe it is rather localized. In North America it is more widely distributed, and it is ubiquitous across the Sun Belt of the USA and in the Neotropics	Throughout Indian subcontinent
14	Ardeidae	*Egretta garzetta*	Little egret	Sch-IV	Nearctic, Palearctic, Oriental, Ethiopian, Neotropical, Australian, Oceanic Islands	Throughout Indian subcontinent
15	Caprimulgidae	*Caprimulgus indicus*	Grey nightjar	Sch-IV	India and Sri Lanka	Dry and moist deciduous jungle. Himalayas, North east hill states, assam and rest of Indian subcontinent
16	Cerylidae	*Ceryle rudis*	Lesser pied kingfisher	Sch-IV	Widely distributed across Africa and Asia	
17	Charadriidae	*Vanellus malabaricus*	Yellow wattled lapwing	Sch-IV	They are found in most parts of India, parts of Pakistan, Nepal, Bangladesh and Sri Lanka	Endemic to the Indian subcontinent. It is found mainly on the dry plains of peninsular India
18	Ciconidae	*Mycteria leucocephala*	Painted stork	Sch-IV	Widely distributed over the plains of Asia	Large marshes of Indian subcontinent

(continued)

Table 5.3 (continued)

S. no.	Family	Scientific name	Common name	IWPA	Distribution in world	Distribution in India
19	Ciconidae	*Ciconia episcopus*	White necked stork	Sch-IV	It is a widespread tropical species which breeds in Asia, from India to Indonesia, and also in Africa	Flooded grasslands, irrigated fields, marshes in forest, and also sometimes in dry grasslands of Indian subcontinent
20	Columbidae	*Streptopelia senegalensis*	Laughing dove	Sch-IV	The species is found in much of Sub-Saharan Africa, Saudi Arabia, Iran, Afghanistan, Pakistan and India. It is also found in Israel, Lebanon, Syria, and Turkey	Open semiarid biotope of Indian subcontinent
21	Columbidae	*Streptopelia chinensis*	Spotted dove	Sch-IV	Pakistan, Nepal and India south to Sri Lanka and east to Bhutan and Assam	Dry and moist-deciduous biotope of Indian subcontinent. Except the arid north west region
22	Corvidae	*Corvus splendens*	House crow	Sch-V	Nepal, Bangladesh, India, Pakistan, Sri Lanka, Maldives and Laccadive Islands, South West Thailand and coastal southern Iran, Africa, Zanzibar, Port Sudan, Hook of Holland	South of Himalayan foothills
23	Cuculidae	*Centropus sinensis*	Greater coucal	Sch-IV	A widespread resident in Asia, from India, east to South China and Indonesia	South of the Himalayas and east of Indus valley except Thar desert
24	Cuculidae	*Eudynamys scolopaceus*	Asian Koel	Sch-IV	It is found in South Asia, China, and Southeast Asia	Lightly wooded country, towns and villages of subcontinent
25	Cuculidae	*Phaenicophaeus viridirostris*	Blue faced malkoha	Sch-IV	It is restricted to Sri Lanka and southern India	Throughout southern part of India
26	Dicruridae	*Dicrurus macrocercus*	Black drongo	Sch-IV	Tropical Southern Asia from southwest Iran through India and Sri Lanka east to southern China and Indonesia	Open deciduous forest and cultivation lands. Most of the Indian subcontinent
27	Estrildidae	*Lonchura malabarica*	Indian silver bill	Sch-IV	Found in the Indian subcontinent and joining regions	Middle East and the Indian subcontinent

(continued)

Table 5.3 (continued)

S. no.	Family	Scientific name	Common name	IWPA	Distribution in world	Distribution in India
28	Estrildidae	*Lonchura malacca*	Black headed munnia	Sch-IV	India, Sri Lanka, Southern China, Taiwan and most of Southeast Asia. The species was also introduced to Guam, Jamaica, Palau, Puerto Rico, Hawaii, Vanuatu and Venezuela	Endemic to Western Ghats of the Palghat gap; North east ghats of Orissa and north Andhra
29	Estrildidae	*Lonchura punctulata*	Spoted munia	Sch-IV	Asia and occurs from India and Sri Lanka east to Indonesia and the Philippines	Open woodland, bush-clad hillsides, grassland, gardens and cultivation lands of subcontinent
30	Halcyonidae	*Halcyon smyrnensis*	White breasted king fisher	Sch-IV	Widely distributed in Eurasia from Bulgaria, Turkey, West Asia east through the Indian subcontinent to the Philippines	Most of the subcontinent
31	Motacillidae	*Anthus cinnamomeus*	African Pipit	Sch-IV	Europe, Asia, Africa	Grassland, marshy ground of Indian subcontinent
32	Motacillidae	*Anthus rufulus*	Paddyfield pipit	Sch-IV	Southern Asia east to the Philippines	Grassland, marshy ground of Indian subcontinent
33	Muscicapidae	*Copsychus saularis*	Oriental Magpie robin	Sch-IV	Tropical southern Asia from Bangladesh, interior India, Sri Lanka, and eastern Pakistan east to Indonesia, Thailand, south China, Malaysia, and Singapore	Subcontinent, south of the Himalayas and east of Indus valley and Kachchh, Andaman
34	Muscicapidae	*Saxicoloides fulicata*	Indian robin	Sch-IV	It is widespread in the Indian subcontinent and ranges across Bangladesh, Bhutan, India, Nepal, Pakistan, and Sri Lanka	The Indian Robin is found throughout the whole of India from the Himalayas southwards to Ceylon
35	Muscicapidae	*Saxicola caprata*	Pied bushchat	Sch-IV	The Pied Bush Chat is a resident breeder in tropical southern Asia from the Greater Middle East through Pakistan, India, and Bangladesh eastwards to Indonesia	Most of the subcontinent, south of the Himalayas
36	Nectariniidae	*Nectarinia asiatica*	Purple sunbird (eclipse male)	Sch-IV	The species is distributed widely from West Asia through the Indian subcontinent and into Southeast Asia	Indian subcontinent, except high Himalayas

(continued)

Table 5.3 (continued)

S. no.	Family	Scientific name	Common name	IWPA	Distribution in world	Distribution in India
37	Nectariniidae	*Cinnyris asiaticus*	Purple sun bird	Sch-IV	The species is distributed widely from West Asia through the Indian subcontinent and into Southeast Asia	Indian subcontinent, except high Himalayas
38	Nectariniidae	*Nectarinia minima*	Crimson backed sunbird	Sch-IV	India	Endemic to the Western Ghats of India
39	Oriolidae	*Oriolus kundoo*	Indian golden oriole	Sch-IV	Indian subcontinent and Central Asia	Throughout India
40	Phalacrocoracidae	*Phalacrocorax fuscicollis*	Indian cormorant	Sch-IV	Peninsular India and northern part of Sri Lanka	Peninsular India
41	Phasianidae	*Francolinus pondicerianus*	Grey francolin	Sch-IV	Southeastern Iran and Pakistan, India to Bangladesh and northern Sri Lanka	Peninsular India
42	Phasianidae	*Pavo cristatus*	Indian peafowl	Sch-I	South Asia	Throughout Indian subcontinent
43	Picidae	*Dinopium benghalense*	Lesser goldenbacked woodpecker	Sch-IV	South Asia	Throughout India
44	Ploceidae	*Ploceus philippinus*	Baya weaver	Sch-IV	Lowlands of southeast Pakistan to India and Sri Lanka	Throughout India
45	Psittaculidae	*Psittacula krameri*	Rose ringed parakeet	Sch-IV	Native to India west Pakistan, Nepal, and central Burma	Throughout India
46	Pycnonotidae	*Pycnonotus cafer*	Red vented bulbul	Sch-IV	Indian subcontinent	Throughout India
47	Rallidae	*Amaurornis phoenicurus*	White-breasted water hen	Sch-IV	South Asia from Pakistan, Maldives, India and Sri Lanka to south China and Indonesia	Throughout India

(continued)

Table 5.3 (continued)

S. no.	Family	Scientific name	Common name	IWPA	Distribution in world	Distribution in India
48	Rallidae	*Fulica atra*	Common coot	Sch-IV	Coots have a wide global distribution, occurring from Europe and northern Africa in the west, through the Middle East, and across Eurasia, east to the Pacific coasts of China and Japan. The species also occurs in India, southeast Asia, and Australasia	Throughout India
49	Rallidae	*Porphyrio porphyrio*	Purple moorhen	Sch-IV	Southern Africa, Southern Asia, Papua new Guinea, Australia, and New Zealand	Throughout India
50	Recurvirostridae	*Himantopus himantopus*	Black winged stilt	Sch-IV	Throughout Asia and south Africa	Throughout India
51	Sternidae	*Sterna aurantia*	River tern	Sch-IV	A wide range in southern Asia, being found in Pakistan, India, Nepal, Bhutan, Bangladesh, Myanmar, Thailand, Laos, Cambodia, Vietnam, and southern China	Throughout India
52	Sturnidae	*Acridotheres tristis*	Common myna	Sch-IV	Throughout Asia, New Zealand, Eastern Australia and Southern Africa and Madagascar	Widely distributed in India
53	Sturnidae	*Sturnus malabaricus*	Chestnut tailed starling	Sch-IV	Indian subcontinent	Throughout India
54	Sturnidae	*Sturnus pagodarum*	Brahminy starling	Sch-IV	Afghanistan to Bangladesh, Nepal, India, and Sri Lanka	Throughout India
55	Sturnidae	*Sturnus roseus*	Rosy Starling (juvenile)	Sch-IV	India, turkey, Kazakhstan, Ukraine, Afghanistan	Southern part of India
56	Threskionithidae	*Plegadis falcinellus*	Glossy ibis	Sch-IV	Warm regions of Europe, Asia, Africa, Australia, and the Atlantic and Caribbean region of the Americas	Warm regions of India
57	Threskionithidae	*Pseudibis papillosa*	Black ibis	Sch-IV	Indian subcontinent	Throughout India

(continued)

Table 5.3 (continued)

S. no.	Family	Scientific name	Common name	IWPA	Distribution in world	Distribution in India
58	Timaliidae	*Turdoides striata*	Jungle babbler	Sch-IV	Indian subcontinent	Endemic to south India
59	Timaliidae	*Turdoides affinis*	White headed babbler	Sch-IV	India and Sri Lanka	South India
Mammals						
1	Canidae	*Vulpes bengalensis* Sharo, 1800	Indian fox	Sch-II	Bangladesh, Nepal, and Pakistan	Throughout the country
2	Cercopithecidae	*Macaca radiata* E.Geoffroy, 1812	Common monkey	Sch-II	Only in India	South India, Gujarat and Maharashtra
3	Muridae	*Mus booduga* Gray, 1837	Little Indian field mouse	Sch-V	Bangladesh, Myanmar, Nepal, Pakistan, and Sri Lanka	Almost throughout the India
4	Muridae	*Mus musculus* Linnaeus, 1758	House mouse	Sch-V	Spread throughout most of world association with human beings	Almost throughout India
5	Suidae	*Sus scrofa* Linnaeus, 1758	Wild Boar	Sch-III	All continents except Antarctica and on many oceanic islands	Throughout the country

Table 5.4 CITES status of all fauna species and their distribution

S. no.	Family	Scientific name	Common name	CITES	Distribution in World	Distribution in India
Reptiles						
1	Colubridae	*Ptyas mucosus* Linnaeus, 1758	Rat snake	Ape-II	The oriental rat snake has an extensive geographical distribution in Asia. From west to east, it occurs in Iran, Turkmenistan, Afghanistan, Pakistan, India (incl. Andaman Isl.), Sri Lanka, Nepal, Bangladesh, Myanmar, China (incl. Hainan and Hong Kong), Thailand, Lao PDR, Cambodia, Vietnam, Malaysia, Singapore and Indonesia	It occurs throughout India
2	Elapidae	*Naja naja* Linnaeus, 1758	Indian cobra	Ape-II	The Indian cobra is native to the Indian subcontinent which includes present day Nepal, Pakistan, India, Bangladesh, and Sri Lanka	It occurs throughout India
3	Varanidae	*Varanus bengalensis* Daudin, 1802	Indian monitoring lizard	Ape-I	Afghanistan; Bangladesh; Cambodia; China; India; Indonesia (Jawa, Sumatera); Iran, Islamic Republic of; Lao People's Democratic Republic; Malaysia; Myanmar; Nepal; Pakistan; Sri Lanka; Thailand; Vietnam	It occurs throughout India
Aves						
1	Accipitridae	*Accipiter badius*	Shikra	Ape-II	Widely distributed in Asia and Africa	Throughout India
2	Accipitridae	*Aquila nipalensis*	Stepped eagle	Ape-II	The Steppe Eagle breeds from Romania east through the south Russian and Central Asian steppes to Mongolia. The European and Central Asian birds winter in Africa, and the eastern birds in India. It lays 1–3 eggs in a stick nest in a tree. Throughout its range, it favors open dry habitats, such as desert, semidesert, steppes, or savannah	Central Part of india, Karnataka and Andhra Pradesh

(continued)

Table 5.4 (continued)

S. no.	Family	Scientific name	Common name	CITES	Distribution in World	Distribution in India
3	Accipitridae	*Haliastur indus*	Brahminy kite	Ape-II	This kite is a familiar sight in the skies of Sri Lanka, Nepal, India, Pakistan, Bangladesh, and southeast Asia and as far south as New South Wales, Australia, through which region it is widespread and resident	Throughout India
Mammals						
1	Canidae	*Vulpes bengalensis* Sharo, 1800	Indian fox	Ape-III	Bangladesh, Nepal, and Pakistan	Throughout the country
2	Cercopithecidae	*Macaca radiata E. Geoffroy*, 1812	Common monkey	Ape-II	Only in India	South India, Gujarat and Maharashtra
3	Herpestidae	*Herpestes edwardsii E. Geoffroy* saint, Hlaire, 1818	Indian gray mongoose	Ape-III	Afghanistan; Bahrain; Bhutan; India; Indonesia; Iran, Islamic Republic of; Kuwait; Malaysia; Mauritius (Mauritius (main island)—Introduced); Nepal; Pakistan; Saudi Arabia; Sri Lanka	Throughout the country

Table 5.5 IUCN status of all fauna species and their distribution

S. no.	Family	Scientific name	Common name	IUCN	Distribution in World	Distribution in India
Fishes						
1	Cyprinidae	*Puntius denisonii* F. Day, 1865	Denison Barb	EN	India	Karnataka, Kerala
2	Cyprinidae	*Cyprinus carpio* Linnaeus, 1758	Common carp	VU	Wide spread freshwater fish of eutrophic waters in lakes and large rivers in Europe and Asia	Throughout the India
3	Siluridae	*Ompok bimaculatus* Bloch, 1794	Indian Butter-catfish	NT	Bangladesh; India; Pakistan; Sri Lanka	Throughout India
4	Siluridae	*Wallago attu* Bloch & Schneider, 1801	Wallago	NT	Bangladesh; India; Indonesia (Jawa); Myanmar; Nepal; Pakistan; Sri Lanka; Thailand; Vietnam	Throughout India
Birds						
1	Ciconidae	*Mycteria leucocephala*	Painted stork	NT	Widely distributed over the plains of Asia	Large marshes of Indian subcontinent
2	Sternidae	*Sterna aurantia*	River tern	NT	A wide range in southern Asia, being found in Pakistan, India, Nepal, Bhutan, Bangladesh, Myanmar, Thailand, Laos, Cambodia, Vietnam, and southern China	Throughout India
Mammals						
1	Pteropodidae	*Acerodon jubatus* Eschscholtz, 1831	Giant Golden-crowned flying fox	EN	India, Sri Lanka, and Burma	Rare in west Rajasthan, Cutch and Sindh. Does not occur in the higher hills

Distributional description for flora and fauna mentioned in Chap. 5 is based on IUCN records available on the Web page of IUCN

References

Convention on Biological Diversity https://www.cbd.int/

Kumar U, Asija MJ (2000) Biodiversity principles and conservation. Agrobios, Jodhpur

Khan TI, Shishodia YS (1998) Biodiversity and sustainable development. Pointer Publishers, Jaipur

IUCN, International Union for Conservation of Nature (https://www.iucn.org/)

IUCN Standards and Petitions Subcommittee (2010) Guidelines for Using the IUCN Red List Categories and Criteria. Version 8.1. Prepared by the Standards and Petitions Subcommittee

The Wildlife Protection Act 1972 India (2012) http://www.environmentabout.com/789/the-wildlife-protection-act-1972-india. Retrieved on 28 Nov 2012

Chapter 6
Remote Sensing and GIS in Understanding the Landscape Dynamics

6.1 Introduction

Information on land use/land cover in the form of maps and statistical data is very vital for spatial planning, management, and utilization of land. Land-use and land-cover (LULC) scenario in India has undergone a radical change since the onset of economic revolution in early 1990s (Roy et al. 2007). The change detection techniques are the process of identifying transformations in same area with respect to temporal term (Singh 1989). Multispectral satellite imagery has been successfully used in sustainable landscape management and in mapping and monitoring of biodiversity from local to global scales (Sandstrom et al. 2003; Lillesand et al. 2004). Change detection as defined by Hoffer (1978) is "temporal effects as variation in spectral response involves situations where the spectral characteristics of the vegetation or other cover type in a given location change over time". Change detection analysis provides the past and present information about the area which helps in understanding the relationship between human activity and natural resource utilization while providing the precautions to be taken to conserve the natural resources (Koppad and Tikhile 2012). The knowledge of land use land cover (LULC) is supportive for proper management, planning, and monitoring of natural resources (Zhu 1997).

Remote sensing (RS) technology and geographical information system (GIS) play an important role in understanding the landscape dynamics. The role of RS and GIS is especially important in cases where there is absence of historical ground data due to any constraint. RS is the science of obtaining information about an object, area, or phenomenon through the analysis of data acquired by a device that is not in contact with the object, area, or phenomena under investigation. The data analysis process involves examining the data using various image processing techniques in digital image processing software. Its application in the field of environmental management is of great importance. As stated the GIS is a system of hardware and software used for storage, retrieval, mapping, and analysis of geographical data.

S. Nautiyal et al., *Biodiversity of Semiarid Landscape*,
Environmental Science, DOI 10.1007/978-3-319-15464-0_6

A detailed study of land-use and land-cover change in the study area was done for four points of time (Year 1973, 1980, 2003, 2008). For the proposed study, Gogi village was selected as the center point which is located at 16° 43′ 35.40″N latitude and 76° 44′ 40.91″E longitude and situated in Yadgir district of Karnataka. The total study area is spread over 2800 km^2 in the semiarid landscape.

6.2 Materials and Methods

Landsat (MSS, TM, and ETM+) and IRS P6 (LISS-IV) data were used for studying the landscape dynamics of the study region.

The toposheet of the study area 56D9SE, 56D10NE, 56D14NW, and 56D13SW on 1:25,000 scales was procured from Survey of India (SOI), Bangalore, and georeferencing of these toposheets was done with Geographic Coordinate Projection system, WGS-84 datum. The vector layers of river, canal, water bodies, and road transportation were prepared for the entire area by digitizing process in the GIS-based software.

The flowchart (Fig. 6.1) illustrates the sequential steps and procedures, which have been executed for the generation of various vector layer and thematic maps of the study region.

Both ArcGIS and ERDAS IMAGINE software were used for raster and vector data analysis and map generation. The methodology applied for the processing of raster and vector data is as follows.

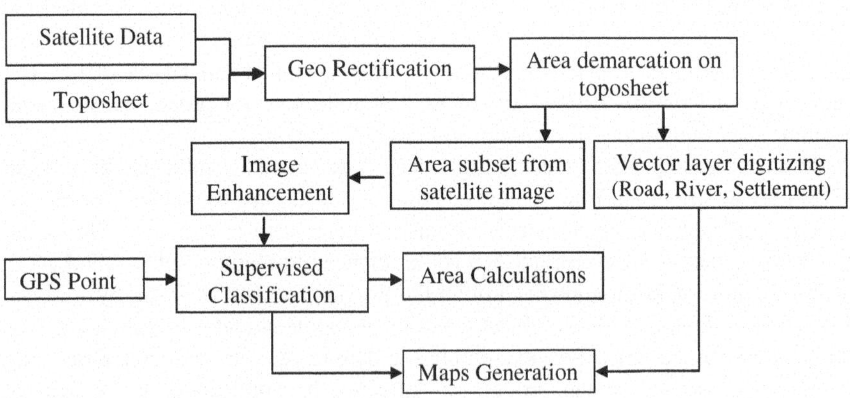

Fig. 6.1 Flowchart illustrating the methodology

6.2.1 Georectification

Geometric correction as mentioned in digital image processing methods, includes correction for geometric distortions due to sensor, earth geometry variations, and conversion of the data to real-world coordinates (e.g., longitude/latitude) on the earth's surface.

The images were checked for occasional shortcomings in the quality of radiometric and line dropouts and this is prerequisite, described in various research methodologies relating to RS GIS in landscape research. These images were then geometrically rectified with reference to the georeferenced Survey of India toposheets and ground control points (GCPs). GCPs are taken in the field using handheld GPS by nearest neighborhood resampling algorithm (Jensen 1996). The scenes were geometrically corrected with SOI toposheets using proper identification of GCPs with a root mean square (RMS) error of 0.0002–0.0003 pixels. Then, they were coregistered with SOI toposheets using Geographic Coordinate Projection system with WGS 84 datum.

6.2.2 Image Subset

The georeferenced toposheet was used for area demarcation with the help of polygon vector layer. The same vector file was used to subset area of interest from toposheet as well as satellite image using ERDAS IMAGINE software. Landsat (TM) image of the study region is given in Fig. 6.2.

Fig. 6.2 Landsat image of study region

6.2.3 *Image Enhancement*

Image enhancement is an important image processing function primarily done to improve the appearance of the imagery and to assist in visual interpretation and analysis. Various options of image enhancement techniques were tried out to get the best image for visual interpretation. Histogram equalization and stretch enhancement techniques were applied to the imagery of the study area for better appearance of different features in the satellite imagery.

6.2.4 *Generation of Vector Layers*

The toposheets was geo-referenced with projected coordinate system (WGS 84) and the same projection system was used for preparing the shape files to indicate the different landscape features for example forest areas, agriculture, settlement etc. Using ArcGIS software, the standard digitization process was used to prepare vector layers for study region to categorise them to the various layers, namely canal (Fig. 6.3), forest boundaries (Fig. 6.4), river and drainage system (Fig. 6.5), road network (Fig. 6.6), water bodies (Fig. 6.7), and village settlements (Fig. 6.8).

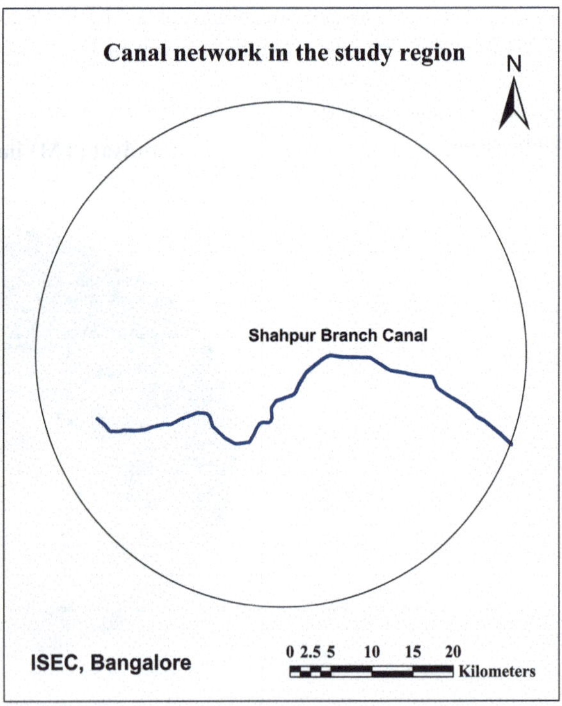

Fig. 6.3 Canal in the study region

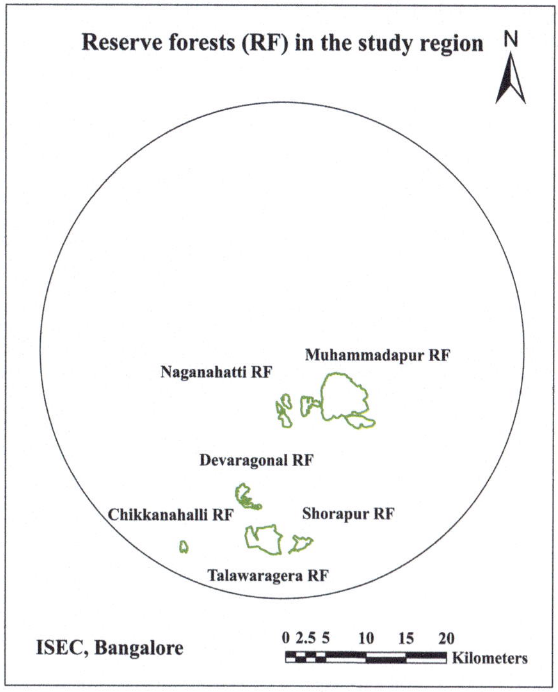

Fig. 6.4 Reserve forests (RF) in the study region

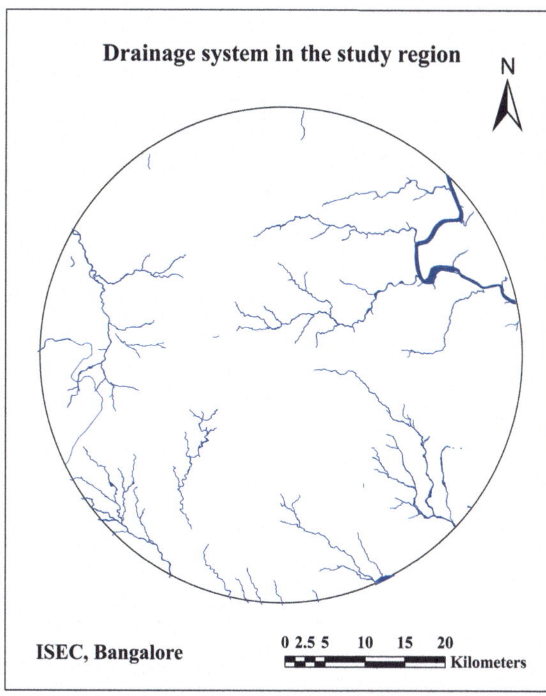

Fig. 6.5 Drainage system in the study region

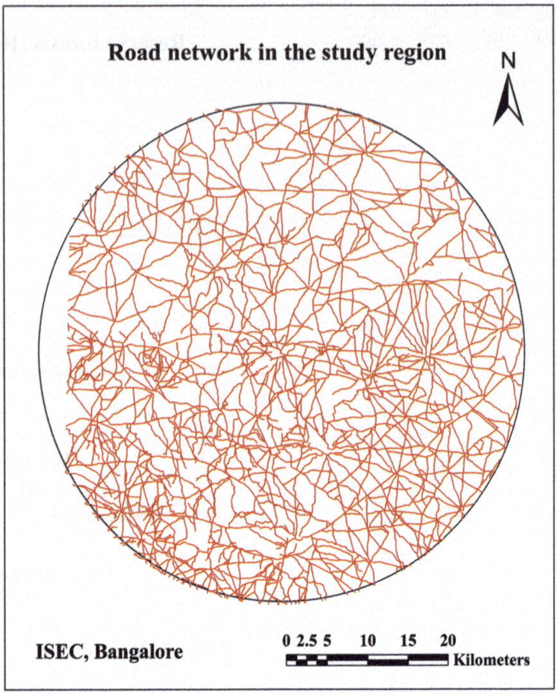

Fig. 6.6 Road network in the study region

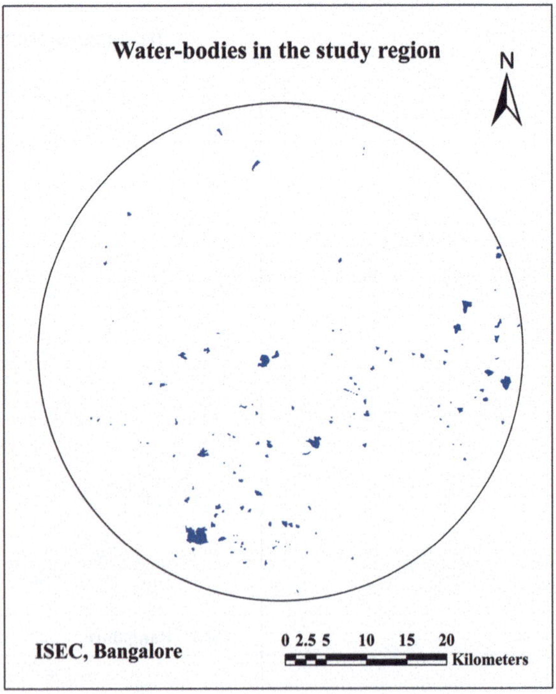

Fig. 6.7 Water bodies in semiarid landscape

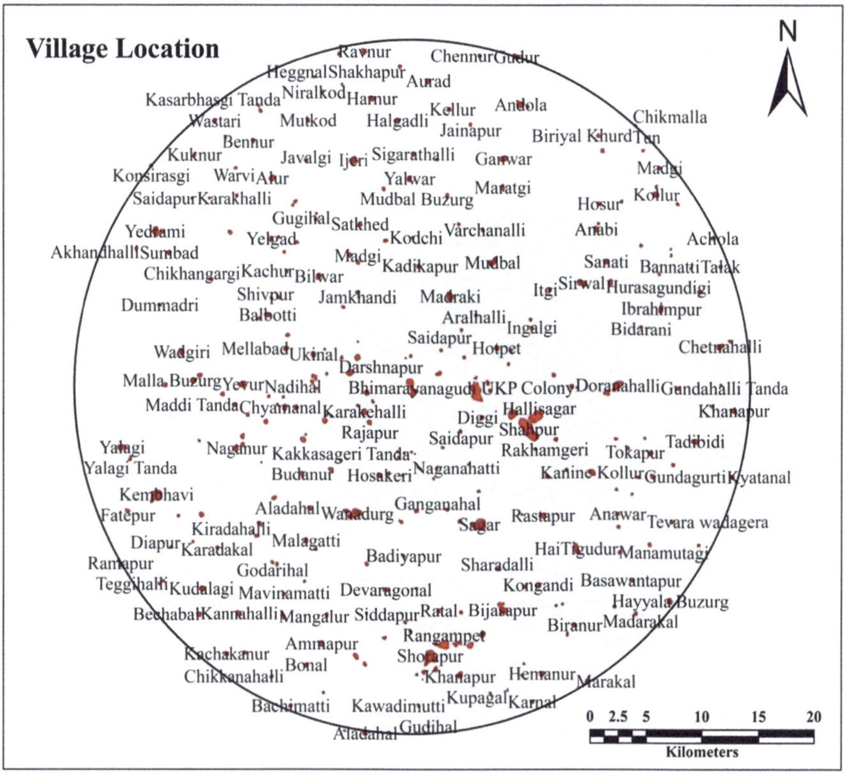

Fig. 6.8 Location of villages

6.2.5 Classification

For the generation of thematic map of study region from 1973 to 2008, the supervised classification method was used in ERDAS software. In the supervised classification the satellite image was classified in three steps—first is selection of training site for each land-cover classes defined by user with the help of GPS point. The second step is converting this training site into the signature file, and the third step is the classification process using different algorithms. The common supervised classification algorithms are maximum-likelihood and minimum distance classification. The classified land cover classes were identified and validated based on the available ancillary ground truth data and in-depth field knowledge. After classification process, water body, settlement, stony land, canal, agriculture, forest land, and fallow land were identified.

6.3 Results

The digital elevation model (DEM), slope map, and aspect map of the study region showing the topographic features are depicted in Figs. 6.9, 6.10 and 6.11, respectively.

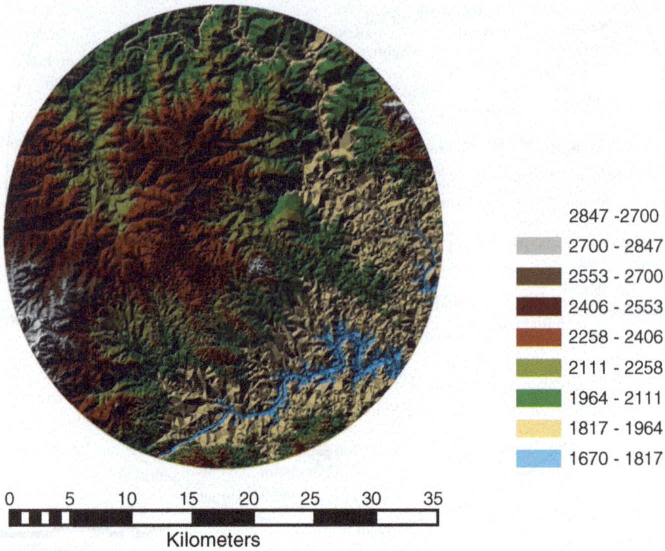

Fig. 6.9 The DEM of the study region

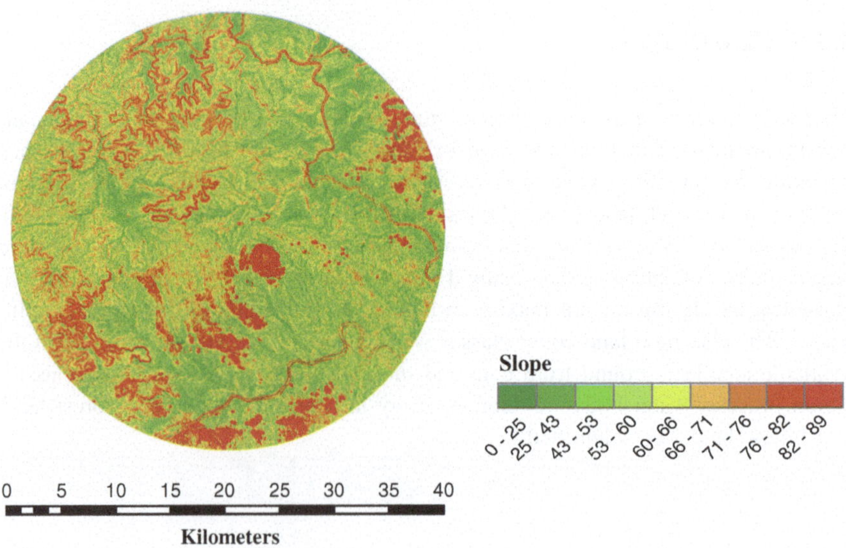

Fig. 6.10 Slope map of the study region

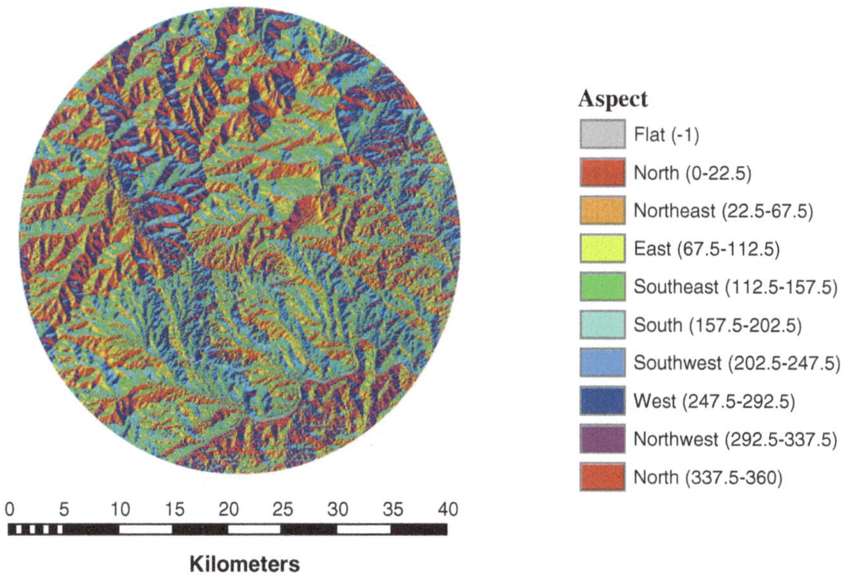

Fig. 6.11 Aspect map of the study region

6.3.1 Land Use/Land Cover from 1973 to 2008

The LULC area from 1973 to 2008 is shown in Tables 6.1 and 6.2 and Figs. 6.12, 6.13, 6.14, and 6.15. In the study region, maximum area is covered by agriculture land, followed by open/grassland and stony land (Table 6.1). The area available under forest cover is very less, i.e., 3.64 % of the total geographical area in the year

Table 6.1 Land-use and land-cover pattern of the study region at four points of time (1973, 1980, 2003, and 2008)

Land use and land cover	1973		1980		2003		2008	
	Area (km^2)	Area (%)	Area (km^2)	Area (%)	Area (km^2)	Area (%)	Area (km^2)	Area (%)
Forest	103.04	3.64	90.11	3.19	84.5	2.99	76.85	2.72
Settlement	30.01	1.06	33.01	1.17	37.01	1.31	39.07	1.38
Stony land	486.92	17.22	470.65	16.65	280.22	9.91	250.22	8.85
Canal	14.20	0.50	14.20	0.50	14.20	0.50	14.2	0.50
River	159.93	5.66	159.93	5.66	159.93	5.66	159.87	5.66
Agriculture	1239.49	43.84	1240.86	43.89	1308.21	46.28	1346.86	47.64
Open/grassland	592.49	20.96	485.98	17.19	444.55	15.73	402.56	14.24
Other infrastructure	200.92	7.11	332.26	11.75	498.38	17.63	537.37	19.01
Total area	*2827*		*2827*		*2827*		*2827*	

1973 (Fig. 6.12). Further, the forest area shows a declining trend over a period of time (Figs. 6.13, 6.14 and 6.15). The decline in forest area was maximum from 1973 to 1980 which was calculated to be about 12.93 km^2. The overall decrease in forest areas was recorded to be 26.19 km^2 from 1973 to 2008 (Table 6.2). In the

Table 6.2 Land use and land cover in the study region: a comparison for change at four points of time

Land use and land cover	1973–1980	1980–2003	2003–2008	1973–2008
	Area change	Area change	Area change	Area change
Forest	−12.93	−5.61	−7.65	−26.19
Settlement	3.00	4.00	2.06	9.06
Stony land	−16.27	−190.43	−30	−236.70
Canal	No change	No change	No change	No change
River	No change	No change	No change	No change
Agriculture	1.37	67.35	38.65	107.37
Open/grassland	−106.51	−41.43	−41.99	−189.93
Other infrastructure	131.34	166.12	38.99	336.45

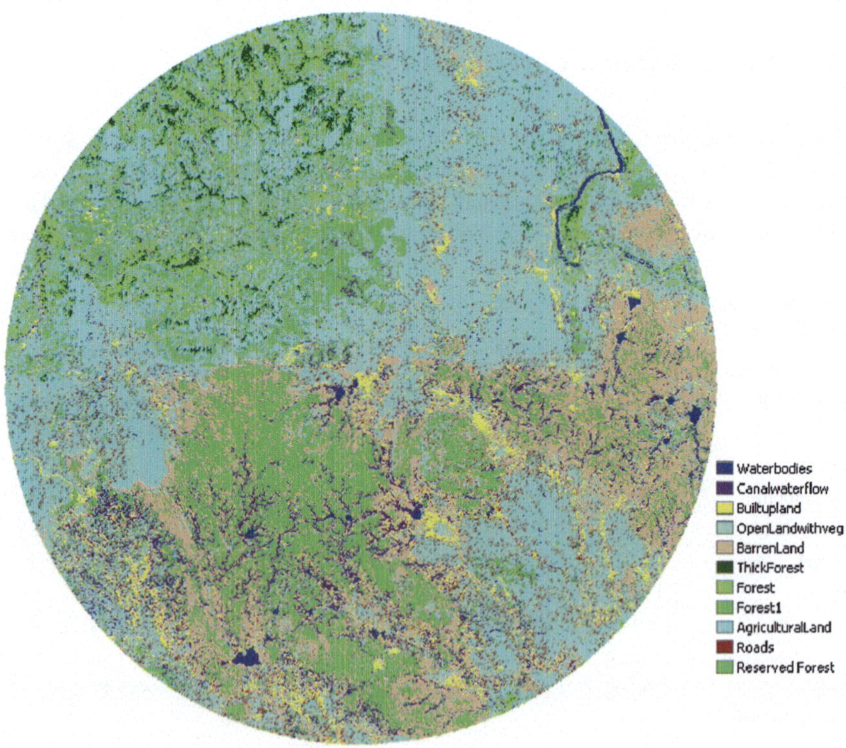

Fig. 6.12 Classified image of the area (1973)

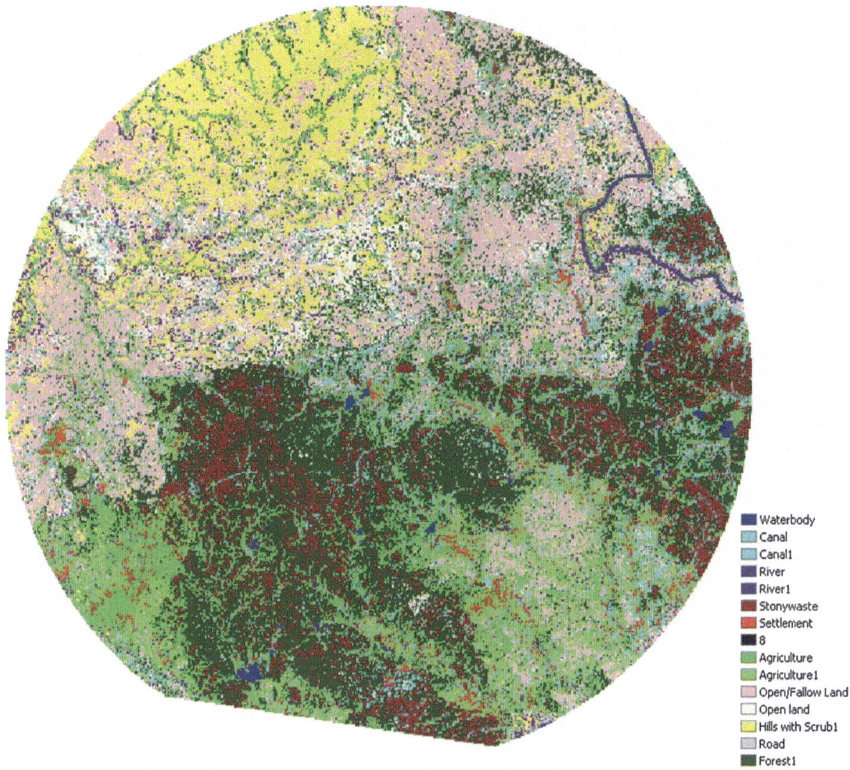

Fig. 6.13 Classified image of the area (1980)

study area, there are four notified reserve forests, namely Mahmmadapur, Naganattigi, Talwaragera, and those located in Shahapur taluk. The local communities access the reserved forests for fuel wood and fodder collection. Vegetation cover in the natural forest ecosystems is very sparse and dominated by *Prosopis juliflora* (Sarkara Jali). This species is occasionally associated with *Acacia nilotica* sub sp. *indica* (Kari jali) and *Ziziphus mauritiana*. The Forest Department has started growing species such as *Eucalyptus globulus*, *Azadirachata indica*, *Pongamia pinnata*, *Chloroxylon swietenia*, *Cassia fistula*, and *Acacia auriculiformis* plantations on barren land, wasteland, and along roadsides in many villages of the study region. The land-use and land-cover change analysis does not show any increase in total vegetation cover. In fact, the forest cover has shrunk by 3.64–1.38 % from 1973 to 2008.

Area under agriculture has had a positive change with an increase of 107.37 km^2 from 1973 to 2008. The maximum expansion of area under agriculture was found from 1980 to 2003. Total settlement area in the study region is 39.07 km^2 in the year 2008. The development of the settlement in the region is 3 km^2 during 1973–1980, 4 km^2 during 1980–2003, and 2.06 km^2 during 2003–2008, respectively.

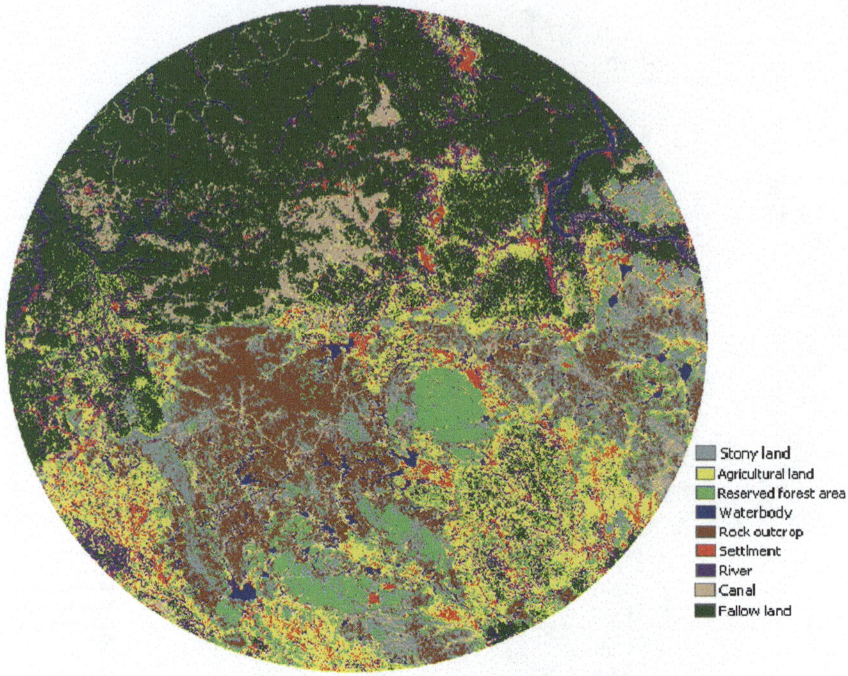

Fig. 6.14 Classified image of the area (2003)

Of special significance is the increase in infrastructure observed in the study region, especially from 1973 to 1980 (Table 6.2). The area under infrastructure was calculated to be 131.34 km^2 during 1973–1980 which was furhter expanded to 166.12 km^2 during the period of 1980–2003. Interestingly, the changed observed during 1973–1980 can be attributed to loss of open/grassland, however, the change in area for infrastructure for the period of 1980–2003 led to reduction in stony land in the region. A reduction of about 106.51 km^2 area under open/grassland was found between two points of time (1973 and 1980). Conversely, a reduction in area of 190.43 km^2 under stony land is observed during the period 1980–2003. While some of the stony land may have come under agricultural area, however, bigger chunk of the land has gone into development of infrastructure such as spreading the road network and construction of canal (Tables 6.1 and 6.2).

On a strong positive note, there has been no change whatsoever in area under wetlands of the region. As can be observed from the table, canals and rivers occupy the same area in 2008 as was in 1973. It is especially heartening to see that canals which occupied a mere 0.5 % of the total land-use and land-cover area in the study region still retain its hold on this area.

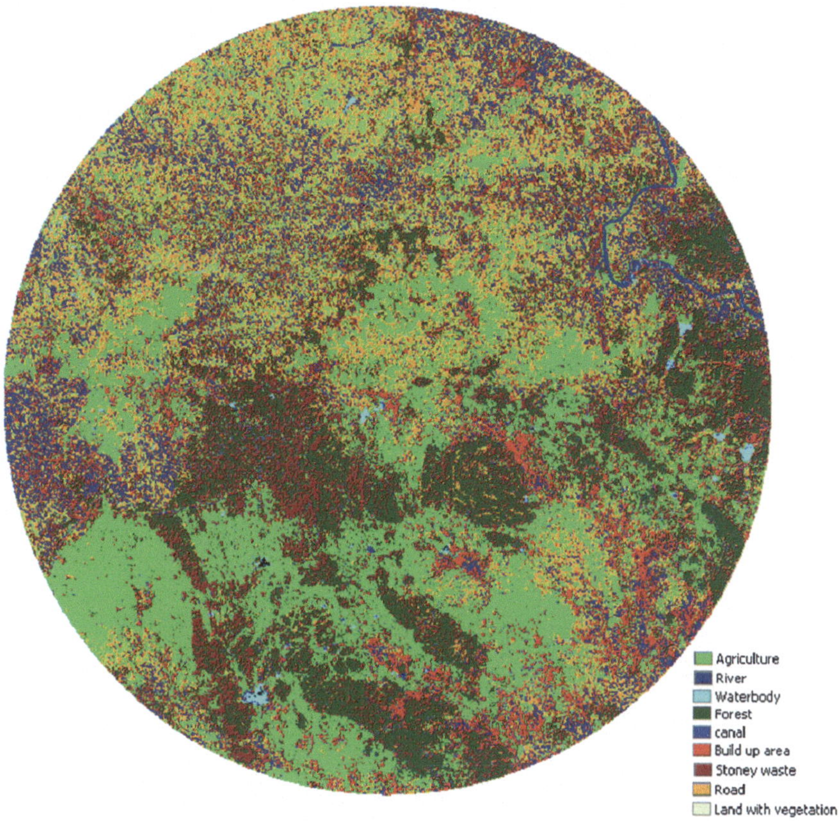

Fig. 6.15 Classified image of the area (2008)

References

Hoffer RM (1978) Biological and physical considerations in applying computer-aided analysis techniques to remote sensor data. In: Swain PH, Davis SM (eds) Remote sensing. The quantitative approach, New York, McGraw-Hill, pp 227–289

Koppad AG, Tikhile P (2012) Anthropogenic impact assessment on forest biodiversity in coastal region of Uttara Kannada district using RS and GIS technique. The Ecoscan 1(1):287–291

Jensen JR (1996) Introductory digital image processing: a remote sensing perspective. Englewood Cliffs, New Jersey: Prentice-Hall

Lillesand TM, Kiefer RW, Chipman JW (2004) Remote sensing and image interpretation, 5th edn. Wiley, New York

Roy SS, Mahmood R, Niyogi D, Lei M, Foster SA, Hubbard KG, Douglas E, Pielke RA Sr (2007) Impacts of the agricultural green revolution-induced land use changes on air temperatures in India. J Geophys Res 112:D21108. doi:10.1029/2007JD008834

Sandstrom P, Pahlen TG, Edenius L, Tommervik H, Hagner O, Hemberg L, Olsson H, Baer K, Stenlund T, Brandt LG, Egberth M (2003) Conflicts resolution by participatory management: remote sensing and GIS as tool for communicating land-use needs for reindeer herding in Nothern Sweden. Ambio 23:557–567

Singh A (1989) Digital change detection techniques using remotely-sensed data. Int J Remote Sens 10(6):989–1003

Zhu AX (1997) Measuring uncertainty in class assignment for natural resource maps under fuzzy logic. Photogram Eng Remote Sens 63(10):1195–1202

Chapter 7
Conclusion and Recommendations

Arid and semiarid regions come under dry lands which cover approximately 40–42 % of land area of planet Earth and support about two billion people of the world and 90 % live in developing countries. A variety of factors influence the process of desertification and land degradation in semiarid region; therefore, the area under this land category is expanding year by year. India occupies only 2.4 % of the world's geographical area, yet supports about 16.7 % of the world's human population; it has only 0.5 % of the world's grazing land, but supports 18 % of the world's cattle population. About 50.8 m ha land area (15.8 % of the country's geographical area) is arid, 123.4 m ha (37.6 %) is semiarid, and 54.1 m ha (16.5 %) area falls in the dry sub-humid region. All put together, about 228 m ha area, i.e., 69 % of the geographic area of the country is dry land (Ajai et al. 2009). In arid and semiarid regions, the livelihood activities of the residing population depend on two sectors—agriculture land use in which the cultivation is restricted to high productive but limited land area, and animal husbandry as availability of rangelands provides fodder to large animal population which depends on typical vegetation of arid and semiarid regions. The arid and semiarid regions are highly complex, thus make a sensitive ecosystem in which disturbances at smaller scale would create threats to the sustainability, which seems to be irreversible keeping current environmental and ecological problems in view. The semiarid region in India is home to very significant population which depends on the available land resources, natural as well as cultivable lands. Therefore, the challenge for the stakeholders is to achieve the goals of economic security and environment sustainability. Landscape of semiarid zone comprises variety of ecosystems, which are highly fragile with large risks that is caused due to factors such as increasing population, resource exploitation, unfavorable climate and weather conditions, climate change, increasing pressure on rangelands during excessive grazing by domesticated animals, high intensification of available arable lands, and land degradation and desertification. During last four decades, the area under semiarid region in India has increased by 3.45 million ha. Therefore, the sustainable management of land and resources is prime concern for the various stakeholders. The study on biodiversity is crucial to understand the potential and prospects for any ecosystems. In the study undertaken in the entire area about 2800 km of semiarid region of Karnataka, India examined both the plant and animal diversity. This rigorous work on the biodiversity covering phytoplankton to higher plants and zooplankton to

S. Nautiyal et al., *Biodiversity of Semiarid Landscape*,
Environmental Science, DOI 10.1007/978-3-319-15464-0_7

invertebrates and vertebrates provides strong baseline study of biodiversity in the area located in semiarid region. With the help of Phytosociological 376 species of angiosperms (Trees, Shrubs, Herbs, and Climbers), 1 bryophyte, 4 pteridophytes, and 21 phytoplankton were recorded. The number of bryophytes and pteridophytes species recorded is low as a geo-climatic condition of the area is not suitable for those species. However, gymnosperm was reported only in cultivated land. The recorded faunal diversity of the area is composed of 164 species of insects, 82 species of spiders, 17 mollusks, 11 fishes, 5 amphibians, 13 reptiles, 71 aves, 11 mammals, and 24 zooplankton. Varieties of butterflies were spotted in and around the study zone. The biodiversity of the region is somewhat typical characteristics of any semiarid climatic zone of the country. The diversity among the floral and faunal species was found to be good in the study region. Moreover, 3 plant species, viz., *Santalum album, Acacia ferruginea, Sphaeranthus amarathoides* were found under red list of IUCN and put the categories of endangered, vulnerable, and least concerned status, respectively. However, there would be degradation of forest and consequent adverse impacts on the existing biota of the area due to various anthropogenic activities. Those impacts in the form of loss or disturbance of natural values such as biodiversity, conservation, and landscape need to be managed. The following measures can be suggested to manage the adverse impacts on biodiversity.

1. Ex-situ conservation of biodiversity is recommended, if directly affected by any human activity. Proper area of suitable environmental parameters should be identified, and methods for management should be developed as a part of Environmental Management Plan.
2. Afforestation Plan and Command Area Development Plan should be formulated following due procedure to ensure the sustainable landscape development.
3. The possible loss of seed bank from the topsoil should be compensated in nearby suitable area for regeneration of vulnerable species, preferably in the buffer zone of the study region.
4. Special care should be taken to conserve the economically important and medicinal plants which are of significant importance for the local people.
5. Proper reclamation of soil should be done and ecosystem rehabilitation should be practiced using only local plant species.
6. Assessment of biodiversity should be done periodically to evaluate the biodiversity of the study area.

7.1 Conservation and Multiplication of Multipurpose Tree Species

Forest conservation is one of the primary steps required for the development of the region. This region is under semiarid zones in southern India where the plant growth is limited as well as sparsely distributed. There are two forests developed in last three decades although need proper maintenance and efforts toward development of forest ecosystems in the region. Selection of multipurpose tree species would support conservation options in the region. There is a need to cultivate bamboo, and species such as *Acacia* spp. and *Casuarina equisetifolia, Melia dubia* for plantation at larger scale, which are fast growing tree species and having potential to fulfill the requirement of people for their fuel, fodder, and timber demand. Apart from this, bioaccumulator species have to be planted for phytoremediation. In the study area, a few bioaccumulator species were found, viz., *Calotropis procera*; *Hibiscus esculentus*; *Chenopodium album; Typha angustifolia; Amaranthus viridis; Cyperus rotundus; Ricinus communis; Solanum nigram; Phyla nodiflora* in terrestrial ecosystems and species such as *Hydrilla; Chara* in aquatic ecosystems.

7.2 Development and Demonstration of Agroforestry Models in the Region

Agroforestry concept needs to be diffused among the people of the region as this provides an opportunity to harvest the annuals with perennials from the same unit of land. Agroforestry is an old concept in that in different parts of the country, agroforestry (different sub-terms) systems have been developed by the people/traditional societies through their own indigenous ecological knowledge base (which is passed on through generations) for maximizing the output from a unit of land. Sustainable use of land use and land cover is one of the most important issues that suffer as a result of deforestation. Therefore, in the current context, all agroforestry systems have positive effects and influences upon land, water, and resource management. Agroforestry technologies have the potential applicability to improve land use and biophysical properties of lands on the one hand and to provide an opportunity for fulfilling the daily needs of rural populace on the other. Plantation of compatible and desirable species of wood perennials on farmlands results in an improvement in soil fertility. In our study region, *Azadirachta indica and Melia dubia* have potential for agroforestry species. Development of proper agroforestry models includes the following benefits.

1. Plantation of compatible and desirable species of wood perennials on farmlands results in an improvement in soil fertility.
2. This will reduce the dependency of the people on forests for their demand for fuel, fodder, fiber, and leaf litter.
3. Agro-forestry increases the organic contents of the soils through the addition of leaf litter and other plant parts.
4. More efficient nutrient cycling within the system leads to more efficient utilization of nutrients that are either inherently present in the soils or externally applied.
5. Inclusion of perennials with the annuals, in due course of time, would certainly help improve the biophysical components of the landscape, increase the carrying

capacity of lands, and maintain different temperature regimes. However, this needs a long-term scientific planning.

6. Nitrogen-fixing tree species enhance the productivity of crops and reduce farmyard manure input. This again reduces the pressure on the forests.
7. Further, in the state of Karnataka whose maximum land comes under the dryzone, agro-forestry will certainly contribute to the improvement of water drainage for recharging ground water in the region.
8. Agroforestry system provide fodder for animals and thus to some extent, the fodder demand can be reduced by the agroforestry system.

7.3 Development of Bird Sanctuary

Many migratory birds come to the study region from far off lands during certain seasons for breeding and as well as for feeding. These areas can be developed as sanctuary so that it can increase their population. Measures can be taken by the forest department to foster the necessary habitats specific to the birds such as maintaining healthy water bodies without human interference. For example, painted stork and purple heron require watery habitats with plenty of feeding and breeding sites; those water bodies can be maintained as a sanctuary. Tree species such as manila tamarind, tamarind tree and neem tree will improve the population of Koels as this is the favorite fruit. Apart from this, the species, viz, *Acacia nilotica* (kari jali) *Aegle Marmelos* (belapatri); *Albizza lebbeck* (baage mara); *Albizzia amara* (chujjulu); *Azadirachta indica* (bevu); *Delonix alata* (kempukenjiga); *Ficus benghalensis* (alada mara); *Ficus religiosa* (arali mara); *Mangifera indica* (maavina mara); *Melia azedarach* (are bevu); *Samanea saman* (bhagya mara); *Tamarindus indica* (hunsina mara); and *Thespesia populnea* (arasi mara) have to be preferred for plantation as these plants provide nesting habitat and food for the birds.

7.4 Conservation of Wild Animals

This can mainly be developed by involving the local people and educating them. The forest departments and the NGOs can have a joint cooperative effort in this regard rather than doing it all alone. Projects for conserving wild animals can be developed along with educating the people about the significance to conserve them. Many species are being killed for their believed medicinal values, which is a myth in most of the cases and thus damaging the biodiversity of the study region. Some animals due to loss of habitat are forced to search for new habitats that create human–animal conflict in agricultural land and human habitats. Many poisonous snakes are being killed out of fear due to lack of awareness that professional snake catchers could be called to catch the animal without killing and transported to its suitable habitat.

7.5 Carrying Capacity of Grazing Areas/Rangelands for Sustainable Landscape Development

There is a need for estimating the carrying capacity of various rangelands/grazing areas located in semiarid region. This work needs coordination with the department of animal husbandry of the state. The use of resources available in various sub-systems of production provides the sustainability to animal husbandry system. The semiarid regions are highly suitable for animal husbandry, and apart from the main land use, the people of the region depend on animal husbandry system for their subsistence. If the livestock population exceeds, the over-exploitation of resources starts thus creates conflicts between villagers and the various other stakeholders for example the forest department. This further aggravates in the area where apart from high livestock density the conservation policies are implemented for example—protected areas, reserve forests, etc. Therefore, it is necessary to analyze the impact of resource availability on animal husbandry sub-system and its ecological and economic consequences on the region under changing environmental, political, and socioeconomic factors. The options for resource management in semiarid regions are not so simple. The suggestions would be maintaining equilibrium between livestock numbers and the carrying capacity of the area on one hand and exploring alternate livelihood options on the other.

If animal husbandry sector of semiarid region which is the subsidiary occupation for sustaining the livelihood of the people is not beneficial, then to replenish the loss from animal husbandry, the pressure on other sectors of the village ecosystem would increase in the circumstances of (i) intensification of agriculture land use in the region and (ii) over-exploitation of the natural resources for economic gains. But, alternate options would not ensure sustainable development thus cannot be recommended. Therefore, a sound region-specific approach needs to be developed. In this direction, the development of land-use and land-cover models, with an emphasis on land-cover change, and bioresources of the semiarid region would help provide feasible solutions. A data analysis at various spatial scales has to be carried out while applying modern tools such as satellite Imagery (remote sensing) and Geographic Information Systems (GIS). This work will be helpful to understand the consequences of the changing scenario on various habitats of the semiarid regions.

7.6 Drinking Water Supply Through Pipeline

The domestic requirement of water for the villages of the study regions is being provided in the form of wells and borewells. Water supply through pipelines is not available to the people and thus the continuous use of borewell water can cause adverse effects on the vegetation and landscape and also deterioration of ground water table. The probable degradation can be checked by banning use of groundwater while providing drinking water through piped water supply from River Krishna and River Bheema.

7.7 Awareness Programs About the Importance of Semiarid Region Under Changing Climate

Climatic variables such as temperature, rainfall, and humidity have the effect on agriculture and rural development. During the study, it was observed that there has been a shift in rainfall seasons that has impact on agriculture thus has resulted in changing the cropping pattern. This is because of irregularities in climate pattern in the last three decades. Educating the local people by conducting workshops on these issues pertaining to climate change will help them to be prepared for the changes in the climatic pattern.

During empirical field studies, it was noticed that the people are unaware about the outcome of various developmental activities. Therefore, there is a need to provide basic information on scientific and technical aspects of the various developmental programmes currently in implementation form in the study region. Thus, a need is to organize various programmes at regular intervals with the involvement of the local people and various other stakeholders. The programme can include workshop, discussion, and debate at gram panchayat (village council) level, showing posters, video clips, etc., to the people about the contribution of various policies and programmes in the overall development perspective of the semiarid region. This can further be linked with current climate debate in view of that the semiarid regions are highly exposed to the climate vulnerability. Therefore, the mitigation and adaptation approaches can be taken into account with help and supports from various agencies with participation at local level. Teaching school-going children by teachers about the importance and sensitivity of semiarid landscape will not only help the children but also help in educating their parents who have not had formal education. Special local dedicated volunteers can be trained by the government to raise the awareness about biodiversity conservation and management and explain the benefits in the vernacular language.

Reference

Ajai AAS, Dhinwa PS, Pathan SK, Ganesh Raj K (2009) Desertification/land degradation status mapping of India. Curr Sci 97:1478–1483

Erratum to: Plant Biodiversity

Erratum to: Chapter 3 in: S. Nautiyal et al., *Biodiversity of Semiarid Landscape*, Environmental Science, DOI 10.1007/978-3-319-15464-0_3

The phytosociology of herbs in the page numbers 57–62 depicts the density of species 10 m × 10 m not the density per hectare (100 m × 100 m) and it was an inadvertent error. The density per hectare for each herb species is calculated and accordingly Table 3.5 phytosociological attributes of herbs (pages 58–62) is revised. The revised text and table is shown below.

Based on the phytosociological study, the 284500 individuals were found in a hectare of land in semiarid region. Species, namely *Parthenium hysterophorus* (39896 individuals per ha) and *Senna sericea* (29041 individuals per ha), were found to be most dominant among the herbaceous vegetation followed by *Croton bonplandianum* (18979), *Bacopa monnieri* (14927), *Euphorbia hirta* (13219), *Tridax procumbens* (13052), *Typha angustifolia* (12812), and *Achyranthes aspera* (10646). The density value for 138 species was calculated to be less than 1000 individuals per ha (Table 3.5).

The online version of the original chapter can be found under
DOI 10.1007/978-3-319-15464-0_3

S. Nautiyal (✉) · K. Bhaskar · Y.D. Imran Khan
Centre for Ecological Economics and Natural Resources, Institute for Social and Economic Change, Bangalore, Karnataka, India
e-mail: nautiyal_sunil@yahoo.com

S. Nautiyal et al., *Biodiversity of Semiarid Landscape*,
Environmental Science, DOI 10.1007/978-3-319-15464-0_8

Table 3.5 Phytosociological attributes of herbs

S. no.	Botanical Name	Density/ha	Frequency	Abundance
1	*Acalypha indica*	3406.25	8.54	3.99
2	*Achyranthes aspera*	10645.83	37.08	2.87
3	*Aerva lanata*	3968.75	7.19	5.52
4	*Ageratum conyzoides*	770.83	0.42	18.50
5	*Aloe vera*	5968.75	3.02	19.76
6	*Alternanthera pungens*	125.00	0.83	1.50
7	*Alternanthera sessilis*	2572.92	2.92	8.82
8	*Alternanthera tenella*	968.75	3.44	2.82
9	*Alysicarpus vaginalis*	333.33	1.46	2.29
10	*Amaranthus spinosus*	791.67	1.98	4.00
11	*Amaranthus viridis*	187.50	0.63	3.00
12	*Amischophacelus axillaris*	166.67	0.21	8.00
13	*Ammannia baccifera*	1562.50	3.54	4.41
14	*Andrographis paniculata*	20.83	0.21	1.00
15	*Anisochilus carnosus*	104.17	0.63	1.67
16	*Argemone mexicana*	697.92	3.54	1.97
17	*Aristida setacea*	395.83	0.42	9.50
18	*Aristolochia bracteolata*	52.08	0.21	2.50
19	*Arundinella setosa*	218.75	1.35	1.62
20	*Arundo donax*	187.50	0.94	2.00
21	*Asteracantha longifolia*	364.58	0.52	7.00
22	*Bacopa monnieri*	14927.08	19.17	7.79
23	*Bambusa arundinacea*	62.50	0.21	3.00
24	*Bergia ammannioides*	93.75	0.31	3.00
25	*Biophytum sensitivum*	114.58	0.42	2.75
26	*Blainvillea acmella*	31.25	0.10	3.00
27	*Blumea eriantha*	52.08	0.31	1.67
28	*Boerhavia diffusa*	1843.75	8.85	2.08
29	*Boerhavia erecta*	20.83	0.21	1.00
30	*Bulbostylis densa*	93.75	0.31	3.00
31	*Canscora diffusa*	83.33	0.52	1.60
32	*Caralluma attenuata*	1968.75	5.73	3.44
33	*Celosia argentea*	31.25	0.10	3.00
34	*Chenopodium album*	31.25	0.10	3.00
35	*Chloris barbata*	125.00	0.52	2.40
36	*Chrozophora plicata*	62.50	0.21	3.00
37	*Cleome viscosa*	395.83	0.52	7.60
38	*Coldenia procumbens*	229.17	0.52	4.40
39	*Commelina benghalensis*	291.67	0.21	14.00
40	*Commelina diffusa*	416.67	0.94	4.44

(continued)

Table 3.5 (continued)

S. no.	Botanical Name	Density/ha	Frequency	Abundance
41	*Corchorus aestuans*	333.33	1.56	2.13
42	*Corchorus trilocularis*	729.17	2.81	2.59
43	*Cressa cretica*	83.33	0.42	2.00
44	*Crossandra infundibuliformis*	41.67	0.31	1.33
45	*Crotalaria juncea*	114.58	0.63	1.83
46	*Crotalaria pallida*	62.50	0.31	2.00
47	*Crotalaria pusilla*	52.08	0.21	2.50
48	*Crotalaria ramosissima*	20.83	0.10	2.00
49	*Crotalaria retusa*	83.33	0.21	4.00
50	*Croton bonplandianum*	18979.17	48.44	3.92
51	*Cullen corylifolium*	83.33	0.42	2.00
52	*Cyanotis sp.*	93.75	0.31	3.00
53	*Cyanotis tuberosa*	229.17	1.25	1.83
54	*Cynodon dactylon*	9406.25	5.42	17.37
55	*Cyperus alopecuroides*	729.17	1.25	5.83
56	*Cyperus compressus*	291.67	1.15	2.55
57	*Cyperus pangorei*	322.92	0.94	3.44
58	*Cyperus rotundus*	2010.42	3.85	5.22
59	*Cyperus sp.*	104.17	0.63	1.67
60	*Cyperus squarrosus*	697.92	1.25	5.58
61	*Dactyloctenium aegyptium*	62.50	0.31	2.00
62	*Datura metel*	2770.83	10.52	2.63
63	*Desmodium triflorum*	125.00	0.83	1.50
64	*Dicoma tomentosa*	166.67	1.15	1.45
65	*Digera muricata*	104.17	0.42	2.50
66	*Digitaria bicornis*	93.75	0.63	1.50
67	*Dinebra retroflexa*	333.33	1.46	2.29
68	*Dipteracanthus patulus*	125.00	0.52	2.40
69	*Echinochloa colona*	114.58	0.52	2.20
70	*Echinops echinatus*	41.67	0.31	1.33
71	*Eclipta alba*	7354.17	21.04	3.50
72	*Enicostemma axillare*	83.33	0.21	4.00
73	*Eragrostis japonica*	166.67	0.73	2.29
74	*Erigeron alpinus*	156.25	0.42	3.75
75	*Eriocaulon quinquangular*	145.83	0.21	7.00
76	*Eulophia epidendraea*	20.83	0.10	2.00
77	*Eulophia ramentacea*	10.42	0.10	1.00
78	*Euphorbia heterophylla*	6239.58	14.90	4.19
79	*Euphorbia hirta*	13218.75	24.27	5.45
80	*Euphorbia hypericifolia*	291.67	1.35	2.15

(continued)

Table 3.5 (continued)

S. no.	Botanical Name	Density/ha	Frequency	Abundance
81	*Euphorbia indica*	197.92	0.63	3.17
82	*Evolvulus alsinoides*	218.75	0.63	3.50
83	*Exacum pedunculatum*	41.67	0.21	2.00
84	*Fagonia arabica*	62.50	0.42	1.50
85	*Fimbristylis dichotoma*	145.83	0.42	3.50
86	*Flaveria trinervia*	72.92	0.31	2.33
87	*Gomphrena celosioides*	6604.17	10.42	6.34
88	*Grangea maderaspatana*	62.50	0.21	3.00
89	*Hedyotis puberula*	177.08	0.52	3.40
90	*Heliotropium indicum*	2718.75	6.67	4.08
91	*Heliotropium ovalifolium*	114.58	0.31	3.67
92	*Hemidesmus indicus*	2968.75	12.29	2.42
93	*Hybanthus enneaspermus*	4458.33	6.88	6.48
94	*Hymenotherum tennifloum*	114.58	0.52	2.20
95	*Hyptis suaveolens*	385.42	0.94	4.11
96	*Indigofera cordifolia*	145.83	0.73	2.00
97	*Indigofera linnaei*	62.50	0.21	3.00
98	*Ipomoea aquatica*	270.83	0.63	4.33
99	*Ischaemum rugosum*	406.25	1.15	3.55
100	*Justicia sp.*	31.25	0.21	1.50
101	*Lagascea mollis*	83.33	0.42	2.00
102	*Launaea pinnatifida*	114.58	0.63	1.83
103	*Lepidagathis cristata*	177.08	0.73	2.43
104	*Leucas aspera*	4270.83	9.79	4.36
105	*Martynia annua*	5208.33	21.88	2.38
106	*Melhania incana*	83.33	0.31	2.67
107	*Merremia emarginata*	31.25	0.21	1.50
108	*Mimosa pudica*	20.83	0.10	2.00
109	*Mollugo lotoides*	187.50	0.42	4.50
110	*Monochoria vaginalis*	52.08	0.31	1.67
111	*Nesaea brevipes*	114.58	0.42	2.75
112	*Nicandra physalodes*	20.83	0.10	2.00
113	*Nymphoides cristata*	166.67	0.31	5.33
114	*Ocimum americanum*	7979.17	27.81	2.87
115	*Orthosiphon glabratus*	41.67	0.21	2.00
116	*Oxalis corniculata*	1479.17	5.52	2.68
117	*Parthenium hysterophorus*	39895.83	45.42	8.78
118	*Pavonia odorata*	62.50	0.42	1.50
119	*Pentanema indicum*	62.50	0.21	3.00
120	*Peristrophe paniculata*	1510.42	4.27	3.54

(continued)

Table 3.5 (continued)

S. no.	Botanical Name	Density/ha	Frequency	Abundance
121	*Persicaria glabra*	83.33	0.10	8.00
122	*Phyla nodiflora*	385.42	0.94	4.11
123	*Phyllanthus amarus*	5843.75	19.79	2.95
124	*Phyllanthus kozhikodianus*	10.42	0.10	1.00
125	*Phyllanthus maderaspatensis*	114.58	0.63	1.83
126	*Phyllanthus virgatus*	20.83	0.10	2.00
127	*Physalis minima*	93.75	0.42	2.25
128	*Plumbago zeylanica*	218.75	0.83	2.63
129	*Polycarpaea aurea*	83.33	0.63	1.33
130	*Polycarpaea corymbosa*	1822.92	2.92	6.25
131	*Polygala chinensis*	125.00	0.83	1.50
132	*Portulaca oleracea*	177.08	0.63	2.83
133	*Priva leptostachya*	83.33	0.42	2.00
134	*Psilotrichum sp.*	10.42	0.10	1.00
135	*Pulicaria sp.*	291.67	0.42	7.00
136	*Pulicaria wightiana*	291.67	0.94	3.11
137	*Pupalia lappacea*	41.67	0.21	2.00
138	*Pycreus flavidus*	229.17	0.83	2.75
139	*Rostellularia simplex*	125.00	0.83	1.50
140	*Ruellia sp.*	41.67	0.10	4.00
141	*Rungia repens*	52.08	0.42	1.25
142	*Saccharum spontaneum*	83.33	0.10	8.00
143	*Salvadora persica*	10.42	0.10	1.00
144	*Scilla hyacinthina*	760.42	2.08	3.65
145	*Senna italica*	20.83	0.10	2.00
146	*Senna sericea*	29041.67	21.46	13.53
147	*Setaria pumila*	895.83	2.29	3.91
148	*Sida acuta*	6406.25	22.92	2.80
149	*Sida cordata*	114.58	0.52	2.20
150	*Sida cordifolia*	322.92	1.77	1.82
151	*Sida spinosa*	239.58	0.94	2.56
152	*Solanum nigrum*	10.42	0.10	1.00
153	*Solanum trilobatum*	62.50	0.21	3.00
154	*Solanum xanthocarpum*	31.25	0.21	1.50
155	*Sonchus oleraceus*	114.58	0.52	2.20
156	*Spermacoce hispida*	31.25	0.31	1.00
157	*Spermacoce stricta*	489.58	1.35	3.62
158	*Sphaeranthus amaranthoides*	93.75	0.10	9.00
159	*Spilanthes calva*	645.83	2.60	2.48
160	*Stachytarpheta indica*	614.58	1.88	3.28

(continued)

Table 3.5 (continued)

S. no.	Botanical Name	Density/ha	Frequency	Abundance
161	*Stemodia viscosa*	145.83	0.94	1.56
162	*Stylosanthes fruticosa*	41.67	0.10	4.00
163	*Trianthema portulacastrum*	62.50	0.31	2.00
164	*Tribulus lanuginosus*	1208.33	3.54	3.41
165	*Trichodesma indicum*	750.00	2.40	3.13
166	*Tridax procumbens*	13052.08	42.81	3.05
167	*Typha angustifolia*	12812.50	12.40	10.34
168	*Vigna sp.*	62.50	0.42	1.50
169	*Vigna trilobata*	62.50	0.21	3.00
170	*Vinca rosea*	52.08	0.21	2.50
171	*Waltheria indica*	104.17	0.21	5.00
172	*Withania somnifera*	1614.58	7.29	2.21
173	*Xanthium indicum*	3125.00	12.08	2.59

Bibliography

Ahmedullah M, Nayar MP (1986) Endemic plants of the Indian region, vol 1. Peninsular India. Botanical Survey of India, Calcutta, India
Bagnoli P, Goeschl T, Kovács E (2008) People and biodiversity policies: impacts, issues and strategies for policy action. OECD (Organisation for Economic Co-operation and Development) Publications, France
Blecher H, Swale E (1976) A beginner's guide to freshwater algae. Cambridge
Bole PV, Yogini V (1986) Field guide to the common trees of India. World wildlife fund India. Oxford University Press, Oxford
Chapin FS, Matson PA, Vitousek PM (2011) Principles of terrestrial ecosystem ecology, 2nd edn. Springer, USA
Charnley S, Fischer AP, Jones ET (2008) Traditional and local ecological knowledge about forest biodiversity in the Pacific Northwest. Pacific Northwest research station general technical report
Cooke T (1901–1908) The flora of the presidency of Bombay, Vol I and II. Taylor & Francis, London; reprinted edition, 1958, BSI, Calcutta
Curci J (2010) The protection of biodiversity and traditional knowledge in international law of intellectual property. Cambridge University Press, UK
Desonie D (2008) Biosphere: ecosystems and biodiversity loss. Chelsea House Publishers, New York
Franklin J (2009) Mapping species distributions: spatial inference and prediction. Cambridge University Press, Cambridge
Gamble JS (1915) Flora of the presidency of Madras. West, Newman and Adlard, London
Gao J (2009) Digital analysis of remotely sensed imagery. The McGraw-Hill Companies, Inc., New York
Hawksworth DL, Bull AT (2007) Vertebrate conservation and biodiversity. Springer, The Netherlands
Henderson PA (2003) Practical methods in ecology. Blackwell Science Ltd, New Jersey
Hill D, Fasham M, Tucker G, Shewry M, Shaw P (2005) Hand book of biodiversity methods, survey, evaluation and monitoring. Cambridge University Press, UK
Hooker JD (1875–1897) The flora of British India, vol I–VII. L. Reeve, London
Lambin EF, Geist H (2006) Land-use and land-cover change: local processes and global impacts. Springer, Berlin
Light SS (2004) The role of biodiversity conservation in the transition to rural sustainability. IOS Press, Netherlands
Lovett GM, Jones CG, Turner MG, Weathers KC (2005) Ecosystem function in heterogeneous landscapes. Springer, USA
Mölders N (2012) Land-use and land-cover changes: impact on climate and air quality. Springer, USA

S. Nautiyal et al., *Biodiversity of Semiarid Landscape*,
Environmental Science, DOI 10.1007/978-3-319-15464-0

Meffe GK, Nielsen LA, Knight RL, Schenborn DA (2009) Ecosystem management: adaptive community-based conservation. Island Press, Washington, DC

Milner-Gulland EJ, Rowcliffe M (2007) Conservation and sustainable use: a handbook of techniques. Oxford University Press Inc., New York

Naeem S, Bunker DE, Hector A, Loreau M, Perrings C (2009) Biodiversity, ecosystem functioning, and human wellbeing: an ecological and economic perspective. Oxford University Press Inc., New York

Nunes ALDP, Bergh CJMJVD, Nijakamp P (2003) The ecological economics of biodiversity-methods and policy applications. Edaward Elgar, UK & USA

Pandey DS, Singh NP (2005) Common flowers of India. Publications division, Ministry of Information and Broadcasting, Government of India

Sánchez-Carrillo S, Angeler DG (2010) Ecology of threatened semi-arid wetlands. Springer, USA

Secretariat of the convention on biological diversity (2006) Global biodiversity outlook 2. Montreal, UNEP, 81 + vii pages

Shachak M, Gosz JR, Pickett STA, Perevolotsky A (2005) Biodiversity in drylands: toward a unified framework. Oxford University Press Inc., New York

Shrader-Frechette KS, McCoy ED (1993) Method in ecology: strategies for conservation. Cambridge University Press, Cambridge

Slootweg R, Rajvanshi A, Mathur VB, Kolhoff A (2010) Biodiversity in environmental assessment: enhancing ecosystem services for human well-being. Cambridge University Press, New York

Spicer JI (2006) Biodiversity: a beginner's guide. Oneworld Publications, Oxford

Stamp LD (1961) A history of land use in arid regions, United Nations Educational, Scientific and Cultural Organization. Berger-Levrault Press, Nancy

Stephan JG, James SS, John CB (2005) Land use problems and conflicts, causes, consequences and solutions. Routledge, Taylor & Francis, London

Websites

www.afri.res.in
www.bsi.gov.in
www.cazri.res.in
www.cazri.res.in
www.cpcb.nic.in
www.envfor.nic.in
www.envis.frlht.org
www.fowghats.bravepages.com
www.gulbarga.nic.in
www.icrisat.org
www.indiabiodiversity.org
www.kbb.kar.nic.in
www.moef.nic.in
www.mospi.nic.in/Mospi_New/site/home.aspx
www.nbaindia.org
www.nmpb.nic.in
www.nmpb-mpdb.nic.in
www.nrsc.gov.in
www.unccd.int/en/Pages/default.aspx
www.wgbis.ces.iisc.ernet.in/biodiversity/sahyadri_enews/newsletter/issue12/index.htm
www.yadgir.nic.in
www.zsi.gov.in/App/index.aspx

The manufacturer's authorised representative in the EU is Springer
Nature Customer Service Centre GmbH, Europaplatz 3, 69115 Heidelberg,
Germany. If you have any concerns regarding our products, please
contact ProductSafety@springernature.com

Printed and bound by CPI Group (UK) Ltd, Croydon, CR0 4YY
25/09/2026
02222851-0006